Towards Coastal Resilience and Sustainability

Coastal zones represent a frontline in the battle for sustainability, as coastal communities face unprecedented economic challenges. Coastal ecosystems are subject to overuse, loss of resilience and increased vulnerability. This book aims to interrogate the multi-scalar complexities in creating a more sustainable coastal zone. Sustainability transitions are geographical processes, which happen in situated, particular places. However, much contemporary discussion of transition is either aspatial or based on implicit assumptions about spatial homogeneity. This book addresses these limitations through an examination of socio-technological transitions with an explicitly spatial focus in the context of the coastal zone.

The book begins by focusing on theoretical understandings of transition processes specific to the coastal zone and includes detailed empirical case studies. The second half of the book appraises governance initiatives in coastal zones and their efficacy. The authors conclude with an implicit theme of social and environmental justice in coastal sustainability transitions.

Research will be of interest to practitioners, academics and decision-makers active in the sphere of coastal sustainability. The multi-disciplinary nature encourages accessibility for individuals working in the fields of Economic Geography, Regional Development, Public Policy and Planning, Environmental Studies, Social Geography and Sociology.

C. Patrick Heidkamp is an Associate Professor in, and chair of, the Department of the Environment, Geography & Marine Sciences at Southern Connecticut State University in the USA and a visiting Lecturer in the School of Natural Sciences and Psychology at Liverpool John Moores University in the UK. He is also the co-director of the Connecticut State University System Center for Environmental Literacy and Sustainability Education and an affiliate of the Economic Rights Research Group at the University of Connecticut. He is an environmental economic geographer with a research focus on sustainability transitions.

John Morrissey is a Lecturer in Geography at Mary Immaculate College, University of Limerick, Ireland. His research is focused on issues of sustainable development, particularly socio-technical transitions, low-carbon development and challenges of low-carbon economy for urban and coastal communities. His work is informed by environmental economic geography with a focus on socio-spatial differentiation of transition processes.

The Dynamics of Economic Space
Series Editor: *Nuri Yavan, Ankara University, Turkey*

The IGU Commission on 'The Dynamics of Economic Space' aims to play a leading international role in the development, promulgation and dissemination of new ideas in economic geography. It has as its goal the development of a strong analytical perspective on the processes, problems and policies associated with the dynamics of local and regional economies as they are incorporated into the globalizing world economy. In recognition of the increasing complexity of the world economy, the Commission's interests include: industrial production; business, professional and financial services, and the broader service economy including e-business; corporations, corporate power, enterprise and entrepreneurship; the changing world of work and intensifying economic interconnectedness.

Making Connections
Technological Learning and Regional Economic Change
Edited by Edward J. Malecki and Päivi Oinas

Regional Change in Industrializing Asia
Regional and Local Responses to Changing Competitiveness
Edited by Leo van Grunsven

Interdependent and Uneven Development
Global–Local Perspectives
Edited by Michael Taylor and Sergio Conti

The Industrial Enterprise and its Environment
Spatial Perspectives
Edited by Sergio Conti, Edward J. Malecki and Päivi Oinas

Towards Coastal Resilience and Sustainability
Edited by C. Patrick Heidkamp and John Morrissey

For more information about this series, please visit: www.routledge.com/The-Dynamics-of-Economic-Space/book-series/ASHSER1030

Towards Coastal Resilience and Sustainability

Edited by C. Patrick Heidkamp
and John Morrissey

LONDON AND NEW YORK

First published 2019
by Routledge
2 Park Square, Milton Park, Abingdon, Oxon OX14 4RN

and by Routledge
711 Third Avenue, New York, NY 10017

Routledge is an imprint of the Taylor & Francis Group, an informa business

British Library Cataloguing-in-Publication Data
A catalogue record for this book is available from the British Library

Library of Congress Cataloging-in-Publication Data
A catalog record has been requested for this book

ISBN: 978-0-815-35863-3 (hbk)
ISBN: 978-0-429-46372-3 (ebk)

Typeset in Times New Roman
by Out of House Publishing

Printed and bound in Great Britain by
TJ International Ltd, Padstow, Cornwall

Contents

Figures

Tables

Contributors

Dr. Karen Alexander is an interdisciplinary Research Fellow in the Centre for Marine Socioecology at the University of Tasmania. She has wide-ranging interests, centring on marine governance. Karen specialises in issues around the transition to a green (blue) economy, and more recently her research has focused on stakeholder engagement and social license for sectors such as offshore renewable energy and aquaculture.

Dr. Will Allen is an independent systems researcher. He primarily works in sustainable development and natural resource management, helping diverse stakeholder groups to collectively plan, act and evaluate. He also manages the Learning for Sustainability (LfS) website –http://learningforsustainability.net – pointing to online resources around collaboration and adaptation.

Dr. Stephen Axon is an Assistant Professor in Geography at Southern Connecticut State University, where he researches and teaches in sustainability, addressing climate change and coastal transitions. Stephen has expertise in the ways in which the public engage with addressing climate change at the community level; indicating the extent to which individuals want (or do not want) to participate in community-based carbon-reduction strategies and coastal sustainability projects.

Dr. Amelie Bernzen is a Postdoctoral Research Fellow of Human and Economic Geography at the University of Cologne. Her research interests include agri-food systems, global value chains, social and environmental standards, social and economic aspects of climate change, smallholder livelihoods and food and nutrition security. She has conducted research in Europe, Australia, Bangladesh and India.

Ekaterina Bezborodko received her bachelor's degree in economics from Bowdoin College and her master's in geography from Rutgers University. While at Rutgers, she worked as a research assistant with Dr. Robin Leichenko and Dr. Melanie McDermott, transcribing and analysing interviews with key stakeholders on the New Jersey coast about their experience with climate-change impacts. She is now a graduate student in the Earth and Environmental Science doctoral program at the Graduate Center at CUNY (City University of New York), and she teaches geography at Hunter College.

Dr. Jason W. Birkett is a senior lecturer in Analytical Chemistry at Liverpool John Moores University, with over 20 years' academic experience involving the research and teaching of chemistry, environmental chemistry and forensic chemistry within the university sector. Jason has over 35 publications in the peer-reviewed literature on a range of topics involving analytical chemistry. He has also written two specialist environmental chemistry textbooks as a co-author and editor in collaboration with recognised world leaders in their respective fields.

Dr. Paula Blackett is a social scientist with the National Institute of Water and Atmosphere, Wellington, New Zealand. Her research interests are the diverse practices of participation and how contested decisions can be made collectively. She is presently focusing on climate change adaptation and natural resource management decisions.

Dr. Gregory Borne is a research fellow at Plymouth University's Marine Institute. He is the author of a number of books directly addressing sustainable development across different sectors including government, industry and the third sector. Most recently, his book *Surfing and Sustainability* (Routledge, 2018) explores transitions to sustainability within the surfing world.

Prof. Dr. Boris Braun is Professor of Human and Economic Geography at the University of Cologne. His research interests include urban and regional development, social and economic aspects of climate change, and environmental management in regions such as South/Southeast Asia (mostly coastal Bangladesh, India, Indonesia), Western Europe, and Australia. Currently he is involved in a larger research programme on Sea Level Change and Society.

Jessica Brewer received her degree in environmental resources engineering from the State University of New York College of Environmental Science and Forestry. She currently works at Leidos Engineering in the field of energy efficiency.

Dr. Catherine Chambers is the director of the Coastal and Marine Management Master's Program at the University Centre of the Westfjords in Ísafjörður, Iceland. She is an interdisciplinary marine social scientist with expertise in small-scale fisheries, local marine food systems, fishermen's knowledge, marine tourism, coastal communities and fisheries governance. Recent publications include Little kings: community, change and conflict in Icelandic fisheries (2017), Tourism governance for the coastal zone: Reynisfjara beach, Iceland (2017) and Viewpoint: the social dimension in Icelandic fisheries governance (2017).

Hemantkumar A. Chouhan is a Project Research Scientist at the Department of Humanities and Social Sciences at the Indian Institute of Technology, Bombay. He has completed his M.Phil. in Development and Planning from the same department. His research area pertains to environmental justice and political ecology, focusing on the impacts of CRZ violations on coastal environments and livelihoods of fishing communities. He has presented papers in various national and international conferences and published articles in journals and as book chapters. Hermantkumar has wider research interests in urban development, environmental sociology, sociology of law and development of marginalised communities.

Dom Clarke is currently Deputy Director at Plymouth Surfing and Sustainability Research Group, with experience in communicating sustainability, conservation education and surf industry marketing and strategy. Dom has experience engaging the public in marine sustainability issues, developing and implementing initiatives to increase awareness and participation in sustainable lifestyles, and collaborating with influential organisations and individuals to foster positive and sustainable future outcomes.

Dr. Kate Davies is a social scientist at New Zealand's National Institute of Water and Atmospheric Research (NIWA). Her research focuses on collaborative and participatory processes in coastal and marine social-ecological systems.

Natalya Dawe is a research assistant in the Department of Geography at Memorial University of Newfoundland, including the Civic Laboratory for Environmental Action Research. Much of this research focused on social aspects of the commercial fishery. She holds a Bachelor of Arts in Geography from Memorial University of Newfoundland.

Prof. Geraint Ellis is a Professor in the School of Natural and Built Environment and Director of Sustainable Built Environment at Queen's University, Belfast, as well as Adjunct Professor at Western University, Canada. His research interests are in planning and sustainability, renewable energy, planning governance and healthy urban planning. He has published widely on these topics and is a Co-Editor of the *Journal of Environmental Policy and Planning* and a member of the editorial board of the new Routledge journal, *Cities and Health*.

Dr. Wesley Flannery is a Lecturer in the School of Natural and Built Environment and Institute of Environmental and Spatial Planning (ISEP) at Queen's University, Belfast. His key research interests are in marine spatial planning, integrated coastal zone planning, stakeholder participation in environmental decision-making, public attitudes and behaviours towards the environment and planning for environmental risk management. He is an Associate Editor of the *ICES Journal of Marine Science*.

Dr. Paul Foley is an Assistant Professor in the Environmental Policy Institute, School of Science and the Environment, at Memorial University of Newfoundland's Grenfell Campus. He received a Ph.D. in Political Science from York University in Toronto, Canada, specialising in political economy and global governance. His work draws on international political economy and political ecology to study relationships between environmental governance and social development, with a current empirical emphasis on fisheries and seafood.

Bruce Glavovic holds the EQC Chair in Resilience and Natural Hazards Planning at Massey University. His research explores the role of governance in reducing disaster risk, and building resilient and sustainable communities. He focuses on coastal communities, and how to translate knowledge into action through anticipatory planning and adaptation, collaboration and conflict resolution.

Dr. Scott M. Graves is Associate Professor in the Department of Environment, Geography and Marine Sciences at Southern Connecticut State University, where he conducts research and teaches geography and environmental science courses. Dr. Graves earned his Ph.D. in Education (University of Idaho), M.S./ABD in Marine Geophysics/Oceanography (Graduate School of Oceanography: University of Rhode Island), B.S. in Geology/Earth Sciences (University of California at Santa Cruz), and has experience as a Marine Geologist/Oceanographer at the US Geological Survey.

Dr. Marcello Graziano is Assistant Professor of Economic Geography in the Department of Geography & Environmental Studies, and Member of the Institute for Great Lakes Research at Central Michigan University. His research focuses on the energy-water-sustainability-food nexus in the context of economic development, as well as on diffusion studies. Marcello holds a *Laurea Triennale* (B.Sc.) in Foreign Trade and a *Laurea Specialistica* (M.Sc.) in International Economics from the University of Turin, and a Ph.D. in Geography from the University of Connecticut.

Dr. Alison Greenaway is a Senior Researcher in Social Science at Landcare Research – Manaaki Whenua. She attends to co-production of sustainable development knowledge and practice, and facilitates and evaluates processes which foster collective deliberation to address complex issues.

Dr. Madeleine Gustavsson is currently working at Teagasc, Ireland. Her research interests are marine and coastal sustainability drawing on social science methods to understand the lifeworlds of marine actors. She completed her Ph.D. titled Examining the socio-cultural contexts of fishing lives on the Llŷn peninsula, UK, at the University of Liverpool in 2016. Madeleine has been awarded an ESRC New Investigator grant to explore the changing lives of women in small-scale fishing families in the UK, starting in spring 2018.

Dr. Samuel P. Hanes is an Assistant Professor of Anthropology at the University of Maine. Dr. Hanes' background is in historical geography and human ecology. He is presently writing a book on the evolution of oyster management in the USA and is in the middle of a major project studying the human dimensions of pollination in the Northeast. Dr Hanes is also part of a team studying the development of aquaculture in Maine as part of the University's new EPSCoR SEANET project.

Dr. C. Patrick Heidkamp is an Associate Professor in, and chair of, the Department of the Environment, Geography & Marine Sciences at Southern Connecticut State University and a visiting Lecturer in the School of Natural Sciences and Psychology at Liverpool John Moores University. He is also the co-director of the Connecticut State University System Center for Environmental Literacy and Sustainability Education and an affiliate of the Economic Rights Research Group at the University of Connecticut. He is an environmental economic geographer with a research focus on sustainability transitions.

Dr. Dan Hikuroa (Ngāti Maniapoto, Waikato-Tainui) is an Earth System Scientist and an established world expert integrating mātauranga Māori and science to realise the dreams and aspirations of the communities he works with. He is a Senior Lecturer in Māori Studies at the University of Auckland.

Dr. Patrick J. Holladay is an Assistant Professor in the School of Hospitality, Sport and Tourism Management at Troy University-Brunswick. He earned his Ph.D. from Clemson University in Parks, Recreation & Tourism Management with an emphasis on travel and tourism. His research focuses on ecotourism, sustainable tourism for development, and community development and uses tools such as social-ecological systems resilience thinking. He has worked in conservation, environmental education, ecotourism, national park management, sustainability and community development in Eastern Europe, the Caribbean, Central and South America and Central Asia as well as across the USA.

Prof. Ian Jenkinson joined Liverpool John Moores University in 1991 having held professional engineering roles in Rank Industries, David Brown, Philips and British Shipbuilders. He is the Head of Maritime and Mechanical Engineering, responsible for maritime and marine engineering education and including simulation-based training. He has led the development of maritime education and training at the university, working on numerous projects with manufacturing industries, international port operators, shipping companies and maritime authorities. In 2015 he led the formation of the Maritime Knowledge Hub with Mersey Maritime and was appointed a Board Member of Mersey Maritime in 2016. He has led six technology transfer projects with SMEs of which one was awarded best Regional Partnerships NW, and two were awarded Business Leader of Tomorrow. His research focus is on manufacturing process engineering, systems modelling, safety and reliability.

Dr. Teresa R. Johnson is an Associate Professor in the School of Marine Sciences at the University of Maine. Dr. Johnson's interdisciplinary, social science research examines the human dimensions of marine systems and coastal communities, with a focus on the role of participation and different forms of expertise in science and management, and responses to social and ecological change in coastal communities. Dr. Johnson's research has focused on these areas in the context of marine fisheries, renewable energy and aquaculture.

Dr. Christina Kelly is based at the School of Natural and Built Environment, Queen's University, Belfast. Her Ph.D. research investigated integrated estuarine and coastal management solutions and the use of Transition Management as a conceptual framework for adopting short-, medium- and long-term governance change. Her research reflects her background and interests in environmental and coastal planning, policy and governance.

Dr. Jason R. Kirby is a coastal sedimentologist with interest in the past and present behaviour of coastal systems. His work has focused on reconstructing sea level change records from coastal marshes and the response of coastal systems to sea level rise and storm events. His interests also include applied research on sustainable coastal management practice, particularly innovation in sediment management in ports and harbours.

Dr. Oliver Klein is a postdoctoral researcher at the Institute for Geography and Geology, University of Greifswald. In his Ph.D. project, finished in 2015 at the University of Vechta, he addressed network configurations in the globalising pork industry. His research interests lie in the fields of economic geography, agrifood geographies and sustainability transitions, with particular focus on rural areas.

Dr. Erena Le Heron is a Research Fellow in Geography in the School of Environment, University of Auckland. Her interests include the contribution made by plants, animals, materials, technologies and the impact of our stories, to human economies and environments from land to sea, and the way the interactions between humans and these contributors can produce outcomes that are different from our assumptions.

Prof. Richard Le Heron is Professor of Geography, School of Environment, University of Auckland, and a former Vice-President of the Royal Society of New Zealand. He is currently focused on igniting knowledge possibilities and new generation collective frameworks and participatory processes for the reorganisation and redirection of interactions of land and sea in conditions of shallow regulation and governance.

Prof. Robin Leichenko is Professor and Chair of Geography at Rutgers University and Co-Director of the Rutgers Climate Institute. Her current research explores economic vulnerability to climate change, equity implications of climate adaptation and the interplay between climate extremes and urban spatial development. Prof. Leichenko has authored or co-authored two books and more than 70 peer-reviewed journal articles and book chapters. Her book, *Environmental Change and Globalization: Double Exposures* (Oxford University Press, 2008), won the Meridian Book Award for Outstanding Scholarly Contribution from the Association of American Geographers.

Dr. Nicolas Lewis is Associate Professor in the School of Environment at the University of Auckland. Dr. Lewis has been involved in the NZ Government-funded Building Research Capability in the Social Sciences initiative, which aims to help build a foundation for a new generation of collaborative social science. Dr. Lewis has played a part in various BRCSS initiatives, particularly those around postgraduate students and early career researchers. Dr. Lewis has extensively contributed to the politics of knowledge production in the New Zealand context, both as a critic of neo-liberalism and as an advocate for critical social theory and the 'emerging researcher'.

Dr. P. Michael Link is senior researcher at the Research Group Climate Change and Security, which is part of the Center for Earth System Research and Sustainability (CEN) at University of Hamburg. He was previously affiliated with the Christian-Albrechts-University, Kiel and the University of California, Berkeley. His research focuses on societal impacts of climate change with an emphasis on water resources and renewable energy. Michael is also speaker of the Working Group Coastal and Marine Geography of the German Geographical Society.

June Logie is a research associate in the School of Environment at the University of Auckland. She is a graduate of Auckland and Massey Universities with a background in geographical education, business administration and coastal and environmental management. She is a Life fellow of the New Zealand Geographical Society, and is currently co-vice-president and secretary of the Auckland branch, New Zealand Geographical Society.

Dr. Carolyn Lundquist has a joint appointment as a Principal Scientist Marine Ecology at the National Institute of Water and Atmosphere and in the Institute of Marine Science at the University of Auckland. She is an applied marine ecologist, providing scientific and social-scientific input to inform decision-making for coast and ocean management.

Prof. Charles Mather is professor in the Department of Geography, Memorial University of Newfoundland. His most recent work has focused on the resource politics of northern shrimp and Atlantic salmon in Eastern Canada.

Dr. Melanie McDermott is Senior Researcher at the Sustainability Institute of the College of New Jersey. Her projects include defining state and local goals, indicators and standards for a municipal sustainability certification program. With degrees in interdisciplinary social science and forestry from University of California, Berkeley (Ph.D.) and Oxford University (M.Sc.), she also draws on decades of experience in the USA, Southeast Asia, southern Africa and the Caribbean conducting applied research on the political ecology of natural resource management. Recent work concerns climate change mitigation and adaptation policy and practice and its impacts on social equity.

Dr. Matthew D. Miller is an Assistant Professor in the Department of the Environment, Geography and Marine Sciences at Southern Connecticut State University. His expertise is in environmental GIS, particularly for temperate forests and coastal zones. His current research examines the bathymetry of New Haven Harbor through the use of historical nautical charts. Other projects include developing GIS exercises for K-12 classrooms using freeware and exploring the relationships between spatial data models.

Dr. John Morrissey is a Lecturer in Geography at Mary Immaculate College, University of Limerick, Ireland. John's research is focused on issues of sustainable development, particularly socio-technical transitions, low-carbon development and challenges of low-carbon economy for urban and coastal communities. His work is informed by environmental economic geography with a focus on socio-spatial differentiation of transition processes.

Dr. Karyn Morrissey is Senior Lecturer in the European Centre for the Environment and Human Health at University of Exeter Medical School. Over the last eight years, Karyn has developed a substantial research profile in marine economics, recently publishing a book entitled '*The Economics of the Marine*' with Rowan and Littlefield. Karyn is interested in the science-policy interface and produced the first economic valuation of Ireland's ocean economy, which provided the baseline estimates for Ireland's current marine strategy, Harnessing our ocean wealth: an integrated marine plan for Ireland.

Ryan Orlowski received a degree in Earth Science from Southern Connecticut State University and now works for Triumvirate Environmental. Mr. Orlowski was responsible for the fieldwork on which the interpretations of sediment transport in the Prospect Beach area are based.

Prof. D. Parthasarathy is Professor in the Deptartment of Humanities and Social Sciences at the Indian Institute of Technology Bombay, Powai, Mumbai, India.

Sarmishtha Pattanaik is Associate Professor in the Department of Humanities and Social Sciences at the Indian Institute of Technology Bombay, Powai, Mumbai, India.

Amélie Polrot is a PhD student in the School of Natural Sciences and Psychology at Liverpool John Moores University. Her current research in the field of marine bioremediation involves microbiology, chemistry and sedimentology. Her work focuses on assessing environmental factors controlling tributyltin biodegradation activity in sediment in order to optimise the implementation of a sustainable method for sediment management in ports and harbours.

Prof. Dr. Jürgen Scheffran is professor of Integrative Geography at the University of Hamburg and head of the Research Group Climate Change and Security (CLISEC) at the Center for Earth System Research and Sustainability. He has had positions at the University of Marburg, Technical University of Darmstadt, University of Illinois and Potsdam Institute for Climate Impact Research. Research fields include: security risks and conflicts of climate change and climate engineering; energy security and energy landscapes; water-food-land nexus, human migration and rural-urban relations; human-environment interactions and sustainability science; technology assessment and international security.

Prof. George P. Sharples is a Reader in Microbiology at Liverpool John Moores University, with extensive academic experience involving the research and teaching of microbiology with particular emphasis on microbial ecology and the biology of filamentous bacteria and fungi. He is a former Vice-President and General Secretary of the British Mycological Society and is Honorary Professor of Electron Microscopy at Kasetsart University, Thailand. His principal area of research relates to the study of the biology of actinomycetes and fungi, particularly their growth, differentiation and secondary metabolite formation. This research has generated over 75 publications. In addition, he has almost 50 years' experience of transmission and scanning electron microscopy of micro-organisms.

Dr. Kesheng Shu is a senior researcher at the Department of Bioeconomy and Systems Analysis of the Institute of Soil Science and Plant Cultivation State Research Institute of Poland. He previously worked at the University of Hamburg, Fudan University, the Institute for Advanced Studies on Science, Technology and Society in Austria and the Chinese Academy of Sciences. His research interests lie in energy transition, energy systems analysis and climate change.

Jamie Snook is the Executive Director of Torngat Wildlife, Plants and Fisheries Secretariat and for the past decade has been engaged in Indigenous fisheries management in Labrador. In 2016, he started a Public Health Ph.D. at the University of Guelph to research the connection between co-management governance and community health and well-being. In 2017, he was named a Pierre Elliott Trudeau Foundation Scholar.

Dr. James Tait received his Ph.D. in Earth Sciences from the University of California at Santa Cruz with a specialisation in coastal processes. He is currently a faculty member in the Department of the Environment, Geography and Marine Sciences at Southern Connecticut State University. He is also co-coordinator of the Werth Center for Coastal and Marine Studies. Recent research includes beach erosion along the Connecticut shoreline, the impacts of large storms on the coastal zone and strategies for improving coastal resilience.

Prof. Dr. Christine Tamásy is Professor for Human Geography at the Institute for Geography and Geology, University of Greifswald. Her research interests are: rural development, agricultural value chains/agri-food geographies and entrepreneurship in a spatial perspective.

Kristin Weis is a master's student in the Coastal and Marine Management Master's Program at the University Centre of the Westfjords in Ísafjörður, Iceland. Her research interests include social-ecological resilience, coastal communities, coastal and marine tourism, human mobility and migration, and environmental peacebuilding. Current projects explore coastal tourism behaviors, public interaction with marine wildlife, and the role of coastal environments in conflict management and peacebuilding.

Foreword

The coast is an extremely dynamic zone, integrating the natural processes of land, water and atmosphere with the impacts of human development. In itself, the coastal region is replete with challenges and conflicts, each worthy of study, analysis and adaptation. Yet the coast is more than the current stage for this interaction. The coast is undergoing vectors of change driven by an accelerating rate of sea level rise, described in the IPCC Fifth Assessment Report (2014) and verified by a variety of satellite observations (Nerem *et al.*, 2018), interacting with negative sediment budgets and increased storminess. Together, these drivers are altering the foundations of human use of the coast and calling for adaptation in recognition of the vectors of change.

With this background, the organisers of the Coastal Transitions Conference, held under the sponsorship of the Commission on the Dynamics of Economic Spaces of the International Geographical Union (IGU) and the Commission on Coastal Systems of the IGU, have contributed to an understanding of the very broad nature of economic impacts associated with the myriad of changes and challenges at the coast. This volume addresses and questions the concepts of sustainability in this dynamic environmental setting with site-specific data sets. It raises issues with management quandaries and conflicts in past, present and future scenarios. It is a valuable contribution to the consideration of decision-making in real-world situations in this dynamic area.

I see this collection of papers as more than a report of observations of ongoing impacts to coastal and estuarine systems. I see the incorporation of messages about the greater need to understand the vectors of change and the inevitable outcomes associated with the delayed or even lack of recognition of this environmental evolution. I see a contribution by the coastal geographical community of scholars in bringing their spatial skill-sets to bear on one of the very important matters of today, concluding with a call for expanded research into the issues of social and environmental justice. Congratulations to the organisers and the contributors.

Norbert P. Psuty
Professor Emeritus
Rutgers University
East Brunswick, NJ

References

Intergovernmental Panel on Climate Change. (2014). *Climate Change 2014: Synthesis Report. Contribution of Working Groups I, II and III to the Fifth Assessment Report of the Intergovernmental Panel on Climate Change.* Core writing team, Pachauri, R.K., & Meyer, L.A. (Eds.). IPCC: Geneva, Switzerland, 151 pp.

Nerem, R.S., Beckley, B.D., Fasullo, J.T., Hamlington, B.D., Masters, D., & Mitchum, G.T. (2018). Climate-change–driven accelerated sea-level rise detected in the altimeter era. *Proceedings of the National Academy of Sciences*. Advance online publication. https://doi.org/10.1073/pnas.1717312115

Editor's acknowledgements

The editors would like to sincerely thank the following people for their input and support to the development of this edited volume. From conception to final delivery, it was a combined team effort and the finished product would not have been possible without the enthusiasm and expertise of these key people.

Many thanks to all contributors and authors for their superb research, prompt delivery and open engagement.

The editorial team have been a joy to work with. We would like to sincerely thank the Routledge team and their associates for their professionalism, patience and consistent support, particularly: Ruth Anderson, Editorial Assistant – Geography & Tourism; Faye Leerink, Commissioning Editor; Ed Robinson, Project manager at Out of House Publishing and Colette Forder, Copyeditor at Coletteforder.eu.

The IGU Commission on Dynamics of Economic Spaces and the IGU Commission on Coastal Systems were key sponsors of the conference Sustainability Transitions In The Coastal Zone, which took place in SCSU, New Haven CT in March 2017 and from which this edited volume emerged. Their important support is much appreciated.

Many planning discussions took place during the workshop sessions of the Network on Sustainability Transitions in the Coastal Zone, an initiative funded by the Regional Studies Association. The support of the RSA has been vital to maintaining regular trans-Atlantic contact and to keeping up vital momentum on Coastal Transitions related collaboration. We acknowledge and thank the RSA.

Finally, this project started as part of the strategic partnership initiatives on coastal resilience and transitions in the coastal zone between Liverpool John Moores University, UK and Southern Connecticut State University, USA. The editors wish to thank those who contributed significantly to developing and nurturing the exchange between our institutions, particularly the international programming staff and university leadership on both sides of the 'pond'. The editors would also like to acknowledge the role of Mary Immaculate College, University of Limerick Ireland for supporting the project.

1 Introduction

Sustainability in the coastal zone

C. Patrick Heidkamp and John Morrissey

1 The coastal context

Coastal areas provide many co-benefits associated with their natural infrastructure, facilitating activities such as fishing, industry, tourism and transportation, for instance (Barragán & de Andrés, 2015; Sutton-Grier, Wowk & Bamford, 2015). Despite occupying a relatively small percentage of the Earth's land surface, the value of the ecosystems' services of coastal areas is more than one-third of the total for the globe (Barbier *et al.*, 2011; Barragán & de Andrés, 2015). However, while the goods and services provided by coastal ecosystems are essential for economic and social well-being (Mavrommati, Bithas & Panayiotidis, 2013), coastal ecosystems are increasingly at risk. As a direct result of intensive resource use and of concentrations of population, coastal ecosystems increasingly demonstrate direct and adverse impacts of human activities (Swaney *et al.*, 2012). Increasing population growth, movement of populations towards the coast and the increase in coastal development have led to an increase in pressure on, and degradation of, coastal ecosystems (Duxbury & Dickinson, 2007). For large coastal cities, interactions between human activities and coastal systems are intensified due to population density effects and associated economic activities (Mavrommati *et al.*, 2013). In many parts of the world, coastal natural habitats are declining and overexploited coastal resources are dwindling, with associated impacts on ecosystems and on the livelihoods dependent on these (Mee, 2012).

From an ecological perspective, the coastal zone represents a unique transition area between terrestrial and marine environments. Because of these interactions, the coast presents a dynamic and challenging environment for adequate management and planning (Duxbury & Dickinson, 2007). Complex management issues are then inherently exacerbated by demands for space and resources, typical of human coastal use (Cummins & McKenna, 2010). For example, the real-estate premium on coastal land has created problems in the ways in which competing uses are managed (Duxbury & Dickinson, 2007). According to Sale *et al.* (2014), the current management of development, habitat destruction, pollution and overfishing is seriously inadequate, with the result that coastal ecosystems are at increasing risk[1]. In addition, current challenges are being amplified by the effects of climate change (Beeharry,

Makoondlall-Chadee & Bokhoree, 2014). The global decline in estuarine and coastal ecosystems is already affecting a number of critical ecosystem services (Barbier *et al.*, 2011). For example, significant erosion of beaches in turn impacts recreational values and tourism (Remoundou, Diaz-simal, Koundouri & Rulleau, 2014). Small islands and coastal communities around the world are particularly vulnerable to storm surges attributed to more frequent and severe coastal storms and to mounting sea level rise, for instance (Mostofi Camare & Lane, 2015). Eustatic sea level changes threaten the viability of many coastal cities (Duxbury & Dickinson, 2007).

Climate impacts such as sea level rise, storminess, intensity of flooding and coastal erosion are likely to become more severe in the coming decades (Falaleeva *et al.*, 2011). Climate change effects such as ocean acidification and introduction of non-indigenous species result in more fragile marine ecosystems (Remoundou *et al.*, 2014). Accelerated sea level rise and increased frequency of strong hurricanes will increase the vulnerability of natural and human systems while making them less sustainable (Day *et al.*, 2014). The socio-economic, ecological and political risks associated with the combined effects of sea-level rise and natural hazards will increase accordingly (Malone *et al.*, 2010). The need for climate adaptation and climate risk management initiatives are therefore greatest in the coastal zones where population and ecosystem services are most concentrated (Malone *et al.*, 2010) and where coastal communities are particularly vulnerable (Bradley, van Putten & Sheaves, 2015). For example, Sale *et al.* (2014) predict that pressures of coastal development will combine with sea level rise and more intense storms to further intrude on and erode natural coastlines, severely reducing mangrove, saltmarsh and seagrass habitats. In turn, the degradation of coastal ecosystems increases the vulnerability of coastal towns and cities and their populations (Duxbury & Dickinson, 2007), to both environmental and social risks[2].

Climate change is expected to have significant impacts on the physical, social, ecological and economic environments of coastal cities and towns (Celliers, Rosendo, Coetzee & Daniels, 2013). For marine and coastal ecosystems, climate change influences temperatures and sea level, ocean circulation, storminess and wave regimes; nutrient availability, biological productivity and predator-prey relationships across the food-web may all be disrupted or altered (Remoundou *et al.*, 2014). Impacts may actively diminish the ability of coastal and marine systems to support the livelihoods of coastal populations, particularly those already under stress due to poverty and marginalisation (Purvaja, Ramesh, Glavovic, Ittekkot & Samseth, 2015). Poor and vulnerable communities within coastal cities and towns may also be disproportionately impacted (Celliers *et al.*, 2013). Uneven socio-economic impacts will be mirrored by uneven geographical, spatial and sectoral impacts. Impacts on the sustainability of natural environments, agriculture and urban areas will manifest differentially from place to place (Day *et al.*, 2014). For example, the severity of localised risk is proportional to the number of people or the value of the assets affected (Elliott *et al.*, 2014).

In Australia, more than A$226 billion in commercial, industrial, road and rail and residential assets is potentially exposed to inundation and erosion from climate change (Bradley *et al.*, 2015). Hurricane surges of up to 10 m represent a significant threat to densely populated areas of the Gulf and Atlantic coasts in the USA (Day *et al.*, 2014). Combined storm surge and sea level rise is projected to pose a risk of flooding to the Mekong Delta, despite extensive damming upriver in China (Hoa Le, Nguyen, Wolanski, Tran & Haruyama, 2007). These local expressions of climate change are threatening the capacity of coastal ecosystems to support goods and services valued by society on a global scale (Malone *et al.*, 2010). Due to the latency of natural systems, changes to the global climate are now 'locked in' for decades to come, regardless of any action taken to mitigate anthropogenic greenhouse gas emissions (Falaleeva *et al.*, 2011).

2 Responding to coastal sustainability and resilience

2.1 Governance challenges

The complexity and dynamism of climate and coastal policy domains calls for an integrated approach to governance in the coastal zone (Falaleeva *et al.*, 2011). However, in the political sphere, as in the economic sphere, human processes display a lack of cognisance of ecological systems, timescales and processes. Governance entities, including government agencies, local communities and other key stakeholder groups in the coastal zone, exhibit influence at overlapping scales, and are subject to hierarchies and gaps in coverage (Swaney *et al.*, 2012). The management of marine and coastal areas falls under a diverse range of institutional arrangements from multiple levels of government (Bradley, van Putten & Sheaves, 2015). Given the inherent diversity of national coastal resources, it is impractical to govern a coastline in its entirety; decisions around interventions need to be context-specific (Lloyd, Peel & Duck, 2013). Local governments are often perceived as the most appropriate level of government to implement adaptation initiatives, being well positioned to understand, interpret and predict the local implications of global climate change (Bradley *et al.*, 2015; Ford, Berrang-Ford & Paterson, 2011). Local governments therefore emerge as important actors in climate change adaptation; facilitating adaptation at the scale they can influence, and applying legislation that reduces vulnerability to climate change, as well as the provision of relevant information, leadership, advice, supervision and funding support (Celliers, Rosendo, Coetzee & Daniels, 2013). However, capacity at the local level can be a challenge, with a risk that lack of knowledge and skills can result in national policy intention being misinterpreted (Bai, Wieczorek, Kaneko, Lisson & Contreras, 2009). The requirement to integrate national, regional and local policy initiatives, to accommodate local stakeholders and take due cognisance of ecological limits, economic pressures and political imperatives, may simply be too great a challenge for local decision-makers.

Sale *et al.* (2014) suggest that the difference between sustainable coastal ecosystems and substantially degraded ones in 2050 will be determined by the effectiveness of local management systems. Effective management systems may mitigate against the worst impacts of increased local environmental impacts, for instance (Sale *et al.*, 2014). For example, the integration of hazard mitigation into land-use policy and evacuation planning is key to the sustainable governance of coastal areas, particularly in view of natural hazards such as hurricanes, tsunamis and flooding (Duxbury & Dickinson, 2007). New strategies need to be devised that will allow coastal communities to continue to live in these regions without further degradation of natural capital (Duxbury & Dickinson, 2007). Co-ordinated, comprehensive and long-term intervention strategies are urgently required to conserve and protect areas increasingly at risk (Elliott, Cutts & Trono, 2014). Falaleeva *et al.* (2011) sum up the governance problem structure as characterised by persistent uncertainty, intergenerational, functional and spatial interdependence of problems, operations and actors, and considerable risks from global environmental change. While in the academic literature there is growing recognition of the inter-related nature of contemporary environmental problems and of the need for new ideas and approaches capable of engaging with complexity (Shove & Walker, 2007), policy-makers frequently retain simplistic, often iconic perceptions of the world (Bridgewater, 2003). Science also needs to engage with policy formulation in more practical and immediate ways (Bridgewater, 2003).

Lags in the feedback effects of excessive resource use and resource stress (e.g., from climate change, eutrophication, biodiversity loss, etc.) on human well-being and resilience represent a critical knowledge gap (Dearing *et al.*, 2014). For coastal ecosystems subject to changes in terrestrial, oceanic or climatic influences, a lag between cause and effects is typical, compounded by lack of warning indicators of sudden ecological responses to change (Swaney *et al.*, 2012). While the time-frames on which decision-makers and practitioners need to respond to current or anticipated changes in ecosystems should be guided by sustainability principles (Malone *et al.*, 2014), human time-scales tend to be short-term in focus and are influenced strongly by political aspects. For example, the global economy does not evolve in accordance with the environment's capacity to produce the resources required for development needs (Marques, Basset, Brey & Elliott, 2009).

2.2 Rationale for a socio-technical transitions perspective on coasts

Current responses to sustainability in the coastal zone, as in other areas, repeatedly meet with a number of deeply embedded and difficult-to-address barriers. These include:

- Inertia and path-dependency of current approaches. The 'tyranny of here and now' (Tilly, 1986) means that economic imperatives prevent a deep structural questioning of the current economic and governance regime, to the detriment of more radical environmental protection initiatives[3].

- Socio-ecological boundary and threshold issues, including mismatches of governance jurisdictions and ecosystem coverage, mean that frequently nobody has responsibility for the full extent of an environmental issue.
- Disconnect between time-horizons of ecological processes and those of human decision-making and governance structures. This is typified by short-term political electoral imperatives, contrasted with long-term impacts of global processes such as climate change.
- Spatial disconnections between causes and effects of ecosystem degradation.
- Inadequate accounting for risk and uncertainty, including critical lack of attention on threshold effects and on positive and negative feedback mechanisms in coupled socio-ecological systems[4].

In this context, there is a need for conceptualisations which are long-term, multi-faceted and which recognise the coupled nature of coastal ecosystems and coastal economies. In view of new and emerging approaches to 'sustainability science', socio-technical transitions ideas are applied with the aim of developing insight into the long-term sustainability and resilience of the coastal zone; this is undertaken with a view to developing a research agenda on 'coastal sustainability transitions'. Transitions insights into regime functioning provide a lens through which to understand complex stakeholder interaction; for instance, interaction between different groups of terrestrial and marine stakeholders can complicate planning and management interventions in coastal areas (Falaleeva *et al.*, 2011). The transitions literature provides a novel way of understanding innovation for sustainability and this is particularly relevant in the coastal management context where new approaches are urgently required; for example, there is widespread agreement that adaptive, ecosystem-based approaches are needed to manage climate risks and to adapt to the impacts of climate change (Malone *et al.*, 2010). The means through which niche developments are mainstreamed is also a key strand of the transitions literature, and this is pertinent to understanding a) how successful coastal sustainability experiments can be mainstreamed and b) how this mainstreaming process can influence, and be influenced by, already existing spatial differences. In this regard, a new spatial perspective is required for full application of socio-technical understanding in the coastal context; for example, it remains that spatial disconnections[5] often complicate governance responses (Swaney *et al.*, 2012). An understanding of coastal sustainability therefore requires understanding of multi-scalar complexities, a relational understanding of space, as well as insights into the heterogeneous stakeholder landscape, institutional context and resource base, differentiated across particular regional, national and international settings.

3 Volume overview, aims and objectives

It is the aim of this volume to interrogate sustainability challenges in the coastal zone from the perspective of the emerging field of socio-technical transitions

(STT) research. Socio-technical transitions are defined as major technological transformations in the way societal functions such as transportation, energy and food provision are fulfilled (Geels, 2002). The socio-technical approach to transitions highlights co-evolution and complex interactions between industry, technology, markets, policy, culture and civil society (Geels, 2012). A considerable body of research has emerged on STT since the 1990s (Geels, 2002; Geels & Schot, 2007); much of this work argues that policy shifts to longer term perspectives and approaches are critical for environmental sustainability (Geels, 2012; Kemp, Rotmans & Loorbach, 2007). While these studies frequently consider local contexts (Geels, 2012), the spatial dimension in general has been poorly elaborated in transitions studies and, according to Murphy (2015), transitions research would benefit from more explicit and refined engagements with concepts such as socio-spatial and socio-technical embeddedness, multi-scalarity and the role of power in transitions. As described by Hansen & Coenen (2015), sustainability transitions are geographical processes, which rather than being universal and pervasive in nature, happen in situated, particular places and, as Truffer & Coenen (2012) point out, should be analysed in a regional context. However, much contemporary discussion of transition is either aspatial or based on implicit assumptions about spatial homogeneity, with comparatively little attention to how policy proposals will influence current patterns of uneven development (Bridge *et al.*, 2013). In much of the transitions literature, place is represented solely as a site for socio-technical processes rather than as a heterogeneous and situated construct (Murphy, 2015). This represents a major limitation to furthering insights into transition processes.

This collection of research essays therefore aims to employ the concept of socio-technical transitions as a guiding framework for a larger discussion of urgent sustainability and resilience issues in the coastal zone. In doing so, the volume showcases not only recent and innovative approaches in the analysis of coastal change, but also provides an example for, and application of, transitions research in a spatially and geographically sensitive context, thus responding to two key criticisms levied against the bourgeoning socio-technical transitions research: spatial naïveté and the apparent lack of consideration of specific geographical contexts (see Morrissey and Heidkamp, this volume, Chapter 2). On the premise of Coenen *et al.*'s (2012) criticism of the literature on socio-technical transitions:

> "In the past decade, the literature on transitions toward sustainable socio-technical systems has made a considerable contribution in understanding the complex and multi-dimensional shifts considered necessary to adapt societies and economies to sustainable modes of production and consumption. However, transition analyses have often neglected where transitions take place…" (1);

this volume is comprised of chapters that provide interdisciplinary and theoretically, methodologically and empirically varied approaches that provide an answer to 'where transitions take place', ranging from case studies situated in southern Asia (e.g., Braun and Bernzen, Chapter 11; Chouhan *et al.*, Chapter 18), New Zealand (Le Heron *et al.,* Chapter 7; Lewis, Chapter 6) and Europe (e.g., Klein and Tamásy, Chapter 8) to North America (e.g., Tait *et al.,* Chapter 12) and the Caribbean (Weis *et al.*, Chapter 20). Barring the research projects by Johnson and Hanes (Chapter 10), Alexander and Graziano (Chapter 14) and Gustavsson and Morrissey (Chapter 15), which were presented at the 2017 Association of American Geographers Meeting in Boston, and the paper by Polrot *et al.* (Chapter 16), the research for the chapters in this issue was first presented during Coastal Transitions 2017 – a conference hosted by Southern Connecticut State University (SCSU) from 30 March to 2 April 2017. This conference was sponsored by the International Geographical Unions (IGU) Commission on the Dynamics of Economic Spaces and the IGU Commission on Coastal Systems, and jointly organised by the Department of the Environment, Geography & Marine Sciences at SCSU and the Geography Faculty at Liverpool John Moores University (LJMU). The paper by Leichenko *et al.* (Chapter 3) is an adaptation of a paper previously published in the *Journal of Extreme Events* (2015, 2(1)) and the research presented therein was presented in the form of a keynote lecture at this conference.

The book is structured as follows:

- Part I is comprised of contributions that focus on theoretical understandings of coastal socio-technical transition/transformation processes towards more sustainable and/or resilient coastal economic-ecological interactions and resource use. The chapters in this section are united by a focus on underlying drivers of coastal-resource-use dynamics and on avenues for reimagining the framing of problems and solutions in local coastal contexts.
- Part II is comprised of a number of detailed empirical analyses that provide a series of detailed quantitative-based enquiries into the practical and real-world aspects of those concepts discussed in Part 1. In particular, these chapters present detailed situated evaluations of practical initiatives to deal with specific, localised coastal issues across a range of topics and diverse geographic locations.
- Part III consists of a number of detailed analyses of governance initiatives in coastal zones and appraises the respective impacts and efficacy of these. Contributions in this section share a common concern for strategic decision-making responses, the challenges posed by scale and the complications posed by the ever-present 'human-factor' of which governance responses need to be cognisant.
- Part IV of the volume concludes with an implicit theme of social and environmental justice in the transition towards more sustainable and resilient coastal systems. The collated chapters in this section share a focus on, and concern for, marginalised and disadvantaged communities

who frequently represent the first groups impacted by coastal change, and who also frequently are not afforded sufficient participation or agency in coastal decision-making.

While the approaches taken by the contributors to this volume are of an interdisciplinary nature and vary considerably in their theoretical foundations and methodological approaches, collectively the research presented highlights that transition dynamics need to be considered within their local and regional contexts and in the context of associated power structures and power relations. In our opinion, a common thread (either explicitly or implicitly) throughout the volume, is a call for the consideration of social, economic and environmental justice issues, which raises the following questions: What can research on socio-technical transitions contribute to these issues? How can we link research on social, economic and environmental justice issues with research on socio-technical transitions? And, how can a transition towards a sustainable and resilient but also just and equitable coastal zone be facilitated?

Acknowledgements

This chapter was produced as part of the strategic partnership initiatives on coastal resilience and transitions in the coastal zone between Liverpool John Moores University, UK and Southern Connecticut State University, USA. The authors wish to thank those who contributed significantly to developing and nurturing the exchange between our institutions, particularly the international programming staff and university leadership on both sides of the 'pond'.

Notes

1 Coastal social and ecological systems are losing their resilience and appear to be ever more vulnerable to human and natural disasters (Mee, 2012).
2 Coastal hazards including inundation, salinisation of the water supply and land erosion all threaten vital infrastructure that supports coastal communities (Mostofi Camare & Lane, 2015).
3 Political responses, resting on the assertion of a neo-liberal market agenda and public expenditure cutbacks, institutional reforms and organisational change, have limited the scope and potential of the land-use planning system to address particular ecological challenges (Lloyd *et al.*, 2013).
4 Ecologically defined criteria are strict and non-negotiable; devising a management plan that takes complex ecosystem processes and their extent in time and space into account typically requires considerable expertise (Borgström *et al.*, 2015).
5 For example, between decisions made upstream and distant environmental consequences which occur as a result downstream.

Bibliography

Bai, X., Wieczorek, A.J., Kaneko, S., Lisson, S., & Contreras, A. (2009). Enabling sustainability transitions in Asia: the importance of vertical and horizontal linkages. *Technological Forecasting and Social Change*, *76*(2), 255–266. http://doi.org/10.1016/j.techfore.2008.03.022

Barbier, E.B., Hacker, S.D., Kennedy, C., Koch, E.W., Stier, A.C., & Silliman, B.R. (2011). The value of estuarine and coastal ecosystem services. *Ecological Monographs*, *81*(2), 169–193. http://doi.org/10.1890/10–1510.1

Barragán, J.M., & de Andrés, M. (2015). Analysis and trends of the world's coastal cities and agglomerations. *Ocean & Coastal Management*, *114*, 11–20. http://doi.org/10.1016/j.ocecoaman.2015.06.004

Beeharry, Y., Makoondlall-Chadee, T., & Bokhoree, C. (2014). Policy analysis for performance assessment of integrated coastal zone management initiatives for coastal sustainability. *APCBEE Procedia*, *9*, 30–35. http://doi.org/10.1016/j.apcbee.2014.01.006

Borgström, S., Bodin, Ö., Sandström, A., & Crona, B. (2015). Developing an analytical framework for assessing progress toward ecosystem-based management. *Ambio*, *44*(3), 357–369.

Bradley, M., van Putten, I., & Sheaves, M. (2015). The pace and progress of adaptation: Marine climate change preparedness in Australia's coastal communities. *Marine Policy*, *53*, 13–20. http://doi.org/10.1016/j.marpol.2014.11.004

Bridge, G., Bouzarovski, S., Bradshaw, M., & Eyre, N. (2013). Geographies of energy transition: space, place and the low-carbon economy. *Energy Policy*, *53*, 331–340.

Bridgewater, P. (2003). Science for sustainability: an estuarine and coastal focus. *Estuarine, Coastal and Shelf Science*, *56*(1), 3–4.

Celliers, L., Rosendo, S., Coetzee, I., & Daniels, G. (2013). Pathways of integrated coastal management from national policy to local implementation: enabling climate change adaptation. *Marine Policy*, *39*(1), 72–86. http://doi.org/10.1016/j.marpol.2012.10.005

Coenen, L., Benneworth, P., & Truffer, B. (2012). Toward a spatial perspective on sustainability transitions. *Research Policy*, *41*(6), 968–979. http://doi.org/http://dx.doi.org/10.1016/j.respol.2012.02.014

Cummins, V., & McKenna, J. (2010). The potential role of sustainability science in coastal zone management. *Ocean & Coastal Management*, *53*(12), 796–804. http://doi.org/10.1016/j.ocecoaman.2010.10.019

Day, J.W., Moerschbaecher, M., Pimentel, D., Hall, C., & Yáñez-Arancibia, A. (2014). Sustainability and place: how emerging mega-trends of the 21st century will affect humans and nature at the landscape level. *Ecological Engineering*, *65*: 33–48.

Dearing, J.A., Wang, R., Zhang, K., Dyke, J.G., Haberl, H., Hossain, M.S., … Poppy, G.M. (2014). Safe and just operating spaces for regional social-ecological systems. *Global Environmental Change*, *28*, 227–238. http://doi.org/10.1016/j.gloenvcha.2014.06.012

Duxbury, J., & Dickinson, S. (2007). Principles for sustainable governance of the coastal zone: in the context of coastal disasters. *Ecological Economics*, *63*(2–3), 319–330. http://doi.org/10.1016/j.ecolecon.2007.01.016

Elliott, M., Cutts, N.D., & Trono, A. (2014). A typology of marine and estuarine hazards and risks as vectors of change: a review for vulnerable coasts and their management. *Ocean and Coastal Management*, *93*, 88–99. http://doi.org/10.1016/j.ocecoaman.2014.03.014

Falaleeva, M., O'Mahony, C., Gray, S., Desmond, M., Gault, J., & Cummins, V. (2011). Towards climate adaptation and coastal governance in Ireland: integrated architecture for effective management? *Marine Policy*, *35*(6), 784–793. http://doi.org/10.1016/j.marpol.2011.01.005

Ford, J.D., Berrang-Ford, L., & Paterson, J. (2011). A systematic review of observed climate change adaptation in developed nations. *Climatic Change*, *106*(2), 327–336. http://doi.org/10.1007/s10584-011-0045-5

Geels, F.W. (2012). A socio-technical analysis of low-carbon transitions: introducing the multi-level perspective into transport studies. *Journal of Transport Geography*, *24*, 471–482. http://doi.org/10.1016/j.jtrangeo.2012.01.021

Geels, F.W. (2002). Technological transitions as evolutionary reconfiguration processes: a multi-level perspective and a case-study. *Research Policy*, *31*, 1257–1274.

Geels, F.W., & Schot, J. (2007). Typology of sociotechnical transition pathways. *Research Policy*, *36*(3), 399–417. http://doi.org/10.1016/j.respol.2007.01.003

Hansen, T., & Coenen, L. (2015). The geography of sustainability transitions: review, synthesis and reflections on an emergent research field. *Environmental Innovation and Societal Transitions*, *17*, 92–109. http://doi.org/10.1016/j.eist.2014.11.001

Hoa Le., T.V., Nguyen, H.N., Wolanski, E., Tran, T.C., & Haruyama, S. (2007). The combined impact on the flooding in Vietnam's Mekong River delta of local man-made structures, sea level rise, and dams upstream in the river catchment. *Estuarine, Coastal and Shelf Science*, *71*(1–2), 110–116.

Lloyd, M.G., Peel, D., & Duck, R.W. (2013). Towards a social-ecological resilience framework for coastal planning. *Land Use Policy*, *30*(1), 925–933. http://doi.org/10.1016/j.landusepol.2012.06.012

Kemp, R., Rotmans, J., & Loorbach, D. (2007). Assessing the Dutch energy transition policy: how does it deal with dilemmas of managing transitions? *Journal of Environmental Policy & Planning*, *9*(3–4), 315–331. http://doi.org/10.1080/15239080701622816

Malone, T., Davidson, M., Digiacomo, P., Gonçalves, E., Knap, T., Muelbert, J., … Yap, H. (2010). Climate change, sustainable development and coastal ocean information needs. *Procedia Environmental Sciences*, *1*(1), 324–341. http://doi.org/10.1016/j.proenv.2010.09.021

Malone, T., DiGiacomo, P.M., Gonçalves, E., Knap, A.H., Talaue-McManus, L., & de Mora, S. (2014). A global ocean observing system framework for sustainable development. *Marine Policy*, *43*, 262–272. http://doi.org/10.1016/j.marpol.2013.06.008

Marques, J.C., Basset, A., Brey, T., & Elliott, M. (2009). The ecological sustainability trigon – a proposed conceptual framework for creating and testing management scenarios. *Marine Pollution Bulletin*, *58*(12), 1773–1779. http://doi.org/10.1016/j.marpolbul.2009.08.020

Mavrommati, G., Bithas, K., & Panayiotidis, P. (2013). Operationalizing sustainability in urban coastal systems: a system dynamics analysis. *Water Research*, *47*(20), 7235–7250. http://doi.org/10.1016/j.watres.2013.10.041

Mee, L. (2012). Between the devil and the deep blue sea: the coastal zone in an era of globalisation. *Estuarine, Coastal and Shelf Science*, *96*(1), 1–8. http://doi.org/10.1016/j.ecss.2010.02.013

Mostofi Camare, H., & Lane, D.E. (2015). Adaptation analysis for environmental change in coastal communities. *Socio-Economic Planning Sciences*, *51*, 34–45. http://doi.org/10.1016/j.seps.2015.06.003

Murphy, J.T. (2015). Human geography and socio-technical transition studies: promising intersections. *Environmental Innovation and Societal Transitions*, *17*, 73–91.

Purvaja, R., Ramesh, R., Glavovic, B., Ittekkot, V., & Samseth, J. (2015). Regional initiatives for interlinking global coastal scientific research projects. *Environmental Development*, *14*, 66–68.

Remoundou, K., Diaz-simal, P., Koundouri, P., & Rulleau, B. (2014). Valuing climate change mitigation: a choice experiment on a coastal and marine ecosystem. *Ecosystem Services*, *11*, 87–94. http://doi.org/10.1016/j.ecoser.2014.11.003

Sale, P.F., Agardy, T., Ainsworth, C.H., Feist, B.E., Bell, J.D., Christie, P., … Sheppard, C.R.C. (2014). Transforming management of tropical coastal seas to cope with challenges of the 21st century. *Marine Pollution Bulletin*, *85*(1), 8–23. http://doi.org/10.1016/j.marpolbul.2014.06.005

Shove, E., & Walker, G. (2007). Caution! Transitions ahead: politics, practice, and sustainable transition management. *Environment and Planning A*, *39*(4), 763–770. http://doi.org/10.1068/a39310

Sutton-Grier, A.E., Wowk, K., & Bamford, H. (2015). Future of our coasts: the potential for natural and hybrid infrastructure to enhance the resilience of our coastal communities, economies and ecosystems. *Environmental Science & Policy*, *51*, 137–148. http://doi.org/10.1016/j.envsci.2015.04.006

Swaney, D.P., Humborg, C., Emeis, K., Kannen, A., Silvert, W., Tett, P., … Nicholls, R. (2012). Five critical questions of scale for the coastal zone. *Estuarine, Coastal and Shelf Science*, *96*(1), 9–21. http://doi.org/10.1016/j.ecss.2011.04.010

Tilly, C. (1986). The tyranny of here and now. *Sociological Forum*, *1*(1), 178–188.

Truffer, B., & Coenen, L. (2012). Environmental innovation and sustainability transitions in regional studies. *Regional Studies*, *46*(1), 1–21.

Part I

Theory and methods

2 A transitions perspective on coastal sustainability

John Morrissey and C. Patrick Heidkamp

1 Introduction

1.1 Socio-technical transitions

Socio-technical transitions (STT) are defined as major technological transformations in the way societal functions such as transportation, energy and food provision are fulfilled (Geels, 2002). The socio-technical approach to transitions highlights co-evolution and complex interactions between industry, technology, markets, policy, culture and civil society (Geels, 2012). A considerable body of research has emerged on STT since the 1990s (Geels, 2002; Kemp, 1994; Geels & Schot, 2007); much of this work argues that policy shifts to longer term perspectives and approaches are critical for environmental sustainability (Geels, 2011; Kemp, Rotmans & Loorbach2007). While the transitions literature is diverse, a general feature is that transitions towards sustainability are framed from a systems perspective (Farla, Markard, Raven & Coenen, 2012)[1]. Here, the multi-level perspective (MLP) as a heuristic framework underpins a widely cited strand of STT theory (Geels, 2002; Kemp, 1994; Rip & Kemp, 1998). The MLP is a hybrid theoretical framework bridging science and technology studies and evolutionary economics, drawing extensively on institutional analysis as a middle ground spanning these disciplines (Coenen, Benneworth & Truffer, 2012)[2]. The MLP distinguishes three levels of heuristic, analytical concepts, which combine as a nested hierarchy to create a socio-technical system; these are landscape, regime, and niche levels (Crabbé, Jacobs, van Hoof, Bergmans & van Acker, 2013). The MLP posits that transitions come about through interactions between processes at three levels: (a) niche innovations afford space for new ideas to be tested and developed[3]; (b) changes at the landscape level create pressure on the regime, typically through large-scale macro factors, including changes brought about by global forces like climate change; and (c) destabilisation of the regime creates windows of opportunity for niche innovations to emerge (Geels & Schot, 2007); see Figure 2.1.

Socio-technical regimes are typically defined as relatively stable configurations of institutions, techniques and artefacts, as well as rules,

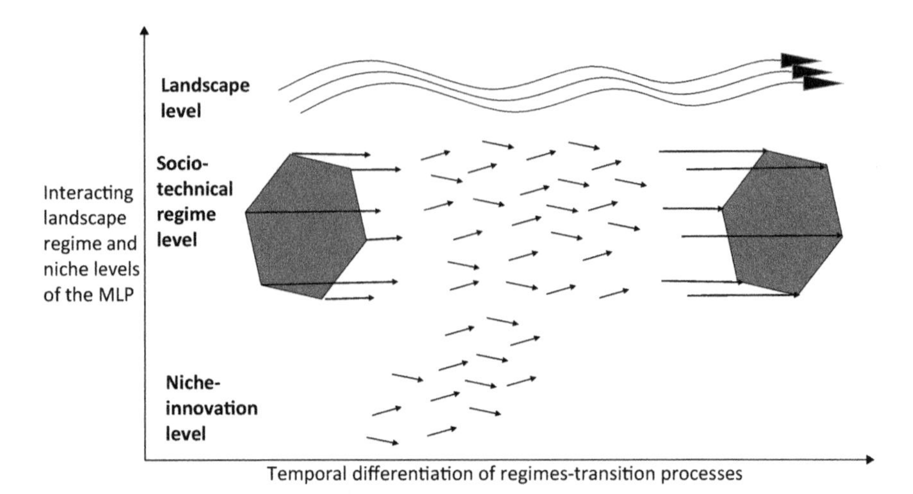

Figure 2.1 The multi-level perspective (MLP) on transition, after the schematic developed by Nykvist & Whitmarsh (2008) and Geels (2012)

practices and networks that determine the development and use of technologies (Rip & Kemp, 1998; Smith *et al.*, 2005). While existing systems are characterised by stability, lock-in and path dependence, giving rise to incremental change along predictable trajectories, radical alternatives can also emerge at the niche level (Geels, 2012). As represented by the MLP heuristic, key to explanations of change over time is the interaction between nested levels of the MLP, that is, niches, regimes and landscapes (Raven, Schot & Berkhout, 2012). As transitions are processes and constituent elements vary over time, it follows that transitions do not follow a stable pathway but are gradual and non-linear (Faller, 2016). Depending on timing and qualitatively different niche-regime-landscape interactions, transitions can evolve following different types of transition pathways (Geels & Schot, 2007).

1.2 Geographies of transitions and need for spatial perspective

While studies of niche innovations frequently consider local contexts (Geels, 2012), the spatial dimension in general has been poorly elaborated to date in the MLP, as well as in transitions studies more broadly. According to Murphy (2015), transitions research would benefit from more explicit and refined engagements with concepts such as socio-spatial and socio-technical embeddedness, multi-scalarity, and the role of power in transitions. As described by Hansen & Coenen (2015), sustainability transitions are geographical processes, which rather than being universal and pervasive in nature, happen in situated, particular places. However, much contemporary discussion of energy transition is either aspatial or based on implicit assumptions about spatial homogeneity, with comparatively little attention to how policy proposals for the low-carbon economy, for instance, will influence current patterns of uneven development (Bridge, Bouzarovski, Bradshaw & Eyre, 2013). In much of the transitions literature, place is represented solely as a site for socio-technical processes rather than as a heterogeneous and situated construct (Murphy, 2015). This represents a major limitation to furthering insights into transition processes. A consideration of space is critical to provide an appropriate understanding and representation of unevenness, heterogeneity and asymmetry in socio-technical systems (Raven *et al.*, 2012). The potential for sustainability transitions differs qualitatively between regions, for instance (Hansen & Coenen, 2015). Such differences are important in not only understanding contemporary transition processes, but also in highlighting future vulnerability and equity implications of transitions. The form of low-carbon transitions is likely to be influenced by existing regional differences (Balta-ozkan, Watson, & Mocca, 2015), potentially leading to further spatial differentiation and uneven development (Bridge *et al.*, 2013), with emerging differences overlaid upon existing heterogeneous socio-economic structures (Balta-ozkan *et al.*, 2015). Balta-ozkan *et al.*(2015) identify five issues through which low carbon transitions might exacerbate spatial differentiation and uneven development:

1) The clustering of low-carbon technologies;
2) The differences in energy demand between urban and rural settings;
3) The economic growth and job creation potential of a low-carbon transition;
4) The trade-offs between agglomeration economies and network constraints;
5) The public good problem involving different actors in liberalised markets.

This chapter forwards a spatially differentiated representation of the MLP, which accounts for uneven development processes and recognises the heterogeneity of different places in terms of potential for low-carbon transition (Figure 2.2).

In this chapter, the ideas represented in Figure 2.2 will be interrogated with respect to a spatial context deemed of critical importance to future sustainability, i.e. the coastal zone. The 100 km wide coastal strip is occupied by over 2.6 billion people at an average density of 97 persons/km^2 (Barbier *et al.*, 2011; Barragán & de Andrés, 2015; Sale *et al.*, 2014); that is, over 40% of the world's population. Eleven of the world's 15 largest cities are located in coastal areas and the number of people living in the coastal zone is expected to double by 2050 (Malone *et al.*, 2014; Small & Cohen, 2004). Coastal areas are resource-rich zones (Mee, 2012); the value of the ecosystems' services of coastal areas is more than one-third of the total for the globe (Barbier *et al.*, 2011; Barragán & de Andrés, 2015). In spatial terms, coastal areas can be deemed to be of strategic priority in any low-carbon transition. For example, the potential of tidal power is larger in coastal regions with large tides providing significant

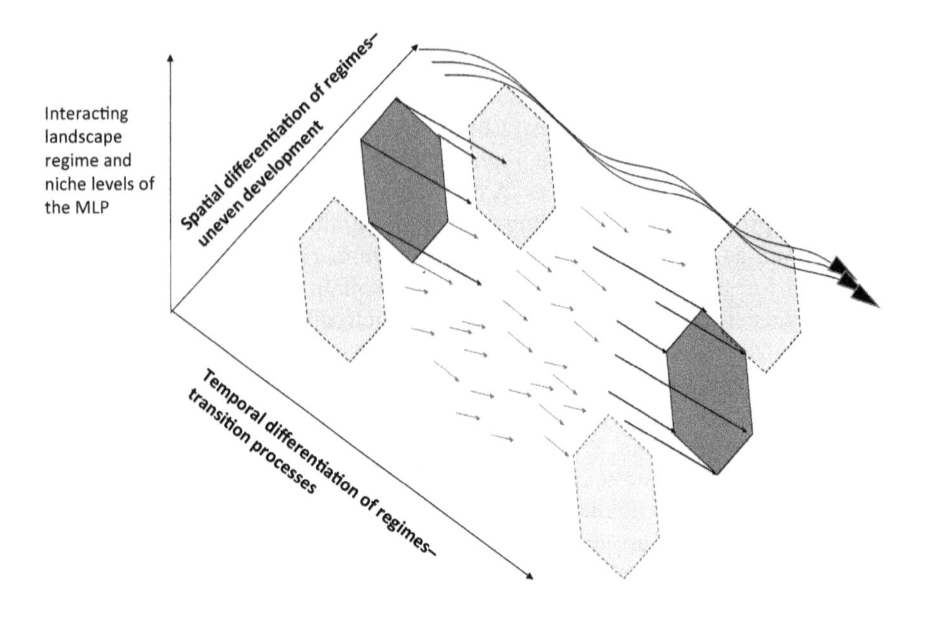

Figure 2.2 Spatially differentiated multi-level perspective (MLP) on transition

local natural resource endowments for sustainability transitions (Hansen & Coenen, 2015). This chapter is structured as follows: Part 2 of the chapter focuses on some general ideas on how to respond to these challenges. Part 3 provides three vignettes aimed at highlighting the geographical complexities related to innovation, and in a coastal context and their relation to the MLP concept of socio-technical transition. The final section of the chapter poses concluding remarks and sets the agenda for further research in coastal sustainability transitions. The aims of the chapter are: (1) to present a review of the multi-scalar complexities of coastal sustainability challenges; (2) to investigate the context for innovation, from social and technical perspectives in the dynamic, multi-issue and stakeholder-contested coastal zone; and (3) to argue that analysing the potential for sustainability transitions in the coastal zone requires an approach rooted in environmental-economic geography, explicitly acknowledging spatial differentiation and both uneven regional development histories and capacities as well as future potential.

2 Potential for coastal transitions

This section of the chapter illustrates how any response to coastal sustainability and resilience is intertwined with and related to the level of maturity of economic process, informed by spatially uneven patterns of development.

In particular, we focus on three discrete vignettes of innovation, which range from a novel and emerging innovation, a maturing innovation and an innovation which has become mainstream and part of the prevailing socio-technical regime. These vignettes aim to provide clear illustrations of complex socio-technical processes active in the coastal zone, and will support our argument that a spatially differentiated conceptualisation of transition is critical to a) understanding and b) realising sustainability in coastal areas. The first vignette focuses on 3D ocean farming off the coast of Connecticut in Long Island Sound, USA, as an example of an emerging niche-level innovation; the second vignette discusses the development of offshore wind energy capacity in Liverpool Bay, UK as an example of a socio-technical innovation that is more mature, and the final vignette discusses containerisation as an example of a fully matured socio-technological change. In addition to a discussion of the level of maturity of an innovation – on the continuum from niche innovation to maturity – the three vignettes provide examples for food systems transition, energy transitions and finally transportation transitions in the coastal zone, thus covering three of the key socio-technical transformations discussed in the transitions literature (Geels, 2012; Kemp, Rotmans & Loorbach, 2007; Morrissey, Mirosa & Abbott, 2014).

2.1 3D ocean farming in Long Island Sound

3D ocean farming is a zero-input farming process in which the entire ocean column – from seafloor to surface – is used to produce and harvest oysters, mussels, scallops, sea salt and kelp (Figure 2.3). In addition, the idea of 3D ocean farming is argued to have further positive environmental effects in terms of carbon sequestration, regulation of the nitrogen cycle (and thus the prevention of dead zones in the ocean) and the protection of the commons through an innovative use of ocean space. 3D ocean farms are also argued to have the potential to mirror natural reef systems, thus contributing to a more resilient coastline in terms of climate change-related weather events. The concept of 3D ocean farming is advanced by project GreenWave – an organisation whose mission it is to "support a new generation of ocean farmers and innovators working to restore ecosystems, mitigate climate change, and build a blue-green economy" – as an open-source concept that has the aim of 3D farm replication. In this vein GreenWave supports potential 3D ocean farmers, from permitting of the farm to harvest of the product[4]. 3D farms have a very small footprint in terms of their overall size or, as Brendan Smith, 3D ocean farmer and executive director of GreenWave, points out: "My [oyster] farm used to be 100 acres [before 3D farming]; now it's down to 20 acres, but it produces much more food than before. If you want 'small is beautiful', here it is" (Smith *et al.*, 2016).

In the context of the MLP to sustainability transitions, 3D ocean farming would be considered a niche innovation. In fact

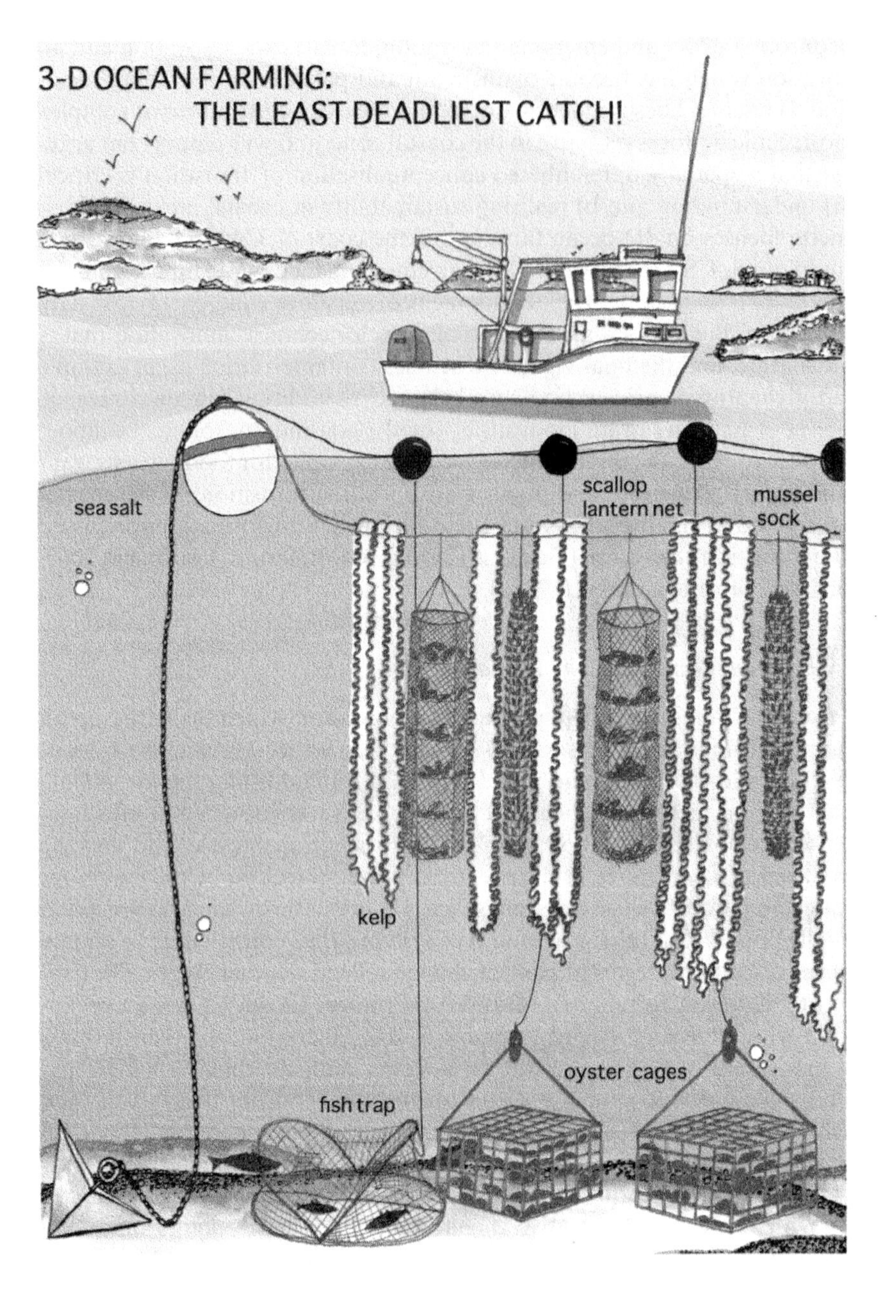

Figure 2.3 3D ocean farm, project by Greenwave.org (illustration by Stephanie Stroud for GreenWave, used by permission)

> *GreenWave's restorative 3D Ocean Farming model was awarded the Buckminster Fuller Prize for ecological design, profiled in the New Yorker (which called our model "the culinary equivalent of the electric car"), and honored by the President Bill Clinton as a keystone of ocean innovation*
> (Smith *et al.*, 2016)

As a niche innovation, 3D farming has not yet received any significant pushback from other stakeholders in Long Island Sound, which may include private entities such as pleasure boaters, the state, which is responsible for regulating land/ocean use in Long Island Sound, and NGOs that may or may not be supportive of the idea depending on their focus of engagement. Thus, 3D Ocean farming has not had a major impact on relevant governance structures – specific regulations for 3D ocean farming still need to be written. Currently the process operates within existing United States Department of Agriculture and Connecticut Shellfish Commission rules that have not been adapted for this innovation. It remains to be seen if local institutions and stakeholders will embrace 3D ocean farming and whether current regulations will be adapted to reflect potential needs for 3D ocean farmers. Insights from economic geography regarding institutional thickness and embeddedness might be useful for an analysis of the transition potential of this niche innovation. The near and longer-term potential of 3D ocean farming to contribute to coastal transitions hinges on the scalability – in terms of scaling up – of the concept. The specific case of 3D farming in Long Island Sound illustrates the multi-scalar complexities of coastal transitions quite well. Not only do existing local (e.g. water quality), regional (e.g. weather patterns) and global (e.g., sea level rise) biophysical conditions have to be taken into consideration, but these need to be examined in terms of their interactions and future changes caused by climate change. In the same vein local, regional and global regulations and governance schemes – and their potential interactions – have to be examined. In the case at hand, governance of Long Island Sound has to consider and reconcile rules and regulations imposed by the states of New York State, Connecticut and Rhode Island, as well as the US federal government. Beyond that the changing patterns of marine transportation (ferries, cargo, pleasure boats), issues of coastal access – who is allowed to farm in the coastal zone and where? – and issues of distribution channels for the produced goods, which are all spatial in nature, have to be examined in detail – not to speak of the socio-cultural aspects that may or may not lead to the adoption of kelp as a common food source. A complete assessment of the scalability of 3D ocean farming thus requires an analysis that is geographical in nature and multi-scalar in its approach. Coenen *et al.*'s (2012) call for a geographically sensitive transition analysis and their argument that the multi-level perspective also needs to be a multi-scalar perspective rings true here.

2.2 Coastal energy transitions: Burbo Bank Wind Farm, Liverpool Bay

The UK has been a world leader in offshore wind since October 2008, with total offshore generating capacity in UK waters of approximately 15 terawatt-hours (TWh) of electricity per annum (RenewableUK, 2016). The sector has developed with a series of licensing 'Rounds' co-ordinated by the Crown Estate (TCE), the landlord of the UK seabed (Barker, 2016; RenewableUK, 2016). Burbo Bank Offshore Wind Farm is situated on the Burbo Flats in Liverpool Bay at the entrance to the River Mersey, approximately 6.4 km (4 miles) from the Sefton coastline and 7.2 km (4.5 miles) from North Wirral. The wind farm has been fully operational since 2007 and consists of 25 turbines with a total capacity of 90MW (developed from Round 1). The location in Liverpool Bay was chosen for a number of reasons (DONG Energy Power (UK), 2015):

- Suitable average wind speed;
- Shallow water depth;
- No perceived environmental issues;
- Good seabed conditions for foundations;
- Close to an onshore electricity connection;
- Within Port Authority jurisdiction (for safety reasons);
- Local wind- power expertise/skills availability.

An extension to the project was awarded consent in 2014, planned to cover an area of 40 km^2 and with an estimated generating capacity of up to 250 MW (Dong Energy, 2016)[5]. The proposed project (Round 2.5) is to be located west of the current operational wind farm in Liverpool Bay (National Infrastructure Planning (UK), 2016). Table 2.1 presents a summary of the original wind farm at Burbo Bank and the new proposed extension.

Despite some local opposition on the basis of disruption to the aesthetics of the area and change to the seascape, plans have been greeted favourably for the most part; for instance, the official UK Government petitions website had received just 32 signatories as of February 2013 (UK Government and

Table 2.1 Burbo Bank Wind Farm, UK

Wind farm phase	*Year constructed*	*Capacity (MW)*	*Total cost (million US$)*	*Depth (m)*	*Turbine size (MW)*	*Number of turbines*	*Distance to shore (km)*
Original Burbo Bank	2007	90	185	5	3.6	25	6.5
Burbo Bank extention	Commissioned 2017 – expected	258	360 (estimated)	3.5–17	Up to 8.0	32	7

Source: data from Dong Energy (2016); Snyder & Kaiser (2009)

Parliament, 2013). For a project of this size, scale and geographic coverage, this is indeed surprising and may point to the maturation and/or increasing public acceptance of certain renewable energy technologies and projects. However, the impact of space and place cannot be ignored. The Burbo Bank project can be contrasted with plans to expand the electricity infrastructure in mid-Wales[6] within a comparable time-frame, from 2011, which resulted in a steady increase in opposition to the construction of pylons and onshore wind turbines in the valleys and on the hills in Powys, much of which was related to the scale of developments proposed. Proposed wind farms in northern Powys, Wales were large-scale, over 50 megawatt capacity and witnessed extensive protests, as well as an online petition signed by over 7,000 people (Dafydd, 2013).

Research by Bush & Hoagland (2016) in the US has shown that residents' opinions changed during the debate over offshore wind in relation to understanding of the project impacts. During ten-plus years of debate over a proposed offshore wind facility off Cape Cod, Massachusetts, the public's understanding of its environmental, economic and visual impacts matured. Massachusetts residents became more supportive of the project, while the gap between scientific and lay knowledge diminished late in the debate (Bush & Hoagland, 2016). In terms of the mainstreaming of innovations into the dominant socio-technical regime, wind farms are more mature and should no longer be considered a niche innovation. Offshore wind thus already has a more pronounced impact on the overall socio-technical system – here the energy system. Regarding governance, there are already regulations in place – though they are still being challenged and are thus regularly changing, as wind farms have already experienced pushback and critiques on a number of levels. The maturation of wind energy technology also poses interesting questions from a sustainability point of view. As discussed by Smil (2016, p. 1), "Wind turbines are the most visible symbols of the quest for renewable electricity generation. And yet, although they exploit the wind, which is as free and as green as energy can be, the machines themselves are pure embodiments of fossil fuels". While the counterargument is that all innovations require initial resource inputs that are initially beyond a sustainable cost-benefit threshold, the question still arises: at what time in the transition process from innovation to maturity does a new technology become sustainable from an environmental cost-benefit point of view? In addition, the case of wind energy demonstrates that there is a certain amount of lag time from the innovation phase to the stage when governments use incentives to foster widespread adoption. This lag time will, as outlined in the previous vignette, depend largely on the institutional context in which the newly emerging industry is situated.

2.3 Containerisation

Our third vignette represents a former innovation that has reached full maturity and incorporation into the prevailing socio-technical regime: the case of containerisation. Developed in the 1960s, containerisation and the related multi-modal transport of goods have developed rapidly, and have not only become commonplace but have revolutionised economic practices (Rodrigue & Notteboom, 2009). Containers do not only make it easy to load and unload ships, but they also allow for goods to be easily transferred from ships to trains, trucks and even airplanes. Containers stacked on top of one another (e.g. on trains or ships) lead to tremendous financial savings in terms of the economies of scale, the reduction in labour costs, and in terms of environmental sustainability through carbon-cost savings. The world shipping council, for example, states:

> Container shipping is the most carbon efficient means of transporting most goods across the world.... Containerization has revolutionized the movement of goods and the increased efficiency of moving goods has produced numerous benefits including lower environmental impacts associated with the movement of products from one point to another.
>
> (World Shipping Council, 2016, p. 1)

While it can certainly be argued that containerisation in its transition towards maturity (in the context of the MLP) has had a positive environmental impact in terms of carbon-cost reduction of shipping, it needs to also be pointed out that containerisation and the related increasing size of container ships have led to increased uneven coastal development pressures. Not all seaports have been able to accommodate large container ships, thus creating an economic landscape where a smaller number of large ports dominate and essentially causing a decline in the importance of many smaller shipping ports – especially those close to city centres. Costly infrastructure development, associated congestion and pollution (McCalla, 1999) as well as the creation (and loss) of job opportunities[7] related to the development as well as the operation of container shipping ports have been highly uneven. In the Northeastern USA, for example, containerisation and load centring, and their related spatial demands, led to the rising dominance of the ports of New York/New Jersey and Boston, and while some smaller city regions – such as Portland Maine, which was able to build a container terminal nearby in Casco Bay – were able to at least partially respond to these imposed changing spatial demands, the majority of port cities in the New England lost most if not all of their waterfront-dependent uses and thus often their economic base. This in turn led to widespread under-utilisation of the port-city interface, local unemployment and often dilapidated waterfronts (Heidkamp & Lucas, 2006). Booming ports in one area and dilapidated urban waterfronts in other cities are clear markers of an unevenly developed economic landscape; "once close infrastructural, economic and institutional ties between ports and their traditional city-regions have been loosened, since major ports now serve producers and consumers in widely dispersed hinterlands" (Hall, 2009, p. 67).

3 Scalar and spatial complexities of transitions

The three vignettes presented illustrate not only that there might be a significant relationship between the maturity of a socio-technical transition process and its impact on uneven development, but also show that the impact is highly dependent on the geographical context and the multi-scale relationships and interactions prevalent in socio-economic/social as well as biophysical conditions and processes. Cities and regions with more autonomy or with strong leadership may be more proactive in pushing a sustainability agenda, for example (Mans, 2014). Place-specific norms and values will have important influences on the geographically uneven landscape of sustainability transition (Hansen & Coenen, 2015). The availability of resources (materials and energy) is limited and distributed heterogeneously over the landscape, and determines current uneven development patterns as well as future development potential. These three vignettes illustrate that it is crucial to pay attention to the multi-scalar complexities of the geographical context of any innovation in order to assess its potential to contribute to sustainability transitions. For 3D ocean farming, for example, at which point in its transition towards maturity will the transition process slow down (or accelerate)? Will stakeholders push back? How does this relate to the thickness of local or regional institutions, or to embeddedness in local practices and routines? At which level of maturity is there a danger that the innovation will take on industrial dimensions and thus possibly set off processes of uneven development – when is 'small is beautiful', as in the words of GreenWave's Brendan Smith, no longer small – no longer beautiful?

The importance of a regional perspective is therefore clear. A regional level is forwarded for a number of reasons. There is evidence that policy at national level often fails to attend to the particular local situation, and that goals and targets set by national government do not reflect the reality of the local situation (Bai *et al.*, 2009). We believe that regional initiatives hold the most promise to foster greater understanding and awareness of the nature of the coast, its processes, problems and values, and thus allow individuals and groups that have vested interests to play more interactive and constructive roles (Gallagher *et al.*, 2004). The end goal should be a collaborative, reflexive and integrative governance network, where difference of opinion and debate are accommodated and encouraged. Scientific evidence, socio-economic values and priorities, and environmental considerations can be weighted differently, depending on political priority and site-specificities (Lloyd *et al.*, 2013). Stakeholder input is critical to appropriately resolving these complexities. Considering the USA, for example, there are marked differences in governance: on the West Coast, county governments do play a valid role, whereas on the East Coast governance for many coastal issues is in fact at the local (town) level and counties do not play any role. Within a regional perspective, local stakeholder-led policy development with emphasis on visioning and resilience building, consideration of MLP dynamics in planning and policy, and a spatial perspective on transition are forwarded as key tactics for realising coastal sustainability.

Best practice for contemporary environmental policy stipulates stakeholder engagement across all levels (Sandström, Bodin & Crona, 2015). In fact, stakeholder commitment is vital to achieving a comprehensive understanding of the range of needs and values attributed to the coastal and marine environments, to bring together all relevant data and to gain compliance with and support of new initiatives (Mason *et al.*, 2015). As coastal-zone sustainability policies are socially constructed, the voluntary co-operation of stakeholders with competing interests and priorities is essential for their effective implementation (Davos, 1998). Stakeholder commitment and co-operation is essential for the success of strategic coastal and marine management initiatives (Mason, Lim-Camacho, Scheepers, & Parr, 2015). Appropriate stakeholder engagement can enable local communities to take a degree of 'ownership' over local development trajectories (Barker, 2005). In this way, effective policy needs to be in tune with local-level reality (Bai *et al.*, 2009). In aggregate, local actors in fact play a significant role in delivering global environmental sustainability through the implementation of cumulative local action (Cummins & McKenna, 2010). However, stakeholder input usually reflects a broad range of interests, with potentially conflicting values and attitudes towards the coastal and marine environments; willingness to co-operate with other stakeholders will depend on the nature of the relationships between stakeholder groups (Mason *et al.*, 2015). Sustained capacity building is critical in this regard (Malone *et al.*, 2014). There is a scale element at work here also. Harrison (2008) and Cox (1998) emphasise that the protection of localised social relations at the community level (community groups) may depend on social and political networks operating at larger scales (national or federal bodies).

A consideration of MLP dynamics in planning and policy potentially delivers a number of benefits. At the landscapes level of the MLP, the future outlook looks to be increasingly volatile and unpredictable. Climate change impacts will be particularly acute in coastal areas where there is a dynamic and unpredictable interplay of natural and social environments (Lloyd, Peel & Duck, 2013). Ideas of social learning emphasise that developing understanding through interaction can be highly beneficial in terms of the exchange of information and ideas and relationship-building (Lloyd *et al.*, 2013). Through careful design, processes for public participation can move beyond public outreach and toward strategic engagement capable of serving multiple purposes through reciprocal knowledge exchange (Lazarus *et al.*, 2016). Regime 'activation' measures could seek to achieve better alignment of regime actors and processes; for example, more ambitious cross-scale linkages of groups across various governance levels can aid knowledge dissemination and increase stakeholder participation (Duxbury & Dickinson, 2007).

Finally, this chapter has argued that a spatial perspective on transition is critical to adequately account for uneven development processes and to recognise the heterogeneity of different places in terms of potential for low-carbon transition. Building on the argument made by Coenen *et al.* (2012) that much transitions research has had either "missing or naïve conceptualizations of space" this chapter argues that:

- While there is a growing realisation that place-specificity matters, there is still little generalisable knowledge about how place-specificity matters for transitions; few studies suggest alternative frameworks to study sustainability transitions (Hansen & Coenen, 2015).
- The study of place and place-making processes can reveal novel insights into the power relations and political processes underlying transition processes, and thus enable transition researchers to better account for the relationalities and context-specific forces determining the pace, scale and direction of socio-technical change (Murphy, 2015).
- Dynamics in socio-technical systems are also explained not only by interactions between modes of structuration and developments over time, but also by interactions between actors and institutions situated across different levels of spatial scale (Raven *et al.*, 2012).

4 A coastal sustainability transitions agenda

We argue that there is a need for a regional approach that enables a detailed analysis of the multi-scalar complexities of the geographical (socio-ecological and institutional context) approach. We posit that future research trajectories on socio-technological transitions need to focus on the following questions:

1) At which level of scale do institutional thickness/embeddedness of local and regional regulations slow down the transition process from innovation to maturity?
2) At which point in the transition process – from innovation to maturity – do processes that lead to uneven development appear?
3) At which level of geographical scale do non-sustainability practices take hold again?

This chapter argues for a regional approach to coastal sustainability transitions, based on the idea of regions as "complex institutions that coordinate and direct development, in detail and strategically, and are both defined by common norms and conventions and, conversely, disputes over what these norms should be" (Hayter 2008, p. 842). A promising example for such a regional approach is currently under way in the Northeastern USA with the United States Northeastern Regional Coastal Plan (NERPB, 2016); a plan which we believe is a promising path forward. The plan, currently in draft form, has been developed by the Northeast Regional Planning Body (RPB) and is a direct response to a National Ocean Policy issued by President Barack Obama on 19 July 2010. While not perfect in its current state, the plan deals conceptually with many of the issues outlined in this chapter, such as: a science-based regional environmental focus; an analysis of the regulatory framework for managing ocean and coastal resources; an inclusive view of regional ocean resources and activities, including the food system, the energy system, the transportation system, etc. It is noteworthy that the terms 'sustainability' and 'transitions' are not widely applied in this document (four and zero mentions respectively). The 'sustainability transitions' terminology is not widely known or adopted in the American context, while, in contrast, terms such as balance and resilience, perceived as being politically more neutral, are used in the plan more often (six and nine mentions respectively). Nevertheless, the way the planning process has been set up – i.e. the ability of the planning process to respond to concerns from the bottom-up – is promising and might lead to broad-based support for the future directions outlined, important factors towards the fulfilment of a transition towards sustainability. A process such as this might lead to ownership of the process by stakeholders with opposing viewpoints; critical from both a plan-delivery perspective as well as a future capacity-building point of view. Ultimately, the social dimension of sustainability may hold the key to transition. Effective, functional local alliances between disparate stakeholders may well represent the most important resource in determining local and regional transitions potential.

Acknowledgements

This chapter was produced as part of the strategic partnership initiatives on coastal resilience and transitions in the coastal zone between Liverpool John Moores University, UK and Southern Connecticut State University, USA. The authors wish to thanks those who contributed significantly to developing and nurturing the exchange between our institutions, particularly the international programming staff and university leadership on both sides of the 'pond'.

Notes

1 A detailed review of transitions approaches can be found in the review paper by Lachman (2013).
2 The MLP was developed in the field of innovation studies, drawing on insights from evolutionary economics (technological trajectories, regimes, niches, speciation), sociology of technology (innovations are socially constructed through interactions between engineers, firms, consumers, policy-makers) and neo-institutional theory (actors are constrained by shared beliefs, norms and regulations) (Geels, 2012)
3 Niches of innovation offer opportunities to experiment with new practices, technologies and organisational models, with subsequent potential for wider social transformation, should these niche innovations be suitable for wider uptake and diffusion (Geels, 2002; Geels & Schot, 2007; Seyfang & Smith, 2007; Seyfang, 2010).

4 For more information visit www.greenwave.org.
5 The extension will consist of 32 MHI Vestas Offshore Wind V164–8.0MW wind turbines with a total capacity of 258MW (Dong Energy, 2016).
6 Straight-line distance from Burbo Bank to Carnedd Wen is 94.44 km.
7 See Herod (2002) for a geographical analysis of the implications of containerisation on the longshore labour force in the USA.

Bibliography

Bai, X., Wieczorek, A.J., Kaneko, S., Lisson, S., & Contreras, A. (2009). Enabling sustainability transitions in Asia: the importance of vertical and horizontal linkages. *Technological Forecasting and Social Change*, *76*(2), 255–266. http://doi.org/10.1016/j.techfore.2008.03.022

Balta-ozkan, N., Watson, T., & Mocca, E. (2015). Spatially uneven development and low carbon transitions: insights from urban and regional planning. *Energy Policy*, *85*, 500–510. http://doi.org/10.1016/j.enpol.2015.05.013

Barbier, E.B., Hacker, S.D., Kennedy, C., Koch, E.W., Stier, A.C., & Silliman, B.R. (2011). The value of estuarine and coastal ecosystem services. *Ecological Monographs*, *81*(2), 169–193. http://doi.org/10.1890/10-1510.1

Barker, P. (2016). UK offshore wind – dodging the cuts. Retrieved 2 June 2016, from www.maritimejournal.com/news101/marine-renewable-energy/uk-offshore-wind-dodging-the-cuts2#sthash.mcO3NvKl.dpuf

Barker, A. (2005). Capacity building for sustainability: towards community development in coastal Scotland. *Journal of Environmental Management*, *75*(1), 11–19.

Barragán, J.M., & de Andrés, M. (2015). Analysis and trends of the world's coastal cities and agglomerations. *Ocean & Coastal Management*, *114*, 11–20. http://doi.org/10.1016/j.ocecoaman.2015.06.004

Bridge, G., Bouzarovski, S., Bradshaw, M., & Eyre, N. (2013). Geographies of energy transition: space, place and the low-carbon economy. *Energy Policy*, *53*, 331–340. http://doi.org/10.1016/j.enpol.2012.10.066

Bush, D., & Hoagland, P. (2016). Ocean & coastal management public opinion and the environmental, economic and aesthetic impacts of offshore wind. *Ocean and Coastal Management*, *120*, 70–79. http://doi.org/10.1016/j.ocecoaman.2015.11.018

Coenen, L., Benneworth, P., & Truffer, B. (2012). Toward a spatial perspective on sustainability transitions. *Research Policy*, *41*(6), 968–979. http://doi.org/http://dx.doi.org/10.1016/j.respol.2012.02.014

Cox, K.R. (1998). Spaces of dependence, spaces of engagement and the politics of scale, or: looking for local politics. *Political Geography*, *17*(1), 1–23.

Crabbé, A., Jacobs, R., van Hoof, V., Bergmans, A., & van Acker, K. (2013). Transition towards sustainable material innovation: evidence and evaluation of the Flemish case. *Journal of Cleaner Production*, *56*, 63–72. https://doi.org/10.1016/j.jclepro.2012.01.023

Cummins, V., & McKenna, J. (2010). The potential role of sustainability science in coastal zone management. *Ocean & Coastal Management*, *53*(12), 796–804. http://doi.org/10.1016/j.ocecoaman.2010.10.019

Dafydd, I. ap. (2013). Wind farm inquiry in Welshpool, Powys, attracts 300 protesters. Retrieved 2 June 2016, from www.bbc.co.uk/news/uk-wales-mid-wales-22758109

Davos, C.A. (1998). Sustaining co-operation for coastal sustainability. *Journal of Environmental Management, 52*(4), 379–387.

Dong Energy. (2016). Burbo Bank extension. Retrieved 2 June 2016, from www.burbobankextension.co.uk/en

DONG Energy Power (UK). (2015). Burbo Bank project summary. Retrieved 22 March 2017 from https://assets.dongenergy.com/DONGEnergyDocuments/uk/Project summaries/Burbo Bank Project Summary July 2015.pdf

Duxbury, J., & Dickinson, S. (2007). Principles for sustainable governance of the coastal zone: in the context of coastal disasters. *Ecological Economics*, *63*(2–3), 319–330. http://doi.org/10.1016/j.ecolecon.2007.01.016)

Faller, F. (2016). A practice approach to study the spatial dimensions of the energy transition. *Environmental Innovation and Societal Transitions*, *19*, 85–95. http://doi.org/10.1016/j.eist.2015.09.004

Farla, J., Markard, J., Raven, R., & Coenen, L. (2012). Sustainability transitions in the making: a closer look at actors, strategies and resources. *Technological Forecasting and Social Change*, *79*(6), 991–998. http://doi.org/10.1016/j.techfore.2012.02.001

Gallagher, A., Johnson, D., Glegg, G., & Trier, C. (2004). Constructs of sustainability in coastal management. *Marine Policy*, *28*(3), 249–255. http://doi.org/10.1016/j.marpol.2003.08.004

Geels, F.W. (2012). A socio-technical analysis of low-carbon transitions: introducing the multi-level perspective into transport studies. *Journal of Transport Geography*, *24*, 471–482. http://doi.org/10.1016/j.jtrangeo.2012.01.021

Geels, F.W. (2011). The multi-level perspective on sustainability transitions: responses to seven criticisms. *Environmental Innovation and Societal Transitions*, *1*(1), 24–40. http://doi.org/10.1016/j.eist.2011.02.002

Geels, F.W. (2002). Technological transitions as evolutionary reconfiguration processes: a multi-level perspective and a case-study. *Research Policy*, *31*, 1257–1274.

Geels, F.W., & Schot, J. (2007). Typology of sociotechnical transition pathways. *Research Policy*, *36*(3), 399–417. http://doi.org/10.1016/j.respol.2007.01.003

Hall, P.V. (2009). Container ports, local benefits and transportation worker earnings. *GeoJournal*, *74*(1), 67–83. http://doi.org/10.1007/s10708-008-9215-z

Hansen, T., & Coenen, L. (2015). The geography of sustainability transitions: review, synthesis and reflections on an emergent research field. *Environmental Innovation and Societal Transitions*, *17*, 92–109. http://doi.org/10.1016/j.eist.2014.11.001

Harrison, J. (2008). The region in political economy. *Geography Compass*, *2*(3), 814–830.

Hayter, R. (2008). Environmental Economic Geography. *Geography Compass*, *2*(3), 831–850.

Heidkamp, C.P., & Lucas, S. (2006). Finding the gentrification frontier using census data: the case of Portland, Maine. *Urban Geography*, *27*(2), 101–125. http://doi.org/10.2747/0272-3638.27.2.101

Herod, A. (2002). Towards a more productive engagement: industrial relations and economic geography meet. *Labour & Industry: A Journal of the Social and Economic Relations of Work*, *13*(2), 5–17. http://doi.org/10.1080/10301763.2002.10669261

Kemp, R., Rotmans, J., & Loorbach, D. (2007). Assessing the Dutch energy transition policy: how does it deal with dilemmas of managing transitions? *Journal of Environmental Policy & Planning*, *9*(3–4), 315–331. http://doi.org/10.1080/15239080701622816

Kemp, R. (1994). Technology and the transition to environmental sustainability: the problem of technological regime shifts. *Futures*, *26*(10), 1023–1046.

Lachman, Daniël A. (2013). A survey and review of approaches to study transitions. *Energy Policy*, *58*, 269–276.

Lazarus, E.D., Ellis, M.A., Murray, A.B., & Hall, D.M. (2016). An evolving research agenda for human–coastal systems. *Geomorphology*, *256*, 81–90.

Lloyd, M.G., Peel, D., & Duck, R.W. (2013). Towards a social-ecological resilience framework for coastal planning. *Land Use Policy*, *30*(1), 925–933. http://doi.org/10.1016/j.landusepol.2012.06.012

Malone, T., DiGiacomo, P.M., Gonçalves, E., Knap, A.H., Talaue-McManus, L., & de Mora, S. (2014). A global ocean observing system framework for sustainable development. *Marine Policy*, *43*, 262–272. http://doi.org/10.1016/j.marpol.2013.06.008

Mans, U. (2014). Tracking geographies of sustainability transitions: relational and territorial aspects of urban policies in Casablanca and Cape Town. *Geoforum*, *57*, 150–161.

Mason, C.M., Lim-Camacho, L., Scheepers, K., & Parr, J.M. (2015). Testing the water: understanding stakeholder readiness for strategic coastal and marine management. *Ocean & Coastal Management*, *104*, 45–56.

McCalla, R.J. (1999). From St. John's to Miami: containerisation at Eastern Seaboard ports. *GeoJournal*, *48*(1), 21–28. http://doi.org/10.1023/A:1007084618624

Mee, L. (2012). Between the devil and the deep blue sea: the coastal zone in an era of globalisation. *Estuarine, Coastal and Shelf Science*, *96*(1), 1–8. http://doi.org/10.1016/j.ecss.2010.02.013

Morrissey, J.E., Mirosa, M., & Abbott, M. (2014). Identifying transition capacity for agri-food regimes: application of the multi-level perspective for strategic mapping. *Journal of Environmental Policy & Planning*, *16*(2), 281–301. http://doi.org/10.1080/1523908X.2013.845521

Murphy, J.T. (2015). Human geography and socio-technical transition studies: promising intersections. *Environmental Innovation and Societal Transitions*, *17*, 73–91.

National Infrastructure Planning (UK). (2016). Burbo Bank extension offshore wind farm. Retrieved 6 February 2016, from http://infrastructure.planningportal.gov.uk/projects/north-west/burbo-bank-extension-offshore-wind-farm/

NERPB. (2016). Ocean planning in the NorthEast. Retrieved 2 June 2016, from http://neoceanplanning.org/

Nykvist, B., & Whitmarsh, L. (2008). A multi-level analysis of sustainable mobility transitions: niche development in the UK and Sweden. *Technological Forecasting and Social Change*, *75*(9), 1373–1387. http://doi.org/10.1016/j.techfore.2008.05.006

Raven, R., Schot, J., & Berkhout, F. (2012). Environmental innovation and societal transitions space and scale in socio-technical transitions. *Environmental Innovation and Societal Transitions*, *4*, 63–78. http://doi.org/10.1016/j.eist.2012.08.001

RenewableUK. (2016). Offshore Wind. Retrieved 2 June 2016, from www.renewableuk.com/en/renewable-energy/wind-energy/offshore-wind/

Rip, A., & Kemp, R. (1998). Technological change. *Human Choice and Climate Change*, 2, 327–399. In Rayner, S., & Malone, E. L. (Eds.), *Human Choice and Climate Change: An International Assessment* (pp. 327–399). Washington DC: Batelle Press.

Rodrigue, J.-P., & Notteboom, T. (2009). The future of containerization: perspectives from maritime and inland freight distribution. *GeoJournal*, *74*(7), 7–22. http://doi.org/10.1007/s10708-008-9211-3

Sale, P.F., Agardy, T., Ainsworth, C.H., Feist, B.E., Bell, J.D., Christie, P., ... Sheppard, C.R.C. (2014). Transforming management of tropical coastal seas to cope with challenges of the 21st century. *Marine Pollution Bulletin*, *85*(1), 8–23. http://doi.org/10.1016/j.marpolbul.2014.06.005

Sandström, A., Bodin, Ö., & Crona, B. (2015). Network governance from the top – the case of ecosystem-based coastal and marine management. *Marine Policy*, *55*, 57–63. http://doi.org/10.1016/j.marpol.2015.01.009

Small, C., & Cohen, J.E. (2004). Continental physiography, climate, and the global distribution of human population. *Current Anthropology*, *45*(2), 269–277. http://doi.org/10.1086/382255

Smil, V. (2016). What I see when I see a wind turbine. *IEEE Spectrum*. Retrieved 27 November 2016 from www.vaclavsmil.com/wp-content/uploads/15.WINDTURBINE.pdf

Smith, A., Stirling, A., & Berkhout, F. (2005). The governance of sustainable socio-technical transitions. *Research Policy*, *34*(10), 1491–1510.

Smith, B., Holmes, L., Coffey, B., Gautreau, R., Lee, J., & Dickerson, A. (2016). GreenWave. Retrieved 2 June 2016, from http://greenwave.org/

Snyder, B., & Kaiser, M.J. (2009). Ecological and economic cost-benefit analysis of offshore wind energy. *Renewable Energy*, *34*(6), 1567–1578. http://doi.org/10.1016/j.renene.2008.11.015

UK Government and Parliament. (2013). Stop Burbo Bank extension wind farm. Retrieved 2 June 2016, from https://petition.parliament.uk/archived/petitions/34283

World Shipping Council. (2016). Industry issues: environment. Retrieved 2 June 2016, from www.worldshipping.org/industry-issues/environment

3 Barriers, limits and limitations to resilience

Robin Leichenko, Melanie McDermott and Ekaterina Bezborodko

1 Introduction

Enhancing resilience has become a key element of preparedness for extreme events and climate change. While much progress has been made in defining and documenting components of resilience at the community level (e.g., Lopez-Marrero & Tschakert, 2011; Amundsen, 2012; Berkes & Ross, 2013; Frazier *et al.*, 2013; Cutter *et al.*, 2014; Burton, 2015), many questions remain about appropriate strategies for building resilience and potential barriers to implementation of these efforts. New questions are also emerging about inherent limits and limitations of resiliency-based approaches, suggesting that such efforts must be coupled with broader transformations of social and political conditions that create and perpetuate vulnerabilities (Tschakert & Dietrich, 2010; Cote & Nightengale, 2012; MacKinnon & Derickson, 2013; Bahadur & Tanner, 2014b; Brown, 2014; Cretney, 2014). Investigation of barriers, limits and limitations to resilience has particular resonance for urbanised coastal communities, many of which face significant hazards from climate extremes while also encountering a suite of technical, political, financial, legal and policy hurdles to adaptation planning (Moser *et al.*, 2012; Leichenko & Thomas, 2012; Douglas *et al.*, 2012; Jeffers, 2013; Kettle & Dow, 2014; Ziervogel & Parnell, 2014). This study explores barriers, limits and limitations to resilience through a case study of the Barnegat Bay region of coastal New Jersey, USA.

Following the damage wrought by Hurricanes Irene in 2011 and Sandy in 2012, the need to enhance resilience has become a front and centre issue in New Jersey and throughout the Northeastern United States. Although New Jersey has always been subject to severe coastal storms, coastal erosion and related stresses, the back-to-back hurricanes brought home the exposure and vulnerability of the region to climate extremes and climate change, and fostered an upwelling of interest in resilience among stakeholders across many sectors (New Jersey Climate Adaptation Alliance, 2014). Hurricane Irene was especially damaging for New Jersey's inland, riverine communities, causing widespread flooding and power outages. In total, Irene caused 11 fatalities in New Jersey and more than $1 billion in property damage (New Jersey Office of Emergency Management, 2012). Hurricane Sandy's impact was far more devastating for coastal communities in New Jersey. In addition to 34 lives lost, Sandy severely damaged or destroyed more than 30,000 homes in the state, and caused nearly $30 billion in property and infrastructure damage (Office of the Governor, 2012; Centers for Disease Control and Prevention, 2013).

Amidst the state's ongoing Sandy recovery efforts, there is a pressing need for investigation of factors that may promote or hinder resilience to a broad suite of climate-related stresses. This type of investigation is especially warranted given the growing emphasis on resilience as a key strategy for adaptation to climate change among policy-makers and practitioners both within New Jersey and throughout the world (e.g., Rockefeller Foundation, 2010; NRC, 2012; NPCC, 2013; New Jersey Climate Adaptation Alliance, 2014; US Climate Resilience Toolkit, 2015; Asian Cities Climate Change Resilience Network, 2014). Drawing upon emerging best practices in climate change adaptation and disaster risk reduction, which emphasise co-production of adaptation information via collaborative engagement between stakeholders and researchers, the study relies upon local knowledge of stakeholders and decision-makers to define barriers, limits and limitations to resilience in the case study region.

2 Barriers, limits and limitations to resilience

The concept of resilience has long been central to the study of natural hazards and disaster risk reduction (Cutter *et al.*, 2008; Gaillard, 2010; Manyena *et al.*, 2011; Weichselgarter & Kelman, 2015). Emphasis on resilience is also increasingly dominating research, policy and practice on community adaptation to climate change (Leichenko, 2011; Davoudi *et al.*, 2012; Bahadur & Tanner, 2014a; Wise *et al.*, 2014). While there is general agreement across these fields that resilience entails the ability to bounce back from climate and hazard-related shocks and stresses, a number of researchers have emphasised the need to define resilience more broadly to include the notion of 'bouncing forward' (Manyena *et al.*, 2011; Mitchell & Harris, 2012; Scott, 2013; Kresge Foundation, 2015). This expanded definition places emphasis on the capacity to anticipate and rebuild from extreme events in ways that reduce vulnerability and enhance preparedness for future shocks and stresses.

In exploring barriers, limits and limitations to resilience, this study draws from several heretofore discrete areas of scholarship, including literature on adaptation barriers, adaptation limits and critical resilience theory. Research on adaptation barriers has identified numerous political, economic, technical, informational and cultural factors that hinder adaptation planning and implementation (Biesbroek, 2013). While only a few studies directly focus on barriers to resilience (e.g., Bahadur & Tanner, 2014b), many of the identified barriers to climate adaptation are relevant to efforts focused on climate resilience. Research on barriers has demonstrated how different governance frameworks and decision-making contexts constrain adaptation planning (Amundsen *et al.*, 2010; Mozumder *et al.*, 2011; Runhaar *et al.*, 2012; McNeeley, 2012; Mukheibir *et al.*, 2013; Ziervogel & Parnell, 2014; Kettle & Dow, 2014; Dilling *et al.*, 2015). Barriers are often rooted in institutional and legal cultures that favour engineering responses to climate hazards (Harries & Penning-Rowsell, 2011; Jeffers, 2013) or presume, as a point of departure, that existing property rights and land uses must be protected (Few *et al.*, 2007). Socio-cultural factors such as notions of personal freedom can also act as barriers to adaptation (Nielsen & Reenberg, 2010; Jones & Boyd, 2011), as can lack of knowledge and information about climate change (Amundsen *et al.*, 2010; Measham *et al.*, 2011; Lata & Nunn, 2012; Pasquini *et al.*, 2013). Research has further shown that specification of barriers is a vital part of adaptation planning at the local level because it reveals areas where interventions might be needed (Moser & Eckstrom, 2010; Eisenack & Steckler, 2012; Ziervogel & Parnell, 2014).

While barriers are typically seen as mutable, *limits* to adaptation are generally understood as absolute or fixed obstacles that undermine the potential effectiveness of adaptation efforts (Islam *et al.*, 2014). As defined by Adger *et al.* (2007), limits are largely insurmountable and render adaptation ineffective. Research on limits to adaptation often focuses on ecological and physical conditions that represent critical thresholds beyond which existing activities, land uses and ecological systems will not be able to adapt without radical alternation of function or state conditions (Moser & Ecsktom, 2010). Limits are also defined in economic and technological terms, whereby the costs of adaptation exceed socially acceptable levels or adaptations are infeasible based on current technology and projected innovations (Adger *et al.*, 2009). Economics can also come into play when policies such as liberalised trade create absolute limits to adaptation responses (Laube *et al.*, 2012). Perceptions and values play an important role in how limits are defined (O'Brien, 2009; O'Brien & Wolf, 2010; Morrison & Pickering, 2013). As such, limits may be better understood as socially and culturally contingent rather than absolute (Adger *et al.*, 2009). Dow *et al.* (2013) further propose that limits to adaptation can be defined based on socially acceptable levels of risk.

Critical resilience literature identifies many *limitations* associated with strategies commonly used to build resilience. This work demonstrates that many approaches to building resilience are only palliative because they do not address underlying political and economic drivers of vulnerability or causes of climate change (Brown, 2014; Wise *et al.*, 2014). Critics also suggest that resiliency approaches can reinforce unequal power relationships and economic inequalities by failing to consider whose environments and livelihoods are being protected and why (Tschakert & Dietrich, 2010; Cote & Nightengale, 2012; Davoudi *et al.*, 2012; MacKinnon & Derickson, 2013; Cretney, 2014). Weichselgarter & Kelman (2015, p. 15) emphasise conceptual limitations of resilience thinking, suggesting that this work relies on "unchallenged assumptions about the social world, effectively imposing a technical-reductionist framework upon more complex webs of knowledge". Resiliency approaches are also thought to reflect neoliberal forms of governance, whereby responsibilities for climate change and disaster planning devolve from the state to individuals and communities, without commensurate support (MacKinnon & Derickson, 2013; Grove, 2014). However, researchers including Cretney & Bond (2014) and Bahadur & Tanner (2014b) have demonstrated that resiliency framings can also offer openings that challenge neoliberal discourses or status quo power relations and empower local actors and communities to enact transformative change.

Building on the above literatures, barriers, limits and limitations to resilience can be distinguished as follows:

Barriers to resilience are factors that hinder or constrain the implementation of resiliency measures. For example, legal objections by local property owners to beach replenishment projects (e.g., failure to grant easement) are an implementation barrier to a resilience action commonly proposed along the Eastern seaboard/East Coast of the USA.

Limits to resilience are factors that influence the effectiveness of resiliency actions, especially in the long run. For example, some limits to beach replenishment include the fact that this strategy only works for a short period of time because sea level rise renders it increasingly less effective.

Limitations to resilience are inherent shortcomings or undesirable outcomes of resiliency actions. A key limitation of beach replenishment is that this strategy tends to encourage additional property development in highly exposed areas, contributing to increased future exposure and potential vulnerability.

While in practice the boundaries between barriers, limits and limitations are somewhat porous, a key distinction is that barriers are perceived as having the potential to be overcome, while both limits and limitations are seen as more fundamental deficiencies of particular resiliency actions that suggest a need for alternative approaches or societal transformations.

3 Methodology

Co-production methods are increasingly recommended for the development of climate risk and adaptation information (Tribbia & Moser, 2008; Weichselgartner & Kasperson, 2010; Rosenzweig *et al.*, 2011; Cornell *et al.*, 2013; Kirchhoff *et al.*, 2013; Wagner *et al.*, 2014). In this study, researchers and stakeholders collaborated in the development of risk and vulnerability information, documentation of exposure to climate change, and identification of resiliency options (Leichenko *et al.*, 2014; Brady *et al.*, 2015). Study data were developed through interviews, discussions and meetings with stakeholders considered to be experts in public, private and non-profit sectors[1]. These interchanges were facilitated by researchers and staff from the Barnegat Bay Partnership (BBP), a part of the National Estuary Program administered by the US Department of Environmental Protection, and the Jacques Cousteau National Estuary Research Reverse (JCNERR), part of the National Estuary Research Reserve System administered by the US National Oceanic and Atmospheric Organization. Both BBP and JCNEER played the role of "boundary organizations" in the study (Cash & Moser, 2000; Corfee-Morlot *et al.*, 2011), helping to identify stakeholders with relevant sectoral expertise (Table 3.1), and initiating and facilitating interactions between researchers and stakeholders. In addition to input from BBP and JCNEER, names of stakeholder participants were also obtained through a "snowball" sampling technique whereby known individuals recommended other key informants with expertise on particular topics.

Research for the study was carried out over two years, beginning in 2011 and ending in 2013. A total of 29 stakeholders participated in the individual and small group meetings during the first phase of data collection in 2011 and 2012. Written transcripts of these meetings, developed from tape-recordings and interview notes, were coded and indexed in order to identify key topics and sub-topics. Summary results tables were developed for each topic and interview transcripts were used to identify quotations that were illustrative of stakeholder views. Initial summary results were presented at a larger group meeting of stakeholders held in 2013. The larger meeting, which included 36 participants (see Table 3.1), helped to corroborate the findings from the individual and small group meetings. During and following the presentation of the results at the larger meeting, participants contributed feedback and suggestions for refinement of the findings, particularly in light of Sandy's impact on the region.

Table 3.1 Number and type of stakeholders included in the study

Stakeholder category	*Individual and small group meetings*	*Large group meeting*
Public sector		
Conservation and land management	3	10
Fishing and fisheries management	-	1
Tourism and recreation	2	-
Planning and economic development	2	6
Emergency management – federal and state	4	5
Emergency management – local	5	-
Environmental regulation	-	4
Education	-	4
Private and non-profit sectors		
Conservation and land management	3	5
Fishing and fisheries management	2	1
Tourism and recreation	4	-
Real estate and insurance	2	-
Infrastructure	2	-
Total	29	36

Source: Adapted from Leichenko *et al.* (2014)

4 Resilience options, limits and limitations

Home to approximately 600,000 people, the Barnegat Bay region is primarily located in Ocean County, New Jersey (Figure 3.1) and falls within the commuter-shed of New York City, northern New Jersey and Philadelphia. Despite rapid suburban development in recent decades, beach tourism,

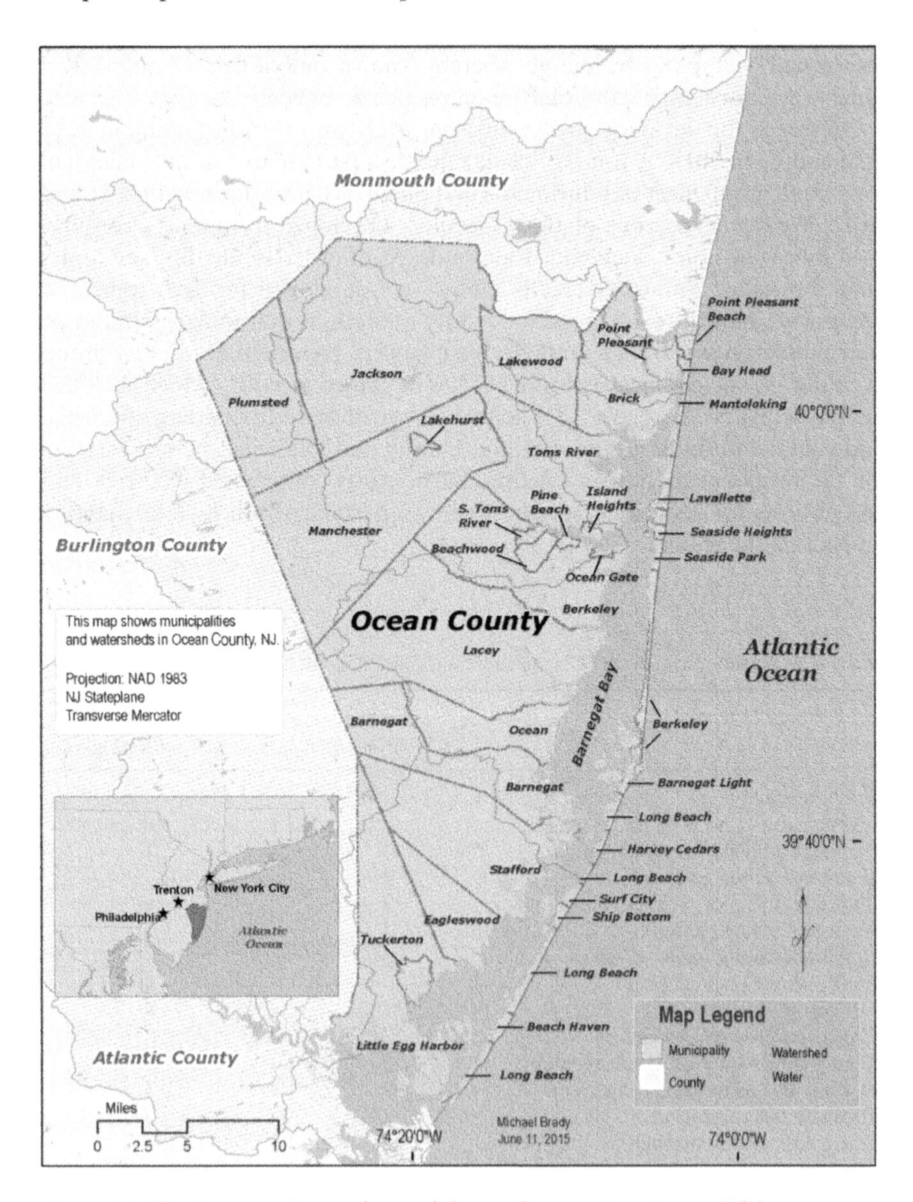

Figure 3.1 The Barnegat Bay region and Ocean County, New Jersey, USA
Source: www.worldscientific.com/doi/abs/10.1142/S2345737615500025

agriculture, and commercial and recreational fishing remain critical parts of the identity of the region and important components of its economy (Leichenko *et al.*, 2014). Like many urbanised coastal watersheds, the Barnegat Bay region is already experiencing significant environmental stresses as the result of population growth and residential and commercial development, including loss of wetlands and natural areas, and contamination and eutrophication of the bay from run-off (Leichenko & Solecki, 2013). Temperature and precipitation records indicate that the region's climate has been gradually warming and that heavy precipitation events have become more frequent. Continued climate change is expected to further exacerbate existing environmental risks in the region and to bring new stresses by altering long-term temperature and precipitation patterns, accelerating rates of sea level rise, and worsening storm surge flooding (NPCC, 2015; Ezer & Atkinson, 2014; Leichenko *et al.*, 2014).

Stakeholders in the study were concerned about numerous ongoing and future climatic stresses in the region. Discussions with stakeholders regarding options for addressing the region's climatic vulnerabilities largely focused on the region's natural and built assets and economic activities. Stakeholders proposed numerous adaptive strategies for increasing climate resilience in the region, but at the same time identified many limits and limitations associated with these efforts. Recommendations generally fell into three broad categories: engineering measures, policy and land use changes, and economic diversification (Table 3.2). As has been documented in other studies (e.g., Penning-Rowsell *et al.*, 2006; Amundsen *et al.*, 2010), stakeholder recommendations generally entailed strategies that responded to known climatic threats, reflected the region's recent experiences of extreme events, particularly Hurricanes Irene and Sandy, and utilised well-known policy and planning mechanisms.

Table 3.2 Options for enhancing resilience

Engineering measures
Replenishment/beach nourishment
Shore armouring
Retrofitting existing infrastructure
Restoration of marshes, dunes and living shorelines
Land-use planning and policy measures
Land-use planning and preservation of open space
Land-use control in flood-prone areas
Strengthening of building codes
Insurance reform
Relocation/retreat
Economic measures
Diversification of tourism
Diversification of fisheries
Expansion of transportation options

4.1 Engineering measures

Coastal engineering and infrastructure-related measures were the most prominent type of resiliency strategy, suggested by approximately half of the stakeholders. These measures also provoked much discussion of both limits and limitations to resilience-based approaches. Beach replenishment or nourishment was mentioned by the majority of stakeholders as an important option to protect coastal assets. While many conceded the short-term effectiveness of this practice in buffering storm impacts and maintaining a recreational space valued by tourists and residents, stakeholders raised concerns about limited effectiveness and mounting costs of these efforts over the longer term.

> ...If the feeling among policy makers in the state is, we want to stay involved in beach nourishment, there's got to be some recognition that it really isn't the long-term answer. It's going to get more expensive over time because the material is going to erode more quickly.
>
> (Real-estate stakeholder)

> ...To the extent that beach nourishment is probably a preferable approach to storm mitigation than building sea walls, right. It's one of those strategies that is available, that you know is obviously designed to help address the impacts of climate change ... Its future is probably fairly limited.
>
> (Conservation stakeholder)

Shore armouring, particularly construction of bulkheads, was mentioned by approximately one-third of the stakeholders as a strategy to increase resilience in the face of climate change. However, many stakeholders pointed out the limits and limitations of this approach, including the tendency for bulkheads and other hard structures to undermine adjacent areas through erosion. A few noted the inequity of the impacts of this approach: many homeowners and small businesses do not have access to bulkheading, but at the same time lose property and/or enjoyment value due to their neighbours' installations. Although there is clearly a private incentive to invest in protective infrastructure, the significant ecological cost of such infrastructure was seen as an important limitation of these efforts. In particular, hard structures pose a direct threat to wetlands and impinge upon their ability to accrete or migrate in response to sea level changes. Stakeholders also noted that shore armouring threatens the long-term viability of ocean and estuarine beaches.

> People are going to want to respond to [more storms] by armoring ... the coastline, which is very, very expensive but also has a huge impact on the geomorphology of the coast, as it accelerates the erosion of the barrier islands in particular.
>
> (Conservation stakeholder)

> We don't know what works, if anything. The [sea walls] don't work. Beach replacement is easy on the oceanfront, but not [marshes]. Bulkheads are nice, but there's nothing for the natural shoreline.
>
> (Natural lands manager)

Other strategies, suggested by stakeholders with expertise in infrastructure, economic development and land management, included improvement of infrastructure through retrofits and restoration efforts such as living shorelines and breakwaters. The perceived imperative to retrofit water and sewer systems to cope with sea level rise and storm surges (e.g., increasing the diameter of storm drains, installing flexible valves, etc.) also raised questions about the technical feasibility, major public cost and scope of such undertakings. Conservation and fisheries stakeholders suggested that interventions that protect and restore ecosystem services in and around Barnegat Bay would have the potential to increase the resilience of the natural systems that directly or indirectly support the region's economic activities. Protecting and restoring dunes and tidal wetlands were seen as possible ways to counteract ongoing loss of habitat as well as provide storm buffers. Stakeholders noted that restoration efforts are in an experimental phase and that demonstration projects might be able to show the benefits of working with nature and possibly inspire others to follow suit.

> Maybe some of these living shorelines and breakwaters are demonstrating that they can have a very positive effect. Maybe you can see some conversion in areas where you don't really need a full bulkhead ... Those solutions are much more effective at trapping sand transport than ... walls.
>
> (Economic development stakeholder)

4.2 Land-use planning and policy measures

Land-use planning and policy, such as open-space protection and other forms of development regulation, were a major area of discussion among stakeholders with expertise in planning, economic development, real estate and conservation. Specific suggestions included use of conservation easements and acquisitions of repetitive-loss properties in order to create and protect dune zones, catch basins and other forms of open space that provide ecosystem services. Implementation of statutes and regulations that proscribe reconstruction in locations vulnerable to continual erosion, flooding and storm damage was also discussed. Stakeholders further recommended that state-level authorities take proactive measures such as the establishment of buffers for salt marshes to allow for their retreat in the face of sea level rise, and that local authorities use zoning to establish setback and elevation provisions and other "platinum standards for building construction".

> Clearly there could be stronger statutes and regulations that don't allow for reconstruction in areas that we know are really going to be vulnerable.
> (Real-estate stakeholder)

> I guess my strategy is if you protect the natural functions of these ecosystems, or better yet, if you enhance and restore them, which is also possible, then the economic activities that come out of them or that are derived from them will both be sustained and enhanced, or made better.
> (Conservation stakeholder)

Insurance reform was suggested to tackle the problem of excessive development in high-risk areas. Several stakeholders mentioned that insurance companies have become eager to withdraw from coastal coverage and suggested that insurance companies may at some point refuse to cover properties in high-risk areas. One stakeholder suggested that insurance companies could send strong signals simply by increasing their premiums, but noted that, under current law, premiums can only be set on the basis of past damages and not on future projections. In addition to changing the laws that regulate insurance premiums, others pointed to the potential for additional reforms in the practices of the private insurance industry to provide incentives to upgrade the storm-worthiness of structures and disincentives to rebuild on repetitive-loss sites.

> There's a lot to do in the world of insurance to deal with this …What other incentives can we provide through insurance that have the dual benefit of increasing or encouraging higher standards and, on the other side of it, reducing potential damage and costs. So it's sort of a win-win for the homeowner and the insurer.
> (Real-estate stakeholder)

In discussing insurance reform, stakeholders were especially vocal about limits and limitations associated with both the US National Flood Insurance Program and the US Federal Emergency Management Agency. In each case, the fact that the bulk of the mounting costs of these two programmes is paid out of the public purse, while the individual communities and property owners reap the benefits, leads to perverse incentives for them to "remain in harm's way" rather than consider strategic retreat. One noted that "complacency among property owners" follows from reliance on insurance.

> The real estate market is so distorted by public subsidies and public interference in the market through flood insurance and through guarantees of evacuation, guarantees of rebuilding, infrastructure support, those types of things….
> (Conservation stakeholder)

Retreat and relocation, either by policy prescription or voluntary action, was also mentioned as a resiliency strategy by stakeholders in conservation, land management, and real estate, but was generally regarded as an option of last resort. Stakeholders mentioned the ecological costs of unplanned retreat from the shoreline (e.g., destruction of marshes). As discussed in the next section, there are also political, cultural and economic barriers to such measures.

> Ten years ago there was a study of adaptation on Long Beach Island. They looked at rollovers ... It's ugly – for example, if you move houses to the back of the beach, you destroy the back marshes. The [sea] will come in from both sides. And retreat? Where? [Sea level rise] is not a nice, linear progression. It's not incremental – it's not moving in a straight line.
>
> (Natural lands manager)

> I just think, and it's heresy to say this stuff in New Jersey ... in the end people are going to have to retreat off these places. They're going to have to move these communities. And it's exceedingly difficult, both economically and politically.
>
> (Conservation stakeholder)

4.3 Economic measures

Stakeholders in several sectors had suggestions for enhancing resilience of the region's economy, particularly via diversification. Within the fisheries sector, for example, stakeholders discussed the benefits of diversifying the range of species caught or expanding aquaculture, but also noted environmental downsides of these efforts. Tourism and economic development stakeholders suggested that tourism operators and investors would do well to explore alternative, non-beach centred ventures, such as environmental tourism, inland kayaking, and nature-orientated, guided trips. Moving to year-round attractions or expanding into the autumn and spring "shoulder seasons" would also increase resilience. The addition of more indoor venues with air-conditioning, such as aquariums, was suggested as a way to attract tourists on rainy days or during periods of extreme heat.

> One of the things that is becoming popular but is really untapped is [environmental tourism]. Rather than a traditional tourism business, if it was focused more on [environmental tourism], the fact that you don't have as many facilities that are endangered by it. There a number of benefits to it, too, because it's environmental education, it's much more low-impact than some other traditional tourist activities.
>
> (Economic development stakeholder)

Concerning the economy of the region overall, stakeholders in economic development, infrastructure and emergency management commented on the need for better transportation in the region, including additional roads and bridges to enhance access to Long Beach Island (the region's main barrier island), and options for public transportation. Improved transportation was seen as necessary both for emergency responsiveness and also for enhancing accessibility in and out of the region for tourists and residents.

> I think the first priority though would be to build extra evacuation routes … You look at our evacuation routes, and they haven't changed since our population [was] 200,000, and we're now approaching 600,000.
>
> (Emergency management stakeholder)

5 Barriers to resilience action

Consideration of resiliency options for the region also generated extensive discussion of barriers to implementation of these efforts. As described earlier, barriers can be distinguished from limits and limitations because barriers have the potential to be overcome, whereas limits and limitations reflect more fundamental deficiencies and suggest a need for alternative options. Barriers include factors that hinder implementation of specific resiliency measures as well as factors that constrain resiliency actions of all types. The barriers identified by stakeholders can be roughly sorted into several categories including: settlement and economic development barriers; political and cultural barriers; and policy and regulatory barriers.

5.1 Settlement and economic development barriers

One important type of perceived barrier to increasing resilience corresponds to the material "facts on the ground" presented by the natural and built

Table 3.3 Barriers to resilience

Settlement and economic development
Density of settlement
Slow progression of sea level rise
High value of coastal property
Political and cultural
Short-term thinking
Lack of political will and denial of climate change
Unequal sharing of burdens
Decreasing exposure of individuals to nature
Policy, regulatory and financial
Inflexible and inappropriate regulations
Expenses and delays entailed in obtaining permits
Perverse incentives and policies
Too lax, or laxly enforced, regulations
Policy inertia and resistance to change

environments, particularly the density, historical pattern and high economic value of settlement in coastal New Jersey. Stakeholders in the real-estate, conservation and land-management sectors noted that options to adapt or enhance resilience such as relocation, rolling easements and the transfer of development rights are all made very difficult when there is no available space to move to or "transfer in".

> And you know, if we had the luxury of having a less densely developed shoreline, I think [setbacks] would happen. If there was more open space, and everything wasn't built and crammed in, and houses weren't a million or two million dollars apiece, it would be nice ... [People in other states talk about] the concept of rolling easements, and the unfortunate thing is that you have to have a little space to roll ... there's nowhere to put people, there's nowhere to set people back.
>
> (Real-estate stakeholder)

Another frequently mentioned barrier to taking adaptive action was the slow progression of climate change, which makes it harder for people to recognise and acknowledge that change is happening. This is especially true of sea level rise. Stakeholders noted that people who are used to storms and flooding, on the barrier islands in particular, do not perceive sea level rise and expect to cope as they always have.

> The thing about the whole sea level rise-climate change thing, it's so gradual, that in a lot of the estuaries and back bays, it's not really perceptible by the public. Property owners don't quite get it. They may notice, boy, the flooding seems worse than it did when we moved here twenty years ago, but because it's such a gradual process in those areas, I don't think it gets people's attention in terms of how the public perceives the risk.
>
> (Real-estate stakeholder)

5.2 Political and cultural barriers

Political and cultural barriers to resiliency action were mentioned by approximately half of the stakeholders. Politicians' tendency to limit their concern to the span of election cycles and the human tendency to think in the short term were seen as major hindrances to action. A number of stakeholders referred to people's "short-term memories" between storm and flood events, and predicted it would take another major catastrophe to provoke serious action.

> And if we go years without a major storm, a major flooding disaster in the state, people have short memories. And mayors especially, [they're] in place for four or six years, or in some places two years, so that's all they're looking for. What are the two-year fixes?
>
> (*Emergency* management stakeholder)

> The decision-making processes are very short-term: what are the costs, what are the benefits over the very short term. And natural systems are very long term.
>
> (Conservation stakeholder)

Stakeholders noted that the lack of political will to undertake adaptive action is exacerbated by the politicisation and denial of climate change. While some stakeholders decried the idea that climate change has become a "political issue" rather than a practical problem, others seemed uneasy acknowledging it as a present reality in light of what they understood to be the scientific uncertainty and political controversy surrounding the subject.

> You have a strong, organized faction … that captures government and tells us to bury our heads in the sand … It's become politically unacceptable to come out and say we must *do* anything!
>
> (Conservation stakeholder)

> [We need to] find some place where government can … do that, without the interferences that you get with all the political garbage … If we need to correct what's going on with global warming, we need to stop everybody making it into a political issue, and start dealing with the issue.
>
> (Fisheries stakeholder)

Several stakeholders also reflected on how decreasing exposure to nature due to cultural shifts and changing lifestyles lessens people's awareness of and concern about climate change. The fact that seasonal residents of the area miss the chance to observe the sea's winter fury and the resultant coastal erosion was also seen as a factor that diminished concern.

> I see the advocates of the future not there … I mean, how did I get into all that I do? … Well, that isn't because I was sitting in a house when I was five years old, playing a video game. That was probably because my grandmother took me out … fishing. You get involved in fishing and you realize the interaction between species.
>
> (Fisheries stakeholder)

5.3 Policy, regulatory and financial barriers

Misguided policies and regulations were seen as another type of barrier to resilience. One-third of the stakeholders provided examples of how inappropriate and inflexible policies and regulations, and expenses and delays in obtaining permits, stymie adaptive responses. Natural lands managers noted, for example, that efforts to put in living-shoreline demonstration projects and other environmental adaptations had to endure multi-year processes for obtaining state permits. Land managers also explained that many types of interventions are currently prohibited in protected areas since they may interfere with natural processes.

> What we're trying is ... to protect our own shoreline in an environmentally friendly way, and this could be a good demonstration project ... The regulatory environment, because there are some modifications needed, anything you do, putting up bio-logs or dune grass, anything you do on the shoreline requires a permit from the [state]. If you have to wait three years for the permit and spend 25,000 dollars, it's [a problem].
>
> (Natural lands manager)

Barriers due to lack of local government funding were also noted by stakeholders in real-estate and emergency management. An "economic climate where everybody is downsizing" was seen as constraining the ability of local government to take action, such as hazard mitigation projects.

Other stakeholders asserted that the problem is that regulations are too lax, or laxly enforced. Real-estate stakeholders pointed to inadequate (re) building standards in construction codes and enforcement. Perverse incentives and misguided policies were also a source of concern. Fisheries stakeholders emphasised ways in which the Magnuson-Stevens Fishery and Conservation and Management Act and the regionally centralised, single-species-driven fisheries management system, along with specific catch requirements, were compromising the economic viability of commercial and recreational fisheries. Conservation and real-estate stakeholders pointed to the fact that provisions of the CAFRA (Coastal Areas Facilities Review Act), as amended in 1993, instituted a "right to rebuild" single family homes that has encouraged owners of repetitive-loss properties to keep on rebuilding after natural disasters. Stakeholders in land management further asserted that private property rights can sometimes become an obstacle to addressing climate risks, as evidenced in cases where beach access has been denied for replenishment projects or homeowners seek to block dune restoration projects that they claim would spoil their ocean views.

> Once bulk heading is in place, it is very difficult to dismantle it, because that would threaten a particular property.
>
> (Natural lands manager)

6 Bouncing forward: openings for transformation

Discussion of limits, limitations and barriers to resilience prompted some consideration of options and strategies to foster more sustainable and equitable adaptation approaches. Stakeholders generally emphasised policy reform as the logical, albeit politically challenging, response to policy-related barriers including perverse incentives, rigid and burdensome regulation, underregulation, uncoordinated planning processes, and haphazard development.

Reforms that were seen as having the greatest potential to promote transformative change in the region included those that would alter how development is regulated and how coastal engineering measures are funded.

> What we need is *better policies.* We can stop continuing to put ourselves at further risk. Most of the impacts are here: we cannot stop them, we can just soften the edge … We need to deploy a range of voluntary and regulatory approaches … There is a role for government at every level. It needs to say to people, 'Here are the things that are no longer acceptable.'
> (Conservation stakeholder)

Stakeholders in economic development, planning, real estate and conservation particularly emphasised the need for implementation of region-wide adaptation and resiliency planning. Examples given of maladaptive outcomes from lack of community and regional co-ordination included shore-armouring strategies that erode neighbouring properties and a lack of geomorphologically strategic sequencing of beach replenishment efforts. One stakeholder noted that because New Jersey is a "home-rule environment", towns do not have to comply with county plans, and while there are incentives to match up with the state plan, there is no consistent and comprehensive co-ordination across scales of governance. The need for regional planning and co-ordination that encompasses watersheds and other natural units that cross political boundaries was also noted. Stakeholders commented that a key element for success of these efforts is communication between levels of government, adjacent local government units, and among all units of government and stakeholder groups.

> We need regional planning, but lack authority at the county level. We have no jurisdiction over land use. The towns do that through their zoning … There is a [county] master plan, but it has no regulatory authority.
> (Natural lands manager)

Explicit recognition of equity issues associated with current policies was seen as another means of opening consideration of alternative adaptation pathways. One economic development stakeholder acknowledged the political necessity of continuing with beach-replenishment projects, but at the same time urged for policy reform to redistribute the burdens and benefits, such that those receiving the latter should pay for the former. The stakeholder advocated that communities receiving beach-replenishment projects should pay a much higher share rather than the current practice of relying on the federal government and ultimately the general public.

> And really, who pays? It should be changed to better reflect who benefits. And that would suggest a much higher burden on the communities … unfortunately the existing policy doesn't provide incentives for

> communities to be more creative in their planning, and how they view what's going to happen in ten years, in twenty years, in fifty years.
>
> (Real-estate stakeholder)

Stakeholders in a range of sectors brought up the importance of education and public awareness about climate change. Education was seen as necessary both in order to make the public understand the threats to their well-being posed by climate change, and to generate interest in taking steps that will enhance resilience. One stakeholder observed that increased resilience will come about if individuals change their behaviour in ways that also mitigate the causes of climate change and reduce their exposure to its inevitable impacts. Civil society and non-governmental organisations were seen as having an important role in efforts to achieve resilience through advocacy, research, demonstration projects and other forms of public outreach.

The notion that future extreme events can provide a political opportunity for more intervention and stronger regulation was yet another type of opening that emerged from stakeholder discussions. In reflecting on the region's past experience with extreme events, stakeholders in a number of sectors mentioned the possibility that large-scale events, such as Hurricanes Irene and Sandy, can act as catalysts for transformation. Stakeholders felt that despite political resistance, fiscal constraints and other impediments to change, policy solutions can be implemented when favourable circumstances align: a confluence of a driving event, leadership and ready-to-go solutions. They noted that ready-to-go practical and policy solutions should be in place so that when a major extreme event occurs it becomes possible to overcome resistance and implement these solutions before the political moment passes.

7 Conclusion

Enhancing resilience has become a dominant strategy for preparedness for extreme events and climate change. Although much research attention has been directed to identification of characteristics of resilient communities, less is known about effective strategies for building community resilience and barriers that may be encountered in these efforts. Drawing from literatures on adaptation barriers, adaptation limits and critical resilience theory, this study explored resilience-building options in the Barnegat Bay region of coastal New Jersey. The collaborative research process uncovered a range of strategies for enhancing resilience, provided valuable insights into context-specific barriers to implementation, and illuminated a range of limits and limitations that influence the effectiveness and viability of these efforts. While specification of barriers is widely accepted as a key element of adaptation planning, these results suggest that explicit consideration of both limits and limitations of various approaches should also be incorporated into resilience-building efforts.

Results of the study reinforced the importance of engaging a broad and diverse set of stakeholders in resilience-building efforts. In addition to their detailed knowledge about the region's history, cultural norms and policy context, stakeholders provided insights into what is feasible politically, economically and technically, what has been tried before, and obstacles that have been encountered the past. One looming challenge for stakeholder-based, resilience-building efforts is to facilitate greater recognition of changing environmental baselines. As illustrated in this study, stakeholder recommendations for responding to extreme events and climate change tend to be reactive rather than anticipatory, emphasising strategies that reflect known threats and recent experiences of extreme events and utilise known policy mechanisms (see Penning-Rowsell *et al.,* 2006; Amundsen *et al.,* 2010). However, as a result of non-stationarity of the climate, future barriers, limits and limitations of resiliency efforts may be difficult to anticipate based on past experience (Solecki & Rosenzweig, 2014).

Collaborative engagement with stakeholders also offered support for the idea that co-production approaches can help to facilitate consideration of alternative adaptation pathways (Eisenhauer, 2016). Coastal regions need to develop a wider array of adaptation responses (Brown *et al.,* 2014), and frank discussion of limits, limitations and barriers to resilience seemed to provide a vehicle for consideration of ways to do things differently. Alternatives that emerged from these discussions included suggestions for significant policy reform, region-wide planning, expanded education on climate risks, and awareness-raising about inequities associated with current strategies. The region's recent experience with an unprecedented extreme event was also seen as a potential opening for policy change. These initial findings suggest a need for further investigation of whether and how co-production methods may facilitate consideration of alternative strategies for responding to extreme events and climate change. One potential avenue for getting stakeholders to think differently about adaptation is through use of new strategies for collaborative research, such as those that draw upon stakeholders' emotional attachments to a region and their personal connections to each other (Ryan, 2015). As efforts gear up to promote community resilience to extreme events and climate change, engagement of a broad suite of stakeholders and application of diverse modes of co-production will be vital to the success of these efforts.

Acknowledgements

This chapter is based on an article that appeared in *Journal of Extreme Events* (2015, 2(1), https://doi.org/10.1142/S2345737615500025.) We thank the journal publisher World Scientific for granting permission to use the article. We also thank the journal editor and reviewers for comments on the original manuscript. The work is the result of research sponsored by the New Jersey Sea Grant Consortium (NJSGC) with funds from the National Oceanic and Atmospheric Administration (NOAA) Office of Sea Grant, US Department of Commerce, under NOAA grant number NA10OAR4170075, and the

NJSGC. The statements, findings, conclusions, and recommendations are those of the author(s) and do not necessarily reflect the views of the NJSGC or the US Department of Commerce. NJSG-15–891. Funding support for the research was also provided by the Barnegat Bay Partnership through a grant from the US Environmental Protection Agency's Climate Ready Estuaries Program. The information herein has not undergone USEPA review and does not necessarily reflect the official views of the USEPA. We thank Michael Brady for cartographic assistance. We also thank two anonymous reviewers for helpful suggestions for improvement.

Note

1 Institutional Review Board (IRB) approval from Rutgers University was obtained for all research on human subjects included in this study under IRB protocol number 11–604M.

Bibliography

Adger, W.N., Agrawala, S., Mirza, M.M.Q., Conde, C., O'Brien, K.L., Pulhin, J., Pulwarty, R., Smit, B., & Takahashi, K. (2007) Assessment of adaptation practices, options, constraints and capacity. In Parry, M.L., Canziani, O.F., Palutikof, J.P., Hanson, C.E., & van der Linden, P.J (Eds.). *Climate Change 2007: Impacts, Adaptation and Vulnerability Contribution of Working Group II to the Fourth Assessment Report of the Intergovernmental Panel on Climate Change.* Cambridge: Cambridge University Press, 717–743.

Adger, W.N., Dessai, S., Goulden, M., Hulme, M., Lorenzoni, I., Nelson, D., Naess, L., Wolf, J., & Wreford, A. (2009). Are there social limits to adaptation to climate change? *Climatic Change, 93*(3), 335–354.

Amundsen, H. (2012). Illusions of resilience? An analysis of community responses to change in northern Norway. *Ecology and Society, 17*(4), 46.

Amundsen, H., Berglund, F., & Westskog, H. (2010). Overcoming barriers to climate change adaptation – a question of multilevel governance? *Environment and Planning C: Government and Policy, 28*(2), 276–289.

Asian Cities Climate Change Resilience Network (2014). About the ACCRN Network. Retrieved 22 May 2015, from http://acccrn.net/about-acccrn

Bagstad, K., Stapleton, K., & D'Agostine, J.R. (2007). Taxes, subsidies, and insurance as drivers of United States coastal development. *Ecological Economics, 63*(2–3), 285–298.

Bahadur, A., & Tanner, T. (2014a). Policy climates and climate policies: analysing the politics of building urban climate change resilience. *Urban Climate, 7*(C),20–32.

Bahadur, A., & Tanner, T. (2014b). Transformational resilience thinking: putting people, power and politics at the heart of urban climate resilience. *Environment and Urbanization, 26*(1), 1–16.

Berkes, F., & Ross, H. (2013). Community resilience: toward an integrated approach. *Society & Natural Resources, 26*(1), 5–20.

Biesbroek, G., Klostermann, J., Termeer, C., & Kabat, P. (2013). On the nature of barriers to climate change adaptation. *Regional Environmental Change, 13*(5), 1119–1129.

Brady, M., Leichenko, R., Auermuller, L., Lathrop, R., & Trimble, J. (2015). Mapping municipal economic exposure to sea level rise in coastal New Jersey. Poster presentation, *New Jersey Sea Grant Consortium Quadrennial Site Review*, Monmouth University: Long Branch, NJ.

Brown, K. (2014). Global environmental change I: a social turn for resilience? *Progress in Human Geography, 38*(1), 107–117.

Brown, S., Nicholls, R., Hanson, S., Brundrit, G., Dearling, J., Dickson, M.E., ... Woodroffe, C.D. (2014). Shifting perspectives on coastal impacts and adaptation. *Nature Climate Change, 4*(Sept), 752–755; available at www.nature.com/articles/nclimate2344

Burton, C. (2015). A validation of metrics for community resilience to natural hazards and disasters using the recovery from Hurricane Katrina as a case study. *Annals of the Association of American Geographers, 105*(1), 67–86.

Cash, D.D. & Moser, S.C. (2000). Linking global and local scales: designing dynamic assessment and management processes. *Global Environmental Change, 10*(2), 109–120. doi.org/10.1016/S0959-3780(00)00017-0

Centers for Disease Control and Prevention (CDC). (2013). Deaths associated with Hurricane Sandy – October–November 2012. *Morbidity and Mortality Report Weekly, 62*(20), 393–397.

Corfee-Morlot, J., Cochran, I., Hallegatte, S., & Teasdale, J.P. (2011). Multilevel risk governance and urban adaptation policy. *Climatic Change, 104*, 169. doi.org/10.1007/s10584-010-9980-9

Cornell, S., Berkhout, F., Tuinstra, W., Tàbara, J., Jäger, J., Chabay, I., ... van Kerkhoff, L. (2013). Opening up knowledge systems for better responses to global environmental change. *Environmental Science and Policy, 28*, 60–70. doi.org/10.1016/j.envsci.2012.11.008

Cote, M., & Nightingale, A. (2012). Resilience thinking meets social theory: situating social change in socio-ecological systems (SES) research. *Progress in Human Geography, 36*(4), 475–489.

Cretney, R. (2014). Resilience for whom? Emerging critical geographies of socioecological resilience. *Geography Compass, 8*(9), 627–640.

Cretney, R., & Bond, S. (2014). 'Bouncing back' to capitalism? Grass-roots autonomous activism in shaping discourses of resilience and transformation following disaster. *Resilience, 2*(1), 18–31.

Cutter, S., Ash, K., & Emrich, C. (2014). The geographies of community disaster resilience. *Global Environmental Change, 29*, 65–77.

Cutter, S., Ahearn, J., Amadei, B., Crawford, P., Eide, E., Galloway, G., ... Zoback, M.L. (2013). Disaster resilience: a national imperative. *Environment: Science and Policy for Sustainable Development, 55*(2), 25–29. doi: 10.1080/00139157.2013.768076

Cutter, S., Barnes, L., Berry, M., Burton, C., Evans, E., Tate, E., & Webb, J. (2008). A place-based model for understanding community resilience to natural disasters. *Global Environmental Change, 18*, 598–606. doi.org/10.1016/j.gloenvcha.2008.07.013

Davoudi, S., Shaw, K., Haider, L., Quinlan, A., Peterson, G., Wilkinson, C., Fünfgeld, F., McEvoy, D., & Porter, L. (2012). Resilience: a bridging concept or a dead end? "Reframing" resilience: challenges for planning theory and practice interacting traps: resilience assessment of a pasture management system. In Davoudi, S., & Porter, L. (Eds.), Northern Afghanistan urban resilience: what does it mean in planning practice? Resilience as a useful concept for climate change adaptation?

The politics of resilience for planning: a cautionary note. *Planning Theory & Practice*, 13(2), 299–333. doi: 10.1080/14649357.2012.677124

Dilling, L., Lackstrom, K., Haywood, B., Dow, K., Lemos, M., Berggren, J., & Kalafati, S. (2015). What stakeholder needs tell us about enabling adaptive capacity: the intersection of context and information provision across regions in the United States. *Weather, Climate, and Society, 7*(1), 5–17. doi.org/10.1175/WCAS-D-14-00001.1

Douglas, E., Kirshen, P., Paolisso, M., Watson, C., Wiggin, J., Enrici, A., & Ruth, M. (2012). Coastal flooding, climate change and environmental justice: identifying obstacles and incentives for adaptation in two metropolitan Boston Massachusetts communities. *Mitigation and Adaptation Strategies For Global Change, 17*(5), 537–562; available at https://link.springer.com/article/10.1007/s11027-011-9340-8

Dow, K., Berkhout, F., Preston, B., Klein, R., Midgley, G., & Shaw, M. (2013). Limits to adaptation. *Nature Climate Change, 3*(4), 305–307.

Eakin, H., & Patt, A. (2011). Are adaptation studies effective, and what can enhance their practical impact? *WIREs Clim Change, 2*(2), 141–153.

Eisenack, K., & Stecker, R. (2012). A framework for analyzing climate change adaptations as actions. *Mitigation and Adaptation Strategies for Global Change, 17*(3), 243–260.

Eisenhauer, D.C. (2016). Pathways to climate change adaptation: making climate change action political. *Geography Compass, 10*(5), 207–221. doi.org/10.1111/gec3.12263.

Ezer, T., & Atkinson, L.P. (2014). Accelerated flooding along the U.S. East Coast: on the impact of sea-level rise, tides, storms, the Gulf Stream, and the North Atlantic Oscillations. *Earth's Future, 2*(8), 362–382. doi: 10.1002/2014EF000252.

Few, R., Brown, K., & Tompkins, E (2.007). Climate change and coastal management decisions: insights from Christchurch Bay, UK. *Coastal Management, 35*(2–3), 255–270.

Frazier, T., Thompson, C., Dezzani, R., & Butsick, D. (2013). Spatial and temporal quantification of resilience at the community scale. *Applied Geography, 42*, 95–107.

Gaillard, J. (2010). Vulnerability, capacity and resilience: perspectives for climate and development policy. *Journal of International Development, 22*(2), 218–232.

Grove, K. (2014). Agency, affect, and the immunological politics of disaster resilience. *Environment and Planning D: Society and Space, 32*(2), 240–256.

Harries, T., & Penning-Rowsell, E. (2011). Victim pressure, institutional inertia and climate change adaptation: the case of flood risk. *Global Environmental Change, 21*(1), 188–197.

Islam, M., Sallu, S., Hubacek, K., & Paavola, J. (2014). Limits and barriers to adaptation to climate variability and change in Bangladeshi coastal fishing communities. *Marine Policy, 43*, 208–216.

Jeffers, J. (2013). Integrating vulnerability analysis and risk assessment in flood loss mitigation: an evaluation of barriers and challenges based on evidence from Ireland. *Applied Geography, 37*, 44–51.

Jones, L., & Boyd, E. (2011). Exploring social barriers to adaptation: insights from Western Nepal. *Global Environmental Change, 21*(4), 1262–1274.

Kettle, N., & Dow, K. (2014). Cross-level differences and similarities in coastal climate change adaptation planning. *Environmental Science and Policy, 44*, 279–290.

Kirchhoff, C., Carmen Lemos, M., & Dessai, S. (2013). Actionable knowledge for environmental decision making: broadening the usability of climate science. *Annual Review of Environment and Resources, 38*(1), 393–414.

Kresge Foundation (2015). Bounce forward: urban resilience in the era of climate change. http://kresge.org/sites/default/files/Bounce-Forward-Urban-Resilience-in-Era-of-Climate-Change-2015.pdf

Lata, S., & Nunn, P. (2012). Misperceptions of climate-change risk as barriers to climate-change adaptation: a case study from the Rewa Delta, Fiji. *Climatic change, 110*(1–2), 169–186.

Laube, W., Schraven, B., & Awo, M. (2012). Smallholder adaptation to climate change: dynamics and limits in Northern Ghana. *Climatic Change, 111*, 753–774.

Leichenko, R. (2011). Climate change and urban resilience. *Current Opinion in Environmental Sustainability*, *3*(3), 164–168.

Leichenko, R., McDermott, M., Bezborodko, E., Brady, M., & Namendorf, E. (2014). Economic vulnerability to climate change in coastal New Jersey: a stakeholder-based assessment. *Journal of Extreme Events, 1*(1), 1450003.

Leichenko, R., & Solecki, W. (2013). Climate change in suburbs: an exploration of key impacts and vulnerabilities. *Urban Climate, 6*, 82–97.

Leichenko, R., & Thomas, A. (2012). Coastal cities and regions in a changing climate: economic impacts, risks and vulnerabilities. *Geography Compass, 6*(6), 327–339.

Lopez-Marrero, T., & Tschakert, P. (2011). From theory to practice: building more resilient communities in flood-prone areas. *Environment and Urbanization, 23*(1), 229–249.

MacKinnon, D., & Derickson, K. (2013). From resilience to resourcefulness: a critique of resilience policy and activism. *Progress in Human Geography*, *37*(2), 253–270.

Manyena, S., O'Brien, G., O'Keefe, P., & Rose, J. (2011). Disaster resilience: a bounce back or bounce forward ability? *Local Environment, 16*(5), 417–424. doi.org/10.1080/13549839.2011.583049.

McNeeley, S. (2012). Examining barriers and opportunities for sustainable adaptation to climate change in Interior Alaska. *Climatic Change, 111*(3–4), 835–857.

Measham, T., Preston, B., Smith, T., Brooke, C., Gorddard, R., Withycombe, G., … Morrison, C. (2011). Adapting to climate change through local municipal planning: barriers and challenges. *Mitigation and Adaptation Strategies for Global Change*, *16*(8), 889–909. Available at https://link.springer.com/article/10.1007/s11027-011-9301-2

Mitchell, T., & Harris, K. (2012). *Resilience: A Risk Management Approach. ODI Background Note, January 2012*. London: Overseas Development Institute.

Morrison, C., & Pickering, C. (2013). Perceptions of climate change impacts, adaptation and limits to adaption in the Australian Alps: the ski-tourism industry and key stakeholders. *Journal of Sustainable Tourism*, *21*(2), 173–191.

Moser, S., & Ekstrom, J. (2010). A framework to diagnose barriers to climate change adaptation. *Proceedings of the National Academy of Sciences, 107*(51), 22026–22031.

Moser, S., Williams, S.J., & Boesch, D. (2012). Wicked challenges at land's end: Managing coastal vulnerability under climate change. *Annual Review of Environment and Resources*, *37*(Nov), 51–78.

Mozumder, P., Flugman, E., & Randhir, T. (2011). Adaptation behavior in the face of global climate change: survey responses from experts and decision makers serving the Florida Keys. *Ocean & Coastal Management, 54*(1), 37–44.

Mukheibir, P., Kuruppu, N., Gero, A., & Herriman, J. (2013). *Cross-scale Barriers to Climate Change Adaptation in Local Government, Australia*. Gold Coast: National Climate Change Adaptation Research Facility.

National Research Council (NRC). (2012). *Disaster Resilience: A National Imperative*. Washington, DC: National Academies Press.

New Jersey Office Of Emergency Management. (2012). Resources for journalists regarding the one-year anniversary of Hurricane Irene. News release, 17 August 2012. Retrieved 26 May 2015, from www.state.nj.us/njoem/media/pr081712.html

New Jersey Climate Adaptation Alliance (NJCAA). (2014). *Resilience. Preparing New Jersey for Climate Change: Policy Considerations from the New Jersey Climate Adaptation Alliance*. Campo, M., Kaplan, M., & Herb, J. (Eds.). New Brunswick, NJ: Rutgers University.

New York City Panel on Climate Change (NPCC). (2015). Building the knowledge base for climate resilience: New York City Panel on Climate Change 2015 Report. *Annals of the New York Academy of Sciences*, *1336*, 1–150.

NPCC (2013). Climate Risk Information 2013: Climate Change Scenarios and Maps. New York: New York City Panel on Climate Change. Retrieved 15 May 2018, from www.nyc.gov/html/planyc2030/downloads/pdf/npcc_climate_risk_information_2013_report.pdf

Nielsen, J., & Reenberg, A. (2010). Cultural barriers to climate change adaptation: a case study from Northern Burkina Faso. *Global Environmental Change, 20*(1), 142–152.

O'Brien, K. (2009). Do values subjectively define the limits to adaptation? In Adger, W.N., Lorenzoni, I., & O'Brien, K. (Eds.). *Adapting to Climate Change: Thresholds, Values, Governance.* Cambridge, UK: Cambridge University Press, 164–180.

O'Brien, K., & Wolf, J. (2010). A values-based approach to vulnerability and adaptation to climate change. *WIREs Clim Change*, *1*(2), 232–242.

Office of the Governor. (2012). Christie administration releases total Hurricane Sandy damage assessment of $36.9 billion. Newsroom press releases, 28 November 2012. Retrieved 24 May 2015, from www.state.nj.us/governor/news/news/552012/approved/20121128e.html

Pasquini, L., Cowling, R., & Ziervogel, G. (2013). Facing the heat: barriers to mainstreaming climate change adaptation in local government in the Western Cape Province, South Africa. *Habitat International, 40*, 225–232.

Pelling, M. (2011) *Adaptation to Climate Change: From Resilience to Transformation*. New York: Routledge Press.

Penning-Rowsell, E., Johnson, C., & Tunstall, S. (2006). 'Signals' from pre-crisis discourse: lessons from UK flooding for global environmental policy change? *Global Environmental Change, 16*(4), 323–339.

Rockefeller Foundation. (2010). *Asian Cities Climate Change Resilience Network*. Rockefeller Foundation, USA.

Rosenzweig, C., Solecki, W., Blake, R., Bowman, M., Faris, C., Gornitz, V., … Zimmerman, R. (2011). Developing coastal adaptation to climate change in the New York City infrastructure-shed: process, approach, tools, and strategies. *Climatic Change, 106*(1), 93–127. Available at https://link.springer.com/article/10.1007/s10584-010-0002-8

Runhaar, H., Mees, H., Wardekker, A., Sluijs, J., & Driessen, P. (2012). Adaptation to climate change-related risks in Dutch urban areas: stimuli and barriers. *Regional Environmental Change, 12*(4), 777–790.

Ryan, K. (2015). Incorporating emotion into climate change research: a case study in Londonderry, Vermont. Theorizing and Communicating Climate Change II: Science, Knowledge and Perceptions Session. Association of American Geographers Annual Meeting. Chicago, Illinois.

Scott, M. (2013). Resilience: a conceptual lens for rural studies? *Geography Compass, 7*(9), 597–610.

Solecki, W., & Rosenzweig, C. (2014) Climate change, extreme events, and Hurricane Sandy: from non-stationary climate to non-stationary policy. *Journal of Extreme Events, 1*(1). doi: 10.1142/S2345737614500080.

Swalheim, S., & Dodman, D. (2008). Building resilience: how the urban poor can drive climate adaptation. *International Institute for Environment and Development, Sustainable Development Opinion* (Nov): 1–2.

Thomas, A., & Leichenko, R. (2011). Adaptation through insurance: lessons from the NFIP. *International Journal of Climate Change Strategies and Management, 3*(3), 250–263.

Tribbia, J., & Moser, S. (2008). More than information: what coastal managers need to plan for climate change. *Environmental Science & Policy, 11*(4), 315–328.

Tschakert, P., & Dietrich, K. (2010). Anticipatory learning for climate change adaptation and resilience. *Ecology and Society, 15*(2). Available at www.jstor.org/stable/26268129?seq=1#page_scan_tab_contents

US Climate Resilience Toolkit. (2015). Retrieved 22 May 2015, from NOAA at toolkit.climate.gov

Wagner, M., Chhetri, N., & Sturm, M. (2014). Adaptive capacity in light of Hurricane Sandy: the need for policy engagement. *Applied Geography, 50*, 15–23.

Weichselgartner, J., & Kasperson, R. (2010). Barriers in the science-policy-practice interface: toward a knowledge-action-system in global environmental change research. *Global Environmental Change, 20*, 266–277.

Weichselgartner, J., & Kelman, I. (2015). Geographies of resilience: challenges and opportunities of a descriptive concept. *Progress in Human Geography, 39*(3), 249–267. First published 7 February 2014.

Wise, R., Fazey, I., Smith, M., Park, S, Eakin, H., Garderen, E., & Campbell, B. (2014). Reconceptualising adaptation to climate change as part of pathways of change and response. *Global Environmental Change, 28*, 325–336. doi.org/10.1016/j.gloenvcha.2013.12.002

World Bank. (2014). Pilot program for climate resilience. Retrieved 25 May 2015, from www.climateinvestmentfunds.org/cif/node/4

Ziervogel, G., & Parnell, S. (2014). Tackling barriers to climate change adaptation in South African coastal cities. In Glavovic, B., & Smith, G.P. (Eds.). *Adapting to Climate Change: Lessons from Natural Hazards Planning*. Dordrecht, Netherlands: Springer Netherlands, 57–73.

4 Exploring transition pathways as an alternative approach for the integrated management of Irish estuaries and coasts

Christina Kelly, Geraint Ellis and Wesley Flannery

1 Introduction

Human reliance on estuarine and coastal resources for a multitude of goods and services is under threat. Due to the impact of anthropogenic activities such as agriculture, urban development, commercial fishing and shipping, these resources are being over-exploited, habitats are being destroyed and important species are being lost (EPA, 2016; Rilo *et al.*, 2013). These adverse effects are evident at global, national and local scales, which makes their management, involving multi-levels of governance, challenging. Existing governance frameworks, where management is fragmented among sectors and institutions, are unable to respond to complex and dynamic ecosystems or build capacity to cope with change (Bigagli, 2015; Boyes & Elliott, 2015; Virapongse *et al.*, 2016). In Ireland, the management of coastal ecosystems is largely carried out in a fragmented, sectoral and uncoordinated manner. There is no overarching national coastal or marine policy and there is no single body with overall responsibility for Irish estuaries and coasts. A more holistic approach is required to deal with the multi-uses, multi-users, multi-scales and multi-effects of estuarine and coastal ecosystems.

The aim of this chapter is to propose an approach for the integrated and sustainable management of estuaries and coasts in Ireland. To do this, it first explores the need for sustainable management, particularly in relation to fragmented governance. Sustainable approaches are analysed and include normative approaches based on the concept of integration. Following an evaluation of a normative approach based on an 'ideal' concept of integration within two Irish case studies, it was evident that long-term sustainable management may not be possible without wider systemic change. Sustainability challenges in the coastal zone are interrogated further from the perspective of the emerging field of transitions research and proposed as an alternative way of moving towards sustainable estuaries and coasts, as policy shifts to longer-term perspectives and approaches are deemed critical for environmental sustainability. The chapter concludes by recommending an 'Integration Transition Pathway' based on the key concept of transition management as a more viable means of facilitating integrated management within short-, medium- and long-term governance change.

2 The need for sustainable management of estuaries and coasts

Estuaries are used for a wide variety of purposes, which are sometimes in conflict. The conflicting functions and overuse of estuaries and catchments have left many in a degraded state (Millennium Ecosystem Assessment, 2005; European Environment Agency, 2015). Typically, however, the management of coastal ecosystems is undertaken in a largely uncoordinated manner, with governance competencies dispersed across a large number of bodies. In order to halt ongoing degradation and achieve sustainable estuaries, better management of these resources is required.

2.1 The effects of fragmented governance in estuarine and coastal management

Fragmented arrangements often emerge from deeply embedded conventions and are constrained by previous decisions that create path dependency. According to Kirk *et al.* (2007, p. 252) "theories of path dependency state that when choices must be made the option most likely to be chosen is that which most closely resembles existing practice or previous choices". Similarly, fragmentation may also materialise from institutional or policy 'layering', which is explained as "gradual institutional transformation through a process in which new elements are attached to existing institutions and so gradually change their status and structure" (van der Heijden, 2011, p. 9). It is imperative, therefore, to recognise that many of the current management or institutional arrangements for estuaries do not necessarily emerge from a grand design, but as a result of *ad hoc* incrementalism.

2.2 Shifting paradigm for integrated environmental management

The legislative and regulatory context relevant to estuary and coastal management is highly complex. As illustrated in Boyes & Elliott's (2014) 'horrendogram', international conventions and European directives initially compelled a mainly sectoral approach to management. However, this has changed with a more integrated approach being actively promoted through holistic policies such as the Water Framework (WFD), Marine Strategy Framework (MSFD) and Maritime Spatial Planning (MSP) directives.

Adding to this level of legislative complexity, current governance arrangements only exacerbate the fragmented management situation. Boyes & Elliott's (2015) 'organogram' highlights a multitude of different government departments and agencies with responsibility for estuary and coastal management. In Ireland this is similar, with 34 different agencies sharing responsibility for coastal and marine management. This *ad hoc* approach results in institutional structures with compartmentalised decision-making processes. This is often denoted by the term 'silo effect', signifying those conditions in which management is fragmented among sectors and institutions with little attention to conflicts or synergies among social, economic and environmental objectives (Holden, 2012; Mitchell, 2005). As a consequence, these arrangements lead to "narrow policy instruments that create incentives for policies and actions that undermine sustainability" (Olsson *et al.*, 2008, p. 9489). Therefore new and more effective governance systems are required.

It has been suggested that, in Ireland, the integrated management of estuarine and coastal ecosystems is occurring within a national policy vacuum (O'Hagan & Ballinger, 2010). This had negative implications for previous attempts at integrated management in Irish coasts. For example, previous management frameworks for Dublin Bay, Bantry Bay and Cork Harbour were never implemented due to a lack of policy commitment, funding and resourcing (Falaleeva *et al.*, 2009; O'Hagan & Ballinger, 2010; Stevens & Associates, 2006).

Despite the lack of an overarching coastal or marine policy, there is the potential to learn from international experiences (Kidd, 1995; Lonsdale *et al.*, 2015; Martin, 2014; Morris, 2008; Stojanovic & Barker, 2008; Elliott, 2013). These experiences can help to conceptualise a move towards the sustainable management of Irish estuaries and coasts.

2.3 The challenge of 'integration'

'Integration' is considered a more appropriate response than current piecemeal governance to address the declining health of marine ecosystems and support more sustainable human activities (Elliott, 2013; Holden, 2012; Kidd & Shaw, 2007; Smith *et al.*, 2011). However, 'integration' is regularly used as a normative concept in environmental management without critical reflection of what this specifically means or how it can be achieved. Consequently, many of the deep-seated challenges to the implementation of integrative approaches are overlooked. From a review of the definition and concept of integration and aspects deemed necessary to address the complexities associated with estuarine and coastal management, a number of key observations were noted:

- The term 'integration' is ambiguous and poorly defined in the literature relating to integrated environmental management.
- Sectoral policy integration is regarded as complex with few examples of best practice (Mullally & Dunphy, 2015; Stead & Meijers, 2009; Velázquez Gomar, 2016).
- With regards to 'integration tools' – i.e., approaches, strategies and instruments – Runhaar (2016) determined that their performance was usually modest and that expectations should be realistic.
- In relation to territorial (i.e., land-sea) integration, there are many issues in the marine environment that differ from land-use planning – i.e., more focus on natural resources and economic development, and less upon social concerns; in addition, poorly developed understanding of the

marine environment, including the scale and significance of land-sea interconnectivities and its 3D nature, is reported by Kidd & Shaw (2013).
- Organisational/institutional integration must have capacity to address the extensive environmental legislation emanating from the EU which has resulted in an excessively complex administrative and organisational structure for the management of estuaries and coasts.
- The integration of stakeholders in environmental governance tends to be tokenistic to date in coastal and marine management (Flannery *et al.*, 2016).
- The management of estuarine and coastal environments requires an extensive range of knowledge, skills and resources (Borja *et al.*, 2016).

These observations highlight a number of challenges for integration. However, there are clear benefits to using integrated environmental management as a potential approach towards the sustainable management of estuaries and coasts. As proposed by Cairns (1991, p. 5) integrated environmental management (IEM) is the "co-ordinated control, direction or influence of all human activities in a defined environmental system to achieve and balance the broadest possible range of short-and long-term objectives". Additional benefits of IEM include improved ecosystem status, potential for sustainable multiple use, cost savings and economic competitiveness, effective compliance with legislation and enhanced environmental awareness.

It is acknowledged, therefore, that IEM is an 'ideal type' rarely achieved in practice because of path dependency and institutional inertia.

3 Methodology

A critical analysis of the concept of integration and IEM was conducted. Following this appraisal, a suite of sustainability principles of IEM was devised specifically for an Irish context. These principles, along with a review of IEM processes and an analysis of international IEM practical examples, helped to guide the development of an integrated Environmental Management and Monitoring System (EMMS) for Irish estuaries and coasts. The proposed EMMS was based on a normative ideal of IEM and comprised a nine-stage step-by-step process for Irish estuaries and coasts which is described in section 4.

The normative model of EMMS was used as a heuristic device in two Irish case studies to evaluate existing management approaches within these sites, and to explore with stakeholders how the EMMS could be used to achieve integration on a national basis. The two Irish case studies selected were the Shannon Estuary and Dublin Bay, which were in the process of progressing their own innovative estuary and bay management initiatives. These case studies represented opportunities to learn about the challenges facing the implementation of the normative model of EMMS. Two one-day workshops were held in each of the case studies and were attended by over 50 stakeholders from a range of organisations including government departments, agencies, local authorities, regulators, research institutes, NGOs, local industries and interest groups. The purpose of presenting the EMMS at the workshops was to instigate a discussion with the stakeholders around existing management initiatives and to consider how the EMMS could be used to achieve sustainability.

Following the workshop events, transcripts were thematically analysed using the software, Atlas TI. The findings revealed that under current management conditions – i.e., without broader institutional change – a normative model of integration was unlikely to achieve the ambition of long-term sustainability in estuaries and coasts. An alternative wider-system approach based on a long-term perspective was deemed necessary. Transition theory presented a more realistic means of conceptualising this change. A proposed 'Integration Transition Pathway' which considers a multi-level and multi-stage transition perspective was explored. This pathway incorporates the relevant stages of the EMMS deemed most effective by the stakeholders and judged to most likely achieve radical, incremental change over short- and medium-term phases. This approach is considered to be a more effective way of thinking about the challenges of sustainable management and provides a means of developing a long-term response to deep-seated problems.

4 Development of a normative EMMS process for Ireland

A normative model of integration can be derived from:

- *Principles of IEM*: a set of 15 principles was devised specifically for Ireland and was based on international agreements and environmental protocols. The principles were evaluated against a set of criteria based on Elliott's (2013) ten interlinked 'tenets' to achieve successful and sustainable marine management. The principles were revised accordingly and are at the core of the proposed EMMS, which aims to ensure an inclusive, fair, integrated and sustainable management process.
- *Integrated processes*: a review of integrated processes related to estuarine and coastal management was conducted and included: environmental management; integrated environmental management (IEM); integrated water resource management (IWRM); integrated coastal zone management (ICZM); ecosystem-based management (EBM); integrated ecosystem assessment (IEA); and marine spatial planning (MSP). This review clarified that there are a number of common and pertinent stages relevant to progressing the proposed EMMS step-by-step process.
- *International practical examples*: the practical implementation of these integrated approaches has been analysed within international examples and included examples of ICZM and estuary management plans in England, coastal and marine planning in Scotland, estuary management in the United States and local community-based initiatives in Ireland. This analysis was to further examine and evaluate the effectiveness of such approaches. These examples help to understand real-world contextual conditions pertinent to the implementation of an integrated approach.

Based on the principles of IEM and the review of integrated processes and international examples, a nine-stage step-by-step normative EMMS has been devised and outlined as follows:

- Stage 1 – Determine vision and objectives
- Stage 2 – Understand ecosystem status
- Stage 3 – Determine ecosystem indicators and reference trends
- Stage 4 – Determine appropriate action
- Stage 5 – Framework planning phase
- Stage 6 – Formal adoption
- Stage 7 – Implementation
- Stage 8 – Monitoring and evaluation
- Stage 9 – Adaptation

To test the robustness and rigour of the proposed EMMS, an evaluation exercise was carried out using a set of criteria to check the intended stages and steps for integrated management. The principles of IEM were used once more in this instance, along with a set of essential and desirable criteria derived from the review of integrated processes and international practical examples. The evaluation process suggested that the EMMS was conceptually fit-for-purpose before being used as a heuristic device in the Shannon Estuary and Dublin Bay. While such normative ideas are present in much policy and institutional discourse, the realities of environmental governance mean that they face substantial challenges in implementation.

5 Identifying and unravelling governance challenges in Irish estuarine and coastal management

As discussed in section 2, there are extensive legislative and regulatory requirements for the management of estuaries and coasts. These represent a magnitude of external pressures from international and European organisations that play an important role in speeding up or slowing down a change in Irish governance. These 'top-down' developments on their own, or simultaneously with 'bottom-up' local initiatives, have the potential to exert pressures on the institutions leading to cracks, tensions and windows of opportunity for change (Kuzemko *et al.*, 2016). These dynamic developments can trigger the evolution of a more integrated system of estuary and coastal management and are explored in these two Irish case studies.

5.1 The Shannon Estuary case study

The Shannon Estuary is the largest estuary in Ireland, located on the west coast where the River Shannon meets the Atlantic Ocean. The estuary area is a multi-functional zone, with the waters and adjoining lands supporting a range of functions, uses and activities.

The Strategic Integrated Framework Plan (SIFP) for the Shannon Estuary is a recent example of a co-ordinated approach to integrated management in Ireland (Clare County Council *et al.*, 2013). The SIFP, developed by a multi-agency steering group comprising 18 organisations, sets out an integrated marine and land-use planning strategy which facilitates economic growth and promotes environmental management within the Shannon Estuary. The SIFP, however, continues to be implemented in accordance with a plethora of international and national legislation, while the management of the estuary is still the responsibility of numerous governmental departments and agencies. Therefore, despite the external legislative pressures, the current governance system is still resisting change and so local initiatives such as the SIFP strive to exert pressure using a 'bottom-up' approach.

5.2 Dublin Bay case study

Dublin Bay is located on the east coast of Ireland, immediately adjacent to the city of Dublin. Dublin Port, the largest in Ireland, is located within Dublin Bay and represents a gateway into the country for millions of visitors each year.

Currently, there is no person or body with sole responsibility for Dublin Bay and there is no integrated plan for the bay area. This is despite numerous plans having been prepared over the years, including city and county development plans, regional strategies and specifically commissioned studies for the bay and surrounding region (O'Hagan, 2010).

The Eastern and Midland Regional Assembly (EMRA) in association with the Celtic Seas Partnership[1] (CSP) recently developed a new strategic management framework (SMF) for Dublin Bay. This has been significant in enhancing stakeholder engagement in the management of the marine resource and has relevance in terms of assisting government in the implementation of the MSFD and MSP in Ireland (EMRA, 2016). The project provided a platform for bringing local, regional and national stakeholders together and found that there was a desire among stakeholders and politically elected representatives to become involved in marine management. The CSP, however, was an EC LIFE+ funded project which finished in December 2016. The follow-up implementation of the SMF for Dublin Bay will be challenging, particularly beyond the lifetime of the project.

The future of the projects in the Shannon Estuary and Dublin Bay is unclear. If there is no political, policy or resource commitment then these initiatives may stall. This type of 'business as usual' approach, or 'status quo' in terms of current management arrangements for estuaries and coasts, may result in a system 'lock-in' as a result of minimal institutional change. To avoid such system outcomes a radical and alternative approach to the current status quo in Irish governance is required. In the two case studies, institutional barriers are examined in more depth as are their potential impacts on future integrated management approaches using the proposed EMMS step-by-step process. These findings will shape how the development of an appropriate approach towards the integrated and sustainable management of Irish estuaries and coasts is ultimately achieved.

5.3 Exploring current and potential management initiatives

At workshops arranged for each case study, each of the stages of the proposed EMMS was analysed with key stakeholders. The positive aspects of the EMMS process were noted and will be incorporated into the final recommended approach. The opportunities and challenges identified as part of the analysis reflected the possibility of implementing the EMMS process within the current governance arrangements in place in the case study sites. A significant number of constraints were highlighted by the workshop participants which will make the implementation of the EMMS difficult within current institutional structures. These relate to: *weak vision, insufficient leadership, uncertainty of resources, lack of policy commitment, inconsistencies between and across government, uncoordinated collection of data and information, inadequate stakeholder engagement, insufficient implementation and communication difficulties.* For example, during the stakeholder workshops it was evident that there was an absence of a political or government vision for estuary and coastal management in Ireland. The ongoing delays in finalising and adopting the Foreshore Bill[2] and transposing the MSP directive by means of regulations rather than through primary legislation emphasise the weak vision, lack of leadership and policy commitment of the government in terms of leading the way on integrated marine and coastal management. Some workshop participants also felt that evolving political regimes and short-termism resulted in a lack of commitment and support from politicians for local initiatives. Instead, local initiatives had to compete with other priorities and objectives as well as opposing views and vested interests. Stakeholders highlighted further aggravating issues such as uncertainty over resources, inadequate engagement with citizens and lack of trust and communication. These barriers reflect a need for fundamental institutional transformation in the areas of legislation, administration, policies, behaviour, culture and practice. In addition to path dependency and institutional inertia, these combined challenges will limit the EMMS in addressing the long-term sustainable management of estuaries and coasts. It is evident, therefore, that a new paradigm is required to deal with these wider political, economic, ecological and environmental problems based on a longer- term perspective. Before identifying solutions, it is necessary to understand these problems in relation to their causes and this will require breaking down the structures in which they are embedded (Grin, 2010).

The findings from the workshops and stakeholder analysis revealed that there were a number of challenges facing estuary and coastal management which are unable to be addressed through more normative approaches to policy development and incremental institutional reform.

In spite of external or 'top-down' pressures from the EU in terms of directives and 'bottom-up' pressures from local initiatives, there has been resistance from the institutions to radically change or transform current governance arrangements in Ireland. This inability to move beyond the existing paradigm is regarded as a 'persistent problem'. Persistent problems have been described as the governance equivalent of Rittel & Webber's 'wicked problems' which are usually associated with policy-making (Rittel & Webber, 1973; Verbong & Loorbach, 2012). Persistent problems cannot be solved by current policies alone; instead they require the application of innovative societal governance approaches (Verbong & Loorbach, 2012).

These institutional challenges at a national level manifest at a local level. The main issues which emerged from the empirical research of the case studies are all typical of persistent problems, as they exhibit similar characteristics to those identified by Schuitmaker (2012), i.e., complex, uncertain, difficult to manage and difficult to grasp. These characteristics of persistent problems in estuarine and coastal management are summarised as follows:

- *Complex:* estuaries and coasts are socio-economically and ecologically important environments. They are sites of competing and conflicting uses and users. They are associated with fragmented governance arrangements. Previous integrated approaches in Ireland have had mixed success. There are inconsistencies with existing policies across and between levels of government.
- *Uncertainty:* as previous attempts at integration in Ireland have proven, no successful solution exists in terms of a co-ordinated approach to managing these environments. Lack of vision, political will, diversity, policy commitment, resourcing and insufficient leadership have contributed to uncertainty in the past. Additionally, despite the availability of more information and data, scientific and technical expertise, the reduction of uncertainty by more knowledge has not always been possible. The inconsistent provision of funding has limited innovative local initiatives.
- *Difficult to manage*: in Ireland, various departments have responsibilities for the management of estuaries and coasts. Many different actors are involved, representing different interests. From the case studies, it was evident that there is a view among stakeholders that some groups have been overlooked during previous attempts at integrated management.
- *Difficult to grasp:* due to the current fragmented governance and policy vacuum, there is no clear structure in terms of responsibility, leadership, guidance, direction, decision-making, information provision and implementation for integrated estuarine management initiatives. From the stakeholder workshops, it is clear that political anchorage is absent. There is a lack of a strong, long-term overriding vision which can guide future reform.

As explored through the Irish case studies, institutional arrangements have a tendency to be constrained by path dependency. This can result in unsustainable practices that become increasingly difficult to change (Schuitmaker, 2012). These institutional norms can have a constraining effect into the future (Greener, 2005) and can occur as a result of institutional inertia. To solve systemic problems, a process of transformation involving both a change in established patterns of action and in the structures in which they take place (system innovation) is needed. An alternative conceptual framework which builds on the positive aspects from the case studies and addresses the key symptoms of persistent problems is explored in the next section.

6 Transitions – an alternative perspective towards integrated sustainability

Transition studies have emerged in the last two decades as a new field of research which aims to foster societal change towards sustainable development and are broadly concerned with the long-term process of radical and structural change (Grin, 2010; Loorbach & Rotmans, 2010; Rotmans *et al.*, 2001; van der Brugge *et al.*, 2005). The motivation for sustainable transitions is a radical diagnosis of 'persistent problems'.

A transition perspective recognises the need for a system-wide approach in dealing with persistent problems, which may be required in addition to an IEM managerial approach for estuarine and coastal management in Ireland. Three key elements of transition theory are particularly relevant in this context: multi-level perspective (MLP), the multi-stage concept and transition management (van der Brugge *et al.*, 2005).

6.1 Multi-level perspective (MLP)

This concept distinguishes between functional scale levels at which transition processes take place. These levels are referred to as niches, regimes and landscapes which have been used to describe technological changes in socio-technical systems (van der Brugge *et al.*, 2005; Grin., 2010).

- *Landscape*: forms a broad external context determined by changes in the macro economy, politics, population dynamics, natural environment, culture and worldviews. This landscape level represents the wider context within which regimes operate and respond to relatively slow trends, long-term changes and large-scale developments or external shocks that play an important role in speeding up or slowing down a transition (Meadowcroft, 2005; van der Brugge *et al.*, 2005).
- *Regimes*: are dominant practices, rules and patterns of institutions, assembled and maintained to perform economic and social activities (Meadowcroft, 2005; van der Brugge *et al.*, 2005).
- *Niches*: are localised areas where innovation can first take place and consists of individual entrepreneurs and innovators willing to take a

chance, alternative technologies and local practices. They are the seeds of transition (Geels & Schot, 2010).

The MLP is a useful analytical concept that helps to explain both how a system of environmental governance in Ireland operates and how it changes. An illustration of the MLP being applied to an Irish estuarine and coastal context is outlined in Figure 4.1. At the regime level, the current legislative and regulatory context for estuarine and coastal management in Ireland is reflected in an adaption of Boyes & Elliott's (2014) 'horrendogram', while current governance arrangements and structures in Ireland are illustrated in an adaption of Boyes & Elliott's (2015) 'organogram'.

The SIFP and the Dublin Bay SMF are types of innovative niches where partnerships were formed and individual actors are involved at the local level within the two case studies. Developing a management framework was a main aim of both projects, which opened up dialogue between key stakeholders and helped them to rethink current and future goals for each of their ecosystems. The SIFP, at a more advanced stage, has already influenced policy change at the county development plan level. By continuing to exert pressure on the marine policy regime, reinforced by European directives from the landscape level, this transition could continue to influence a system shift towards governance across the land-sea interface. As Olsson *et al.*, (2008) discovered, it is important to identify any political or policy window, as they offer opportunities for large-scale change. For example, the forthcoming Foreshore Bill in Ireland may present an opportunity for policy innovation in relation to estuary and coastal management. Similarly, it is important to create the right links, at an opportune time, around the right issues.

6.2 Multi-stage concept

The multi-stage or multi-phase concept (Rotmans *et al.*, 2001) relates to "the systemic change of a dominant regime in four main phases" (Verbong & Loorbach, 2012, p. 10) which is usually illustrated by an S-shaped pattern. These phases include: 1) pre-development; 2) take-off; 3) acceleration; and 4) stabilisation. It is also accepted that the pathway towards stabilisation represents an 'ideal' transition. In this context, the system "adjusts itself successfully to the changing internal and external circumstances, while achieving a higher order of organisation and complexity" (Rotmans & Loorbach, 2010, p. 127).

In an Irish context, previous unimplemented plans and strategies which had been prepared for coastal areas around Ireland can be conceptualised as never having made it beyond the pre-development stage of the multi-stage concept. There are a number of examples where 'non-ideal' transitions can occur. These manifestations, according to Rotmans & Loorbach (2010) include the following:

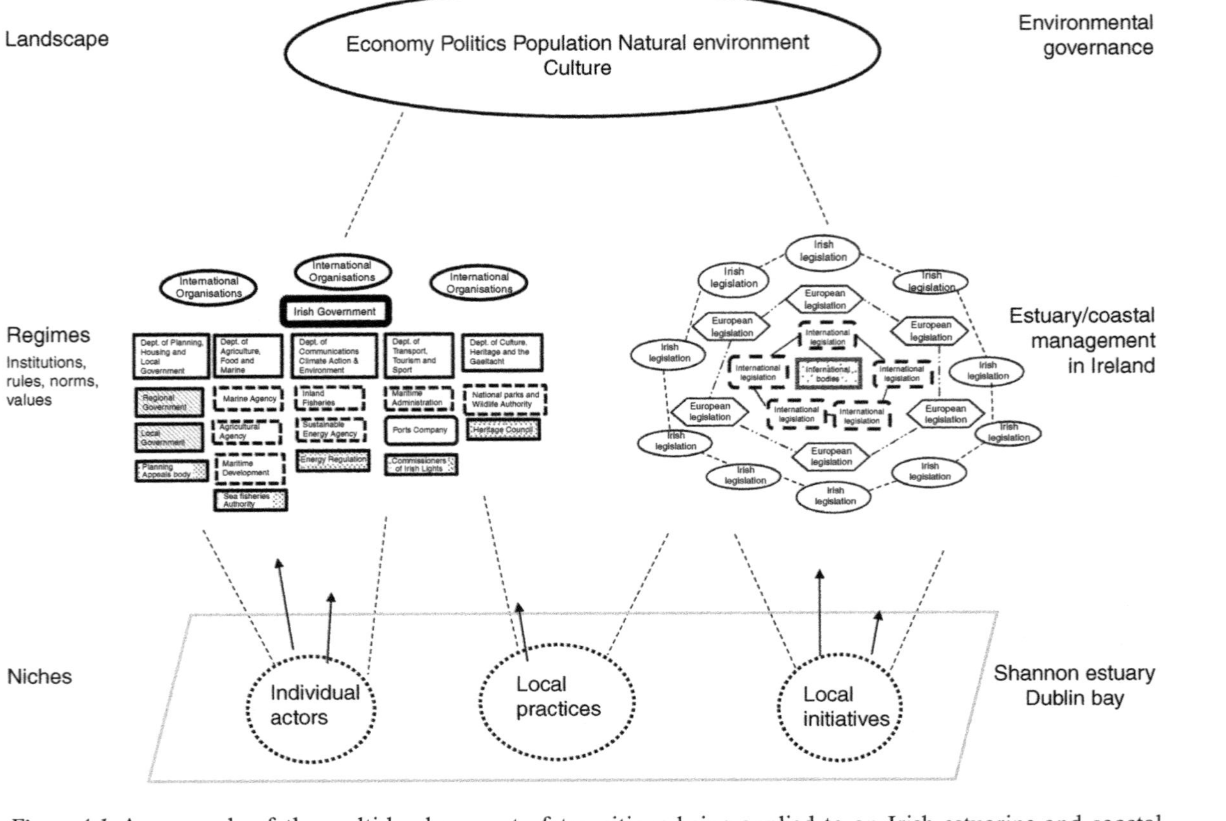

Figure 4.1 An example of the multi-level concept of transitions being applied to an Irish estuarine and coastal context

- Increasing path dependence: i.e., choices made in the past exclude different opportunities in the present, e.g., by ingrained behaviour or ideas that get stuck, so that a 'lock-in' situation emerges.
- Choices made early on can also reduce the necessary diversity, causing a niche 'burn-out' or system 'backlash'. This trajectory can be in response to either top-down or landscape pressures or from bottom-up or niche pressures.
- Due to insufficient knowledge, support or embedding in the system, this can lead to so much resistance that the system innovation path will be blocked.
- An overshoot collapse situation may occur. In this case a reverse transition takes place and the system collapses and eventually dies. This could occur due to external shocks and lack of system resilience.

These multi-stages and pathways suggest specific generic examples such as path dependency which could be recognised and understood, particularly in an Irish context.

6.3 Transition management

Transition management (TM) was developed as a new governance approach to facilitate and accelerate transitions – i.e., to influence the direction and speed of the change dynamics in contributing to end goals such as sustainability or integration. It ultimately co-ordinates multi-actor processes at different levels through joint problem-framing, a long-term vision, innovation networks and experimentation, with the aim of achieving long-term sustainability (van der Brugge *et al.*, 2005).

TM is a cyclical and holistic governance framework. The TM cycle comprises different activities coinciding with four stages (Loorbach & Rotmans, 2010; van der Brugge *et al.*, 2005; Wittmayer *et al.*, 2014) at which policy and negotiation processes take place. Stage 1 focuses on establishing a transition arena and organising a multi-actor network. Within this arena, problem structuring, envisioning and the development of new paradigms and pathways can be developed. Stage 2 involves setting an agenda, coalition-forming and policy deliberation. Stage 3 is the main experimental phase and involves mobilising actors and executing projects and experiments. The fourth and final stage is the reflexive phase which encompasses monitoring, evaluation and learning. These stages will be important considerations in the development of potential pathways towards integrated sustainability at the coast and, in particular, the setting up of a transition arena. A transition arena is an informal and creative space for a group of innovative actors, from various backgrounds, to meet and construct new networks, discourses, agendas and experiments, with the aim of influencing policy and practice. This will be a significant deliberation, particularly when building on the progress of existing partnerships in local management niches.

It is clear that large-scale institutional change over the long term will be necessary to achieve integrated sustainability. Many of the IEM approaches, central to current research in estuarine and marine management (Elliott, 2013; Lonsdale *et al.*, 2015; Sardà *et al.*, 2014) focus only on the intervention, leaving out the context and underlying systemic challenges and problems. The TM approach is more likely to deal with the complexities associated with estuary and coastal management which requires a more inter-disciplinary, flexible and adaptive approach.

7 Proposed 'Integration Transition Pathway'

Based on the key research outcomes, a potential 'Integration Transition Pathway' has been devised for estuarine and coastal management in Ireland. This approach builds on the findings from the analysis of persistent problems, complexities of estuary management, governance challenges and institutional barriers to change. The pathway represents an amalgamation of the MLP, multi-stage and the TM approach and incorporates the effective IEM instruments within the proposed EMMS. The actions set out within the pathway are ultimately aimed towards system improvement and system innovation over the short, medium and long term. A summary of the recommended stages and actions is included in Table 4.1.

The persistent problems associated with system-wide weaknesses in environmental governance and regime impediments have, for many years in Ireland, hindered institutional, legislative, economic and socio-ecological systems change needed for long-term estuary and coastal management. To address these barriers to change, a TM pathway, centred on a shared, long-term, radical vision of sustainability, is recommended. This pathway should build up innovation networks of frontrunners with an ambitious agenda of reform for estuary and coastal management, starting with tangible breakthrough projects, similar to the SIFP for the Shannon Estuary, that illuminate the longer-term sustainability vision (Loorbach & Rotmans, 2010).

The 'Integration Transition Pathway' therefore recognises individual estuary management initiatives as feeding into a wider transition arena and a network of diverse actors that share the debate, thinking and experimenting. This creates conditions for upscaling and breakthrough of innovations (Loorbach & Rotmans, 2010).

The pathway provides guidance for achieving a desirable outcome such as the sustainable management of estuaries and coasts and a desirable direction for the transition as a whole. The suggested stages and actions within the pathway provide a structure to build on the dynamics of the Irish initiatives. It is not a universal blueprint and will have to be adapted to the specific and spatial circumstances and variations emerging throughout the process. As noted by Hansen & Coenen (2015) sustainability transitions are geographical processes which should be analysed in a regional context having regard to place as a heterogeneous and situated construct (Murphy, 2015), as evidenced from the Irish case studies.

Table 4.1 Summary of the recommended stages and actions included in the 'Integration Transition Pathway' for estuary and coastal management in Ireland

Short term (0–3 years)		
Multi-stage	*Recommended steps*	*Key actions*
1. Establishing a transition arena	Forming a transition team Setting the scene System and actor analysis Selection of frontrunners Establish transition arena Problem-framing Framing challenges	An initiating agency forms transition team Specifies issues to be addressed Agree on short-term goals Draft a process plan and explore appropriate resources Delineate the system boundaries in space, time, themes Structure the system by defining the relevant stocks covering the social, environmental and economic domains Collect and analyse data Identify frontrunners or change agents through desktop research, interviews and participative workshops Selection of frontrunners for transition arena Arrange regular meetings to share information, explore challenges and frame problems
2. Envisioning	Sharing a vision Agenda building Transition pathways Transition agenda	Exchanging perspectives on the future Formulating guiding integration principles Creating visionary images Elaborating the vision Backcasting Identifying short-term actions Consolidating the transition agenda

(*continued*)

Table 4.1 (Cont.)

Medium term (3–10 years)		
Multi-stage	*Recommended steps*	*Key actions*
3. Steering process and experimentation	Engaging and anchoring Transition experiments Examples: Record and map uses; Scope pressures and impacts; Examine interconnections; Develop indicators and determine trends; Consider risk and decision analysis Radical short-medium term actions Developing an estuary/coastal planning support system Consultation Management strategies Assist plan-making process Pilot changes to influence legislative and governance change	Kickstart the process with an event Undertake activities to support initiatives Setting up initiatives: specific actions of the transition agenda are selected and progressed Establish working groups where required to work on initiatives Supporting initiatives: a co-ordinating team can assist or facilitate the work of the working groups and other sustainability initiatives by mapping relevant actors, creating a network, searching for funds and playing an active role in the project themselves.
Longer term (10–25yrs)		
4. Monitoring, evaluation and reflexivity	Implementation Learning Reflexive governance Visioning of system transition Large-scale institutional change	Examine achievements Monitor and evaluate transition processes and actors Reflect on lessons learned Radical suggestions for large-scale institutional change Major shifts in dominant structures, cultures and practices Achieve vision of environmental governance and sustainability

Source: adapted from Roorda *et al.* (2014)

8 Conclusions

Estuaries and coasts are highly dynamic and complex ecosystems under significant development pressures which require anticipatory and adaptive management systems that go beyond the existing ways in which policies and strategies are developed (Virapongse *et al.*, 2016). A new kind of governance is needed that develops large-scale institutional change based on a long-term perspective. A process of long-term transformation which incorporates innovative short-term actions based on good practice of integration and environmental sustainability provides a potential solution to this.

As the motivation for sustainable transitions is a radical diagnosis of persistent problems, transition theory presented a more realistic means of conceptualising the long-term change required. Transitions are presented as a new way of realistically progressing towards sustainability. The key concepts discussed in relation to the MLP and the multi-stage pathways have been most effective in contemplating the problems relating to environmental governance and understanding how potential transitions can materialise in estuary and coastal management from innovative niches and external developments. TM is a new governance approach which promotes collective choices and learning at different levels through lengthening time horizons, building networks among innovative stakeholders, focusing on sectoral and spatial dynamics and integrating economic, social and environmental considerations in process and policy design. By exploring different options, flexibility is maintained, which is necessary when dealing with great uncertainty and complexity in environmental governance.

An 'Integration Transition Pathway' is proposed based on an amalgamation of the MLP, multi-stage and TM concepts. The actions set out within the pathway are ultimately aimed towards system improvement and system innovation over the short, medium and long term. This approach is considered a more viable means of facilitating sustainable integrated management of estuaries and coasts within long-term governance change. These recommendations align with the overall aims of this edited volume – to address the limitations of transitions research by spatially focusing on two Irish coastal case studies.

It is acknowledged that further research is necessary to advance this conceptual framework in estuarine and coastal management. In particular, future investigation would contribute to a better understanding of barriers to change and therefore help in the design of different pathway options towards sustainability. A particular emphasis would be on establishing transition arenas, championing change agents and encouraging further experimentation, innovation, learning and change. The piloting of coastal case studies and evaluating a range of TM pathways with interested stakeholders would promote further discussion and debate among environmental and marine researchers, policy-makers and decision-makers, with the potential to address unveven development and power imbalances.

A focus on the learning-by-doing and doing-by-learning philosophy can help to guide flexible and adaptive governance strategies that reflect on and draw conclusions from activities and lessons learned, making a valuable and practical contribution to this field of research in the future.

Acknowledgements

This research was funded by the Environmental Protection Agency Ireland (EPA) and Department of Environment, Community and Local Government (DECLG) through the Science, Technology, Research & Innovation for the Environment (STRIVE) Research Programme 2007–2013.

Abbreviations

CSP	Celtic Seas Partnership
EBM	Ecosystem-based management
EMMS	Environmental Management and Monitoring System
EMRA	Eastern and Midland Regional Assembly
ICZM	Integrated coastal zone management
IEA	Integrated ecosystem assessment
IEM	Integrated environmental management
IWRM	Integrated water resource management
MLP	Multi-level perspective
MSFD	Marine Strategy Framework Directive
MSP	Maritime/Marine Spatial Planning
SIFP	Strategic Integrated Framework Plan for the Shannon Estuary
SMF	Strategic Management Framework for Dublin Bay
TM	Transition management
WFD	Water Framework Directive

Notes

1 Celtic Seas Partnership was a funded EC LIFE+ project (more information at: www.celticseaspartnership.eu/about-us/)
2 The Irish Government is currently proposing legislative changes under the General Scheme of Maritime Area and Foreshore (Amendment) Bill 2013. The changes have yet to be finalised.

Bibliography

Bigagli, E. (2015). The EU legal framework for the management of marine complex social–ecological systems. *Marine Policy, 54*, 44–51. doi:10.1016/j.marpol.2014.11.025

Borja, A., Elliott, M., Snelgrove, P.V.R., Austen, M.C., Berg, T., Cochrane, S., ... Wilson, C. (2016). Bridging the gap between policy and science in assessing the health status of marine ecosystems. *Frontiers in Marine Science, 3* (175). doi:10.3389/fmars.2016.00175

Boyes, S.J., & Elliott, M. (2014). Marine legislation – the ultimate 'horrendogram': international law, European directives and national implementation. *Marine Pollution Bulletin, 86*(1–2), 39–47. doi:10.1016/j.marpolbul.2014.06.055

Boyes, S.J., & Elliott, M. (2015). The excessive complexity of national marine governance systems – has this decreased in England since the introduction of the Marine and Coastal Access Act 2009? *Marine Policy, 51*(1), 57–65. doi:10.1016/j.marpol.2014.07.019

Cairns Jr., J., & Crawford, T.V. (Eds.). (1991). *Integrated Environmental Management*. Michigan: Lewis Publishers.

Clare County Council, Kerry County Council, Limerick City and County Council, and Shannon Development and Shannon Foynes Port Company. (2013). *Strategic Integrated Framework Plan for the Shannon Estuary*. (Volume 1: Written Statement).

Eastern and Midland Regional Assembly. (2016). *Celtic Seas Partnership. Strategic Management Framework for Dublin Bay.* Celtic Seas Partnership.

Elliott, M. (2013). The 10-tenets for integrated, successful and sustainable marine management. *Marine Pollution Bulletin, 74*(1), 1–5. doi:10.1016/j.marpolbul.2013.08.001

EPA. (2016). *Ireland's Environment 2016 – An Assessment*. Wexford: EPA.

European Environment Agency. (2015). SOER 2015 – The European Environment – State and Outlook 2015. *Synthesis Report*. European Environment Agency.

Falaleeva, M., Gray, S., Desmond, M., Gault, J., & Cummins, V. (2009). The role of ICZM in informing the development of climate adaptation policy in Ireland. Paper presented at the 2009 Amsterdam Conference on the Human Dimensions of Global Environmental Change, Earth System Governance: People, Place and the Planet, Amsterdam, 2–4 December 2009.

Flannery, W., Ellis, G., Ellis, G., Flannery, W., Nursey-Bray, M., van Tatenhove, J.P. M, …O'Hagan, A.M. (2016) Exploring the winners and losers of marine environmental governance/Marine spatial planning: Cui bono?/ "More than fishy business": epistemology, integration and conflict in marine spatial planning/ Marine spatial planning: power and scaping/Surely not all planning is evil?/ Marine spatial planning: a Canadian perspective/Maritime spatial planning – "ad utilitatem omnium"/Marine spatial planning: "it is better to be on the train than being hit by it"/Reflections from the perspective of recreational anglers and boats for hire/Maritime spatial planning and marine renewable energy. *Planning Theory & Practice, 17* (1), 121–151. doi: 10.1080/14649357.2015.1131482

Geels, F.W., & Schot, J. (2010). The dynamics of transitions: a socio-technical perspective. In Grin, J., Rotmans, J. & Schot, J. (Eds.), *Transitions to Sustainable Development. New directions in the study of long term transformative change.* New York: Routledge, pp. 11.

Greener, I. (2005). The potential of path dependence in political studies. *Politics, 25*(1), 62–72. doi:10.1111/j.1467-9256.2005.00230

Grin, J. (2010). Understanding transitions from a governance perspective. In: Grin, J., Rotmans, J., & Schot, J. (Eds.), *Transitions to Sustainable Development. New Directions in the Study of Long Term Transformative Change*. New York: Routledge, pp. 223.

Hansen, T., & Coenen, L. (2015). The geography of sustainability transitions: review, synthesis and reflections on an emergent research field. *Environmental Innovation and Societal Transitions, 17,* 92–109.

Holden, M. (2012). Is integrated planning any more than the sum of its parts?: Considerations for planning sustainable cities. *Journal of Planning Education and Research, 32*(3), 305–318. doi:10.1177/0739456X12449483

Kidd, S. (1995). Planning for estuary resources: the Mersey Estuary Management Plan. *Journal of Environmental Planning and Management, 38*(3), 435–442. doi:10.1080/09640569512968

Kidd, S., & Shaw, D. (2007). Integrated water resource management and institutional integration: realising the potential of spatial planning in England. *The Geographical Journal, 173*(4), (*Critical Perspectives on Integrated Water Management*), 312–329. doi:10.1111/j.1475-4959.2007.00260

Kidd, S., & Shaw, D. (2013). Reconceptualising territoriality and spatial planning: insights from the sea. *Planning Theory and Practice, 14*(2), 180–197. doi:10.1080/14649357.2013.784348

Kirk, E., Reeves, A., & Blackstock, K. (2007). Path dependency and the implementation of environmental regulation. *Environment and Planning C Government and Policy, 25* (2), 250–268. doi:10.1068/c0512j

Kuzemko, C., Lockwood, M., Mitchell, C., & Hoggett, R. (2016). Governing for sustainable energy system change: politics, contexts and contingency. *Energy Research and Social Science, 12*, 96–105. doi:10.1016/j.erss.2015.12.022

Lonsdale, J., Weston, K., Barnard, S., Boyes, S.J., & Elliott, M. (2015). Integrating management tools and concepts to develop an estuarine planning support system: a case study of the Humber Estuary, Eastern England. *Marine Pollution Bulletin, 100* (1), 393–405. doi:10.1016/j.marpolbul.2015.08.017

Loorbach, D., & Rotmans, J. (2010). The practice of transition management: examples and lessons from four distinct cases. *Futures, 42*(3), 237–246. doi:10.1016/j.futures.2009.11.009

Martin, L. (2014). The use of ecosystem services information by the US national estuary programs. *Ecosystem Services, 9*(0), 139–154. doi:10.1016/j.ecoser.2014.05.004

Meadowcroft, J. (2005). Environmental political economy, technological transitions and the state. New *Political Economy, 10*(4), 479–498. doi:10.1080/13563460500344419

Millennium Ecosystem Assessment. (2005). *Ecosystems and Human Well-being: Wetlands and Water Synthesis*. Washington, DC: World Resources Institute.

Mitchell, B. (2005). Integrated water resource management, institutional arrangements, and land-use planning. *Environment and Planning A, 37*(8), 1335–1352. doi:10.1068/a37224

Morris, R.K.A. (2008). English Nature's Estuaries Initiative: a review of its contribution to ICZM. *Ocean and Coastal Management, 51*(1), 25–42. doi:10.1016/j.ocecoaman.2007.05.001

Mullally, G., & Dunphy, N. (2015). *State of Play Review of Environmental Policy Integration Literature.* Report for the National Economic and Social Council (NESC). (Research Series No. Paper No.7). NESC. Retrieved 8 July 2015, from: www.nesc.ie/en/publications/publications/nesc-research-series/envpolicy-integration/

Murphy, J.T. (2015). Human geography and socio-technical transition studies: promising intersections. *Environmental Innovation and Societal Transitions, 17,* 73–91. doi:10.1016/j.eist.2015.03.002

O'Hagan, A.M. (2010). New structures for Integrated Coastal Management and Monitoring around Dublin Bay. Retrieved 15 January 2016, from: http://ec.europa.eu/ourcoast/index.cfm?menuID=6andarticleID=291

O'Hagan, A.M., & Ballinger, R.C. (2010). Implementing Integrated Coastal Zone Management in a national policy vacuum: local case studies from Ireland. *Ocean and Coastal Management, 53*(12), 750–759. doi:10.1016/j.ocecoaman.2010.10.014

Olsson, P., Folke, C., & Hughes, T.P. (2008). Navigating the transition to ecosystem-based management of the Great Barrier Reef, Australia. *Proceedings of the National Academy of Sciences of the United States of America, 105* (28), 9489–9494. doi:10.1073/pnas.0706905105

Rilo, A., Freire, P., Guerreiro, M., Fortunato, A.B., & Taborda, R. (2013). Estuarine margins vulnerability to floods for different sea level rise and human occupation scenarios. *Journal of Coastal Research, 65*, 820–825. doi:10.2112/SI65-139

Rittel, H.W.J., & Webber, M.M. (1973). Dilemmas in a general theory of planning. *Policy Sciences, 4*(2), 155–169. doi:10.1007/BF01405730

Roorda, C., Wittmayer, J., Henneman, P., van Steenbergen, F., Frantzeskaki, N., & Loorbach, D. (2014). *Transition Management in the Urban Context. Guidance Manual.* Rotterdam: DRIFT. Retrieved 20 October 2016, from https://drift.eur.nl/publications/transition-management-urban-context-guidance-manual/

Rotmans, J., & Loorbach, D. (2010). Towards a better understanding of transitions and their governance. A systemic and reflexive approach. In: Grin, J., Rotmans, J., & Schot, J. (Eds.), *Transitions to Sustainable Development. New Directions in the Study of Long Term Transformative Change.* London: Routledge, pp. 105.

Rotmans, J., Kemp, R., & van Asselt, M. (2001). More evolution than revolution: transition management in public policy. *Foresight, 3*(1), 15–31. doi:10.1108/14636680110803003

Runhaar, H. (2016). Tools for integrating environmental objectives into policy and practice: what works where? *Environmental Impact Assessment Review, 59*, 1–9. doi:10.1016/j.eiar.2016.03.003

Sardà, R., O'Higgins, T., Cormier, R., Diedrich, A., & Tintoré, J. (2014). A proposed ecosystem-based management system for marine waters: linking the theory of environmental policy to the practice of environmental management. *Ecology and Society, 19*(4). doi:/10.5751/ES-07055-190451

Schuitmaker, T.J. (2012). Identifying and unravelling persistent problems. *Technological Forecasting and Social Change, 79*(6), 1021–1031. doi:10.1016/j.techfore.2011.11.008

Smith, H., Maes, F., Stojanovic, T., & Ballinger, R. (2011). The integration of land and marine spatial planning. *Journal of Coastal Conservation, 15*(2), 291–303. doi:10.1007/s11852-010-0098-z

Stead, D., & Meijers, E. (2009). Spatial planning and policy integration: concepts, facilitators and inhibitors. *Planning Theory and Practice, 10*(3), 317–332.

Stevens & Associates. (2006). Scottish Natural Heritage. A review of relevant experience in sustainable tourism in the coastal and marine environment. Case Studies – Level 1. The Bantry Bay Charter. Retrieved 7 July 2016, from www.snh.org.uk/pdfs/strategy/natparks/CMNP/advice/Final%20report%20-%20CMNP%20sustainable%20tourism%20-%20February%202006.pdf

Stojanovic, T., & Barker, N. (2008). Improving governance through local coastal partnerships in the UK. *Geographical Journal, 174*(4), 344–360. doi:10.1111/j.1475-4959.2008.00303.

Van der Brugge, R., Rotmans, J., & Loorbach, D. (2005). The transition in Dutch water management. *Regional Environmental Change, 5*(4), 164–176. doi:10.1007/s10113-004-0086-7

Van der Heijden, J. (2011). Institutional layering: a review of the use of the concept: institutional layering. *Politics, 31*(1), 9–18. doi:10.1111/j.1467-9256.2010.01397.

Velázquez Gomar, J. (2016). Environmental policy integration among multilateral environmental agreements: the case of biodiversity. *International Environmental Agreements: Politics, Law and Economics, 16*(4), 525–541. doi: 10.1007/s10784-014-9263-4

Verbong, G., & Loorbach, D. (2012). *Governing the Energy Transition: Reality, Illusion or Necessity?* New York: Routledge.

Virapongse, A., Brooks, S., Metcalf, E.C., Zedalis, M., Gosz, J., Kliskey, A., & Alessa, L. (2016). A social-ecological systems approach for environmental management. *Journal of Environmental Management, 178*, 83–91. doi: 10.1016/j.jenvman.2016.02.028

Wittmayer, J., Roorda, C., & van Steenbergen, F. (2014). Governing urban sustainability transitions – inspiring examples. DRIFT. Retrieved 7 June 2016, from https://drift.eur.nl/publications/governing-urban-sustainability-transitions-inspiring-examples/

5 Catalysing transitions through the informational governance of climate change advocacy

Using Web 2.0 in the surfing world

Gregory Borne and Dom Clarke

1 Introduction

This chapter explores transitions to sustainability through the medium of a small non-profit organisation within the surfing world. Surfing as an activity is not only a uniquely coastal activity with multiple impacts, but is also described here as being embedded in a broader systemic network providing unique insights into coastal transitions. This approach is related to a theoretical framework that combines elements of Ulrich Beck's risk society with the notion of socio-technical transitions, which is then applied to the concept of informational governance. The increasing use of Web 2.0 as a medium that changes the dynamics of informational flows provides opportunities for non-governmental organisations to play a more central role in the distribution of information about a range of contemporary environmental concerns, including climate change. This altered process of information distribution is explored through a visible process of transition within the multimillion-dollar surfing industry and directed at an organisation that has been successful in facilitating that transition. Surfing, surfers and surfing organisations provide a unique opportunity to explore information governance, climate change advocacy and the role these play in sustainability transitions for the following reasons. Surfing provides unique opportunities to explore coastal and broader systemic transitions. It is an activity that combines direct and repeated engagement with nature. This is an activity that is impacted by the consequences of climate change through sea level rise and increased extreme weather events. Surfing presents a tension between cultural and subcultural norms, as it simultaneously grapples with notions of freedom, spirituality and radicalised behaviours, whilst being intimately embedded in broader global and corporate realities. Surfing also provides access to a large, globally disparate and heterogeneous, yet technologically connected community, as well as engaging communities and sectors beyond those of surfing. Transitions are a "fundamental change in structure, culture and practice. Structure can include physical infrastructure, economic infrastructure or institution. Culture refers to the collective set of values, norms and perspectives and paradigm in terms of defining problems and solutions. And practice refers to the ensemble of production routines, behaviour, ways of handling and implementing at the individual level" (Borne, 2015, p. 25).

It has been argued that the proliferation of sustainable development discourse on the global stage represents nothing less than an epochal transition from a modern to a reflexive modern society (Borne, 2010). This argument is based on the establishment of a symbiotic relationship between the central tenets of a risk society and that of sustainable development, and has been explored within the context of both local sustainability implementation and global governance processes. Beck's recent work has explored transitions within the context of *The Metamorphosis of the World* (Beck, 2016). For Beck, both climate change and the rise of digital technologies and communications form a central part of the architecture of metamorphosis in a risk society. Beck (2016) considers that if climate change is a fundamental threat to all of humanity, then it might bring about a transformative turn in contemporary life and potentially metamorphosise the world for the better; he calls this "emancipatory catastrophism". Taking these insights forward, Borne has synthesised these theoretical discussions with broader debates on transitions that exist within the socio-technical and transition management literature (Borne, 2018; Grin *et al.*, 2010).

To date, Grin *et al.* (2010) have offered the most comprehensive, extensive and critical analysis of the application of socio-technical and transitional management literature in the context of sustainable development (Geels & Schot, 2010; Rotmans & Loorbach, 2010; Grin, 2010). Grin *et al.* argue that whilst the topic of transitions has been studied and debated in multiple disciplines over a prolonged period of time, none of them "is applicable to the complex nature and multiple dimensions of societal transformations implicated in sustainable development" (2010, p. xvii). Four common themes emerge from the work of Grin *et al.* First, co-evolution where economic, cultural, technological, ecological and institutional subsystems reinforce and co-determine a transition. Secondly, the multi-level perspective forms a central analytical frame. The multi-level perspective explores the interaction of the niche, regime and landscape levels of analysis. What is important to recognise is that the three levels identified "…are levels of analysis rather than levels located at specific geographic, administrative of other types of real world locals" (Grin *et al.*, 2010, p. 324). Transitions result from the alignment of trajectories within and between these levels. Thirdly, the idea of multi-phase, which explores transitions over time with a transition occurring in four alternate phases. These are the pre-development, take-off, acceleration and stabilisation phases. And finally there is the co-design and learning, where knowledge is developed in an interactive and complex way with multiple societal stakeholders. However, it is also recognised that this is not learning "…in the sense of the transfer of knowledge, but more to learning in terms of developing in interaction with other viewpoints of reality" (Grin *et al.*, 2010, p. 5).

With the above in mind Grin *et al.* accept that theoretical bolstering is necessary; for example: "We acknowledge that the role of consumers and grassroots initiatives is underrated and under-conceptualised, therefore we welcome new perspectives which theorise changes in demand-side practices as motors for transition" (Grin *et al.*, 2010, p. 331). To this end Grin (2010) suggests that introducing governance into this framework has a number of advantages. First, integrating governance into the transitions literature provides an historical contextualisation of transitions to sustainability. Secondly, the governance perspective emphasises not only the nature of transition as profound changes in both established patterns of action and the structure in which they are embedded, but also why these changes and practices in a particular domain are influenced by long-term societal trends outside that particular domain. Thirdly, a governance perspective addresses the politics intrinsic to transitions innovation. Taking these observations forward Ehnert *et al.* (2017) introduce the term 'multi-level governance' in an attempt to draw to the fore the notion of agency within the transitions literature. They suggest that "... while multi-level governance has been a long-standing theme in political science research, it has remained under-explored in the study of sustainability transitions" (2017, p. 2).

With the above in mind, this chapter adds an important dynamic to these observations by introducing informational governance and then exploring how this has facilitated a transition catalysed by a particular organisation. Moreover, there is little engagement with the especially disruptive effects of new communication technologies. The following section introduces the notion of informational governance into this framework, providing a further and more sophisticated understanding of transitions within a reflexive modern world.

2 Informational governance

With its foundations in the work of Castells (2009), informational governance is a recent incarnation that focuses directly on the role that information and information technology play in steering society. Mol (2008) explains that perspectives on information governance build upon and share key concerns of the wider literature on new modes of governance, notably the growing involvement of non-governmental actors, the diversification of modes of governing from a monopoly of law-based regulatory intervention, and the complex interdependencies of different levels of governance, ranging from local to global. Within the broader framework, the concept of informational governance emphasises the key importance of information in fundamentally restructuring the processes, institutions and practices of environmental governance.

Information is increasingly regarded as a resource with transformative potential. Consequently, contemporary environmental struggles and movements have orientated towards shaping state governance, and increasingly have defined their agenda through the rights of access to information, production of information, verification of information, and control over information (Mol, 2009; Toonen & Mol, 2013). For Soma *et al.,*:"Informational governance reflects thus on how increased information sharing and interaction transform societies towards more cooperation, empowerment, self-organisation, private governing and interconnectedness" (2016, p. 96).

The rapid development of many digital technologies is constantly reshaping the field of environmental governance; one such innovation is the production and dissemination of information via Web 2.0 (O'Reilly, 2007). New Web 2.0 technologies hosting social media have introduced new interactive communication channels for environmental organisations (Goldkind, 2015), at the heart of which contemporary online communicative tools such as hashtags (denoted with a #) provide organisations with a means of branding campaigns, and spreading information about core values and goals, to like-minded individuals (Sexton *et al.*, 2015). The dramatic rise of social media has led many to argue that it is an ideal tool to improve communication, enhance user engagement and amplify individual participation on such issues as climate change (Hestres, 2014, 2015; Schafer, 2012).

As Hestres (2014) points out, despite the impact of internet-mediated advocacy organisations, little is understood about how they work, even though non-governmental organisations (NGOs) are investing considerable effort into these new modes of communication, searching for new opportunities within nation-state governance structures. The internet is playing an increasingly important role in the efforts of environmental organisations and broader NGOs to limit climate change risks (Schafer, 2012; Shapiro & Park, 2018). Social and environmental movements addressing climate change are transnational in scope and have a dependence on digital technologies for the dissemination and diffusion of information as well as for communication and co-ordination. Internet-mediated social networks are a key ingredient of the environmental movement in the global network society (Ackland & O'Neil, 2011; Castells, 2009). Using digital technologies, NGOs have been engaging in practices that pose a challenge to conventional governance structures.

2.1 Web 2.0 technologies

Since the early 1990s websites containing information about environmental issues accessed via internet connections have dominated the communication of sustainability. However, during the early 2000s communication pathways developed considerably. The interactive nature of the internet changed to incorporate a many-to- many communication model, whereby much of the information became user-generated, with the differentiation between 'senders' and 'receivers' of information becoming ever more blurred. This platform of internet is commonly referred to as Web 2.0 or social media (Schafer, 2012). At present social media is generally taken to consist of internet-based applications in which individuals create, share or exchange information and ideas in virtual communities (Stevens *et al.*, 2016).

Since the inception of Web 2.0, social media platforms have significantly increased the ability of NGOs to communicate with a variety of actors, including clients, volunteers, the media, the general public and policy-makers (Goldkind, 2015). The impacts of social media on the communication of sustainability are significant; as a tool it can improve the communication of scientific issues to a broader public. It highlights previously hidden information using creative new means such as audio-visual and interactive features that enhance user engagement and understanding. Critically, there can be a fragmentation of public debate, as information in Web 2.0 is easily manipulated to fit the agendas of certain parties. Moreover, the lay audience may find it difficult to distinguish reliable information from that emanating from less reputable sources (Schafer, 2012).

Costs and benefits of these types of platforms aside, the increased capacity of digital technologies has enabled NGOs to emerge as legitimate non-governmental communicators of environmental concerns and policies, as well as enabling an adaptive style of governance outside established state channels (Nulman & Özkula, 2015).

The borderless and truly global nature of the internet thus provides an arena for organisations and individuals endeavouring to activate change. NGOs interested in creating and sustaining social change have discovered social networking sites as an efficient means of identifying groups and allied organisations and empathetic individuals with common concerns and agendas. It has been argued that these connections form communities that are the nucleus of social change (Goldkind, 2015). The primary communication tool on many social media platforms is the series of brief messages (i.e., tweets, statuses, comments, updates and photos) that are sent to an organisation's followers on such platforms as Twitter, Facebook and Instagram. As a result, social media research has generally focused on the nature of the messages sent. However, examining the relationship between an organisation's social media and messages in conjunction with the reaction of users in the form of likes, comments, sharing and 'favouriting', allows the measurement of public reaction to an organisation's message and facilitates a shift from the perceptual to the behavioural realm. This type of analysis provides a quantitative and comparable gauge to measure the relative effectiveness of an organisation's messaging strategy (Saxton *et al.*, 2015).

The social media communicatory toolbox is ever-evolving, with new tools for communicating and interacting with the public emerging all the time (likes, comments etc.), and some tools such as the hashtag have become increasingly popular over a range of social media platforms including Facebook, Instagram and YouTube. Hashtags indicate topics or themes, and are particularly engaging due to their participatory nature. Hashtags are not predetermined by a set of users, giving rise to a hashtagging system that is a decentralised, user-generated, organising and classification system (Saxton *et al.*, 2015). Also, using hashtags to classify messages allows organisations to link messages to existing communities.

It is this ability of the hashtag to form *ad hoc* publics (Bruns & Burgess, 2011) or communities that is particularly meaningful. These hashtag communities can be fleeting in reaction to catastrophes or enduring communities of practice (Wenger, 2009) that evolve and spread information on given topics (Saxton *et al.*, 2015). Furthermore, hashtags allow these communities to become more involved in the decision-making process, and there is potential for NGOs to take advantage of this heightened participation in environmental action by reaching far beyond the limited scope of an organisation's physical membership (Roose, 2012). Empirically, therefore, the involvement of an organisation that utilises social media to form communities of practice may suggest that there are opportunities for inclusion of the online community in the decision-making process. The following section introduces the context of, and the specific organisation investigated for, this research.

3 Surfing, and Sustainable Surf's Deep Blue Life

Surfing is an aspirational activity that engages millions of people across the world, whether they surf or not. Surfers themselves have been described as oceanic stewards who are well aware of "sustainability issues such as water quality and pollution, impacts of tourism and local conflicts over coastal development" (Gibson & Warren, 2017, p. 87). An increasing body of work explores surfing's ability to catalyse change not only within the surfing community but across broader society (Borne, 2018; Borne & Ponting, 2017; Lazarow & Olive, 2017). Furthermore the conceptualisation of surfing as intimately embedded in broader networks, "whereby individuals are concerned with using natural areas in ways that sustain them for current and future generations of human beings and other forms of life" (Martin & O'Brien, 2017, p. 25). This also encompasses stakeholder involvement and the development of management plans to forward sustainability scenarios that include historical and cultural dimensions of the system.

As such, surfing's intimate connection with nature in the coastal zone provides a fertile space to explore transitions to sustainability (Borne, 2015, 2018; Borne & Ponting, 2015, 2017). Furthermore, the constructed space of surfing encompasses a broad range of local and global dynamics that are both heterogeneous and constantly changing. With regard to surfing organisations catalysing change, the surfing community has a rich tradition of attempting to draw attention to key environmental issues. Over the past three decades, environmental and social activism and engagement have increased on the part of the surfing community (Hales *et al.*, 2017; Ware *et al.*, 2017). The following section highlights a single organisation whose central mission is to transition the surfing world to a more sustainable operating model.

3.1 Sustainable Surf

Sustainable Surf was launched in 2011. The organisation's primary concern related to market transformation within the surfing industry, and facilitation of efforts of the community to change towards a more sustainable model of doing business (Whilden & Stewart, 2015). Sustainable Surf's initiatives include setting environmental standards for eco-friendly surfboards (ECOBOARD Project), managing waste collection for incorporation into recycled surfboard blanks (Waste to Waves), and working to reduce the environmental impact of surf contests (Deep Blue Surfing Events). In a short time, Sustainable Surf has embedded itself in popular surf culture, collaborating with contest organisers such as at the Volcom Pipe Pro, accrediting major surfboard brands with ECOBOARD project labelling, and recruiting high-profile professional surfers as ambassadors for the organisation. During 2014, Sustainable Surf were awarded the 'Agents of Change' award at the SURFER poll awards, arguably elevating the status, visibility and credibility of the organisation within the surfing community.

Therefore, Sustainable Surf provides an exemplary case study for the examination of the dynamics of a climate change advocacy campaign mediated on social media within the broader context of sustainability (Borne, 2018). Specifically, Sustainable Surf's Deep Blue Life (DBL) campaign, an initiative aimed at lessening the environmental impact of the surfing community while increasing well-being within the surfing population, is explored. The DBL campaign fits into the overall scheme of Sustainable Surf's mission to "... identify key barriers in culture and business that are preventing transformation to sustainability" and "... develop programmes that educate and enable actions that directly break through these barriers" (Whilden & Stewart, 2015, p. 131). Sustainable Surf have used a strategy of engaging both individuals and businesses to solve environmental problems, focusing much of their efforts on climate change and its related impacts.

The Deep Blue Life initiative is the fourth programme by Sustainable Surf, designed to reduce an individual's environmental impact and address the effects of climate change (see Borne, 2018). The programme follows six broad categories of environmental impact, reducing the impact of ecologically damaging behaviours in ways that participants can implement in their own lives. The intention of the programme is to be simple, flexible and to appeal to a large demographic, which is generally speaking the surfing community

The DBL initiative was launched in 2014 and was assigned the corresponding social media hashtag (#deepbluelife), which first appeared on the photo- and video-sharing social networking site Instagram, on 28 January 2014. Data from Sustainable Surf's Instagram account were harvested for a 66-week period from the launch date up to 5 July 2015. An initial search for the term "#deepbluelife" was performed using the Instagram search facility. Data from all the posts containing "#deepbluelife" were then collected, including Sustainable Surf's posts, Advocates of Sustainable Surf's posts and unknown user's posts. The research follows the approach given by Saxton *et al.* (2014) in making sense of the data. An inductive categorisation analysis was performed on the data collected, enabling the research to identify strategies unique to social media.

3.2 The Deep Blue Life initiative Instagram activity

During the period 28 January 2014 through to 5 July 2014, Sustainable Surf uploaded a total of 64 posts to the organisation's Instagram account. These posts consisted of 63 images and one video post. On average, Sustainable Surf posted approximately one item per week. An important measure of public engagement on social media is the number of followers an organisation attracts (Saxton & Waters, 2014). The number of followers is an indication of the size of audience Sustainable Surf is capable of reaching on a given social media platform, as well as a reflection of users that have made a conscious decision to connect with Sustainable Surf (Saxton *et al.*, 2015). During the sampling period, Sustainable Surf had an average of 4,705 followers, ranging between 4,672 and 4,738. At the same time, the organisation itself was following an average of 322 other Instagram users, with a range of 290 to 354.

A useful proxy of engagement and popularity on Instagram is the amount of likes and comments each post can gain (Sheldon & Bryant, 2016). The mean amount of likes that each Sustainable Surf post gained was 108 (STD 39), with a range of 47–230, and the total number of likes during the sample period was 6,937. The mean number of comments per Sustainable Surf post was 3.6 (STD 3.3) with a range of 0–14, with a total amount of comments for the sample period being 233.

The amount of likes remained above 60 per post with the exception of six posts, whereas more interactive engagement in regard to comments was considerably less with only 14 comments made two occasions. The mean number of likes that each advocate of the DBL gained was 88.61 (STD 137.85), with a larger range of 5–486, and the total number of likes during the sample period was 1,861. The mean number of comments per Sustainable Surf post was 4.3 (STD 4.5) Three 'spikes' within the data set were evident; these posts were created by Rob Machado (@robmachadofoundation), Alison Teal (@ alisonadventures) and Greg Long (@gerglong) respectively, all three of whom are members of the ambassador team at Sustainable Surf. Interestingly, there is a positive correlation between the spike in the likes and comments, suggesting that the ambassadors of Sustainable Surf foster greater engagement with their posts than a 'normal' advocate.

3.3 Analysis of the hashtag #deepbluelife

Sustainable Surf used hashtags sparingly during the sample period. Of the total 64 posts, 105 hashtags were included. The mean number of hashtags per post was 1.68 (STD 0.79) with a range of 1–4 hashtags. All the posts associated with the DBL initiative included #deepbluelife, and a further 31 hashtags were used in conjunction with this. The hashtag #deepbluelife was used 64 times, and was the most-used hashtag during the sample period. Other hashtags that were used more than a single time were #ecoboardproject (six times), #ECOBOARD (five times), #wastetowaves (four times) and #deepbluesurfingevent (twice).

During the sample period, Sustainable Surf were not the only users of the hashtag #deepbluelife. Sustainable Surf was the largest user, while other users have been categorised into DBL advocates and unknowing users. Sustainable Surf had the highest percentage of #deepbluelife use with 67.4%, Deep Blue Life advocates had 22.1 % of the use, and unknowing users of the hashtag made up 10.5% of the total use.

3.4 Effectiveness

To gauge the effectiveness of the dissemination of climate change and sustainable lifestyle advocacy, this research used the hashtag #deepbluelife to chart the course of Sustainable Surf's online social media activity. In line with Saxton & Waters (2014), Sustainable Surf's campaign points to the opportunities within smaller-scale social media-based campaigns that can be more meaningful to those receiving them, and have the ability to engage with key influencers rather than an inattentive larger audience.

Also, by exploring the activity of DBL advocates of Sustainable Surf on Instagram, it was evident that the supporting community played an important role as 'co-communicators' of the hashtag #deepbluelife, endorsing the DBL campaign, and having approximately one-fifth of the total hashtag usage. Although engagement in terms of likes and comments with followers was fewer than that of Sustainable Surf, its significance is no less important. Within the DBL advocate community there lie potentially notable influencers to the wider online surfing community. Champions of the DBL such as Greg Long (@gerglong), two-time World Surf League Big Wave World Champion (Surfing Magazine, 2016), and Rob Machado (@robmachadofoundation) (Machado & Toth, 2015) represent a group of professional surfers, which within the surfing community at large have significant power of influence and ability to inspire (Thompson, 2015). Many surfing celebrities also advocate environmental and sustainability-related causes which intensify this connection between chosen celebrity and the message projected (Machado & Toth, 2015).

3.5 Classification of Sustainable Surf Instagram posts

Table 5.1 presents the six categories of the BDL initiative goals, along with two additional categories added by means of inductive analytical processing. To develop the supplementary classifications, the images and text from

Table 5.1 Categorisation of Sustainable Surf Instagram posts

Category/initiative goals	*Frequency of use*	*% use*	*Total % use*
Health management	15	23.44	64.06
Waste management	12	18.75	
Renewable energy	2	3.13	
Cleaner transport	3	4.69	
Community outreach	5	7.81	
Climate impact	4	6.25	
Values	12	18.75	35.94
Arts	11	17.19	

Sustainable Surf's Instagram data set that did not obviously align with the existing initiative goals were arranged into themes. The themes that emerged were 'values' (general organisational values of Sustainable Surf) and 'arts' (defined as a visual painting, sculpture or event).

As an outcome of reviewing the Instagram data, Sustainable Surf's Instagram posts can be categorised into their respective initiative goal categories. Categorisation casts a light as to where the emphasis of an NGO's efforts are being made in terms of prerequisite goals. It emerged that there was general alignment of posts to the DBL initiative goals, with 64.04% of posts matching a DBL category goal. However, it was necessary to introduce two additional categories to fill in the gap of 36.94%; these categories were 'values' and 'arts'. The categorisation of posts is critical as it allows for further understanding of these posts within climate change advocacy (Saxton & Waters, 2014).

The inclusion of 'values' and 'arts' emerged as an important discovery, as these categories represent the organisation's unique and individual approach to communicating issues of climate change advocacy, and differentiates the BDL campaign from other NGO initiatives, along with developing the understanding of Sustainable Surf's social media strategy (Saxton *et al.*, 2015). Differentiation of the BDL campaign is critical within an already crowded surfing-based NGO market (Lazarow *et al.*, 2008). It can be argued that the unique characteristics of a campaign such as DBL can give legitimacy and leverage within decision-making circles (Thorpe & Rinehart, 2010).

Furthermore, the distinctive identity of the DBL campaign provides an alternative vocabulary of climate change advocacy, broadcasting alternative or ignored issues and providing a new position of congruence for the community involved, in turn strengthening the movement (Buttel, 2000). While it could be argued that the promotion of these other categories distracts from the fundamental goals of the DBL initiative, Auer *et al.* (2014) suggest that as long as an organisation's communication flow is actively engaging the community with the issue, and voicing their core values, it will provide a favourable condition to enable societal action on climate change (Moser & Dilling,

2011). While the degree of engagement will vary from organisation to organisation and campaign to campaign, Borne (2018) suggests that a balance should be maintained between campaign goals and the ability to be flexible and adaptable to the interactive social media environment. For Sustainable Surf the inclusion of 'arts' and 'values' to the DBL initiative adds extra merit as an emergent strategy.

3.6 Associating media

Hashtags, when used in conjunction with an event, catastrophe or scandal, can potentially heighten user engagement with the movement. An example of where Sustainable Surf used this to some effect featured an individual post that attained high levels of engagement with the most likes (230), and a high number of comments (11) in comparison with other posts during the data collection period. The post was based around news that the 11-time World Surf League world champion Kelly Slater was competing at the prestigious Bells Beach event in Australia, riding an ECOBOARD construction surfboard. Boykoff & Goodman (2009) explain that celebrities are becoming increasingly important non-state actors in the dissemination of climate change issues across the media. The coupling of a celebrity with a news event (as seen in the example above) arguably has the potential to enhance further engagement with an online campaign.

Stevens *et al.* (2016) note that events that are amplified and particularly problematic within the media attract further public attention, and consequently these occurrences can become news items; it is in these 'moments' when NGOs are provided a platform, whereby association with a news item can amplify the organisation's message (Anderson, 1997). Although within the DBL campaign there is a noticeable effort to align posts with news items, the regional nature of these news events critically affects the potential global reach of the post. During this research data collection period, there were many climate change-related news items that elicited global media hype.

4 Conclusion

The case study of Sustainable Surf and the DBL social media campaign offers relevant insights into the current conversation about new and shifting modes of environmental governance. Sustainable Surf's transition from physical activities, such as 'ECOBOARD Project' and 'Waste to Waves' programmes, to online activities is helping to shape a new form of informational governance, one that situates greater authority on information, and places the organisation at the fulcrum of information distribution about sustainability issues, and reduces the reliance on information coming from existing decision-making institutions (Nulman & Özkula, 2015). Sustainable Surf's acceptance and operational use of digital technologies to campaign for environmental advocacy on such mediums as Instagram, as discussed above, has already contributed to progressions within informational governance. This is in line with multiple and burgeoning environmental groups as well as those more specifically located in the surf zone (Surfers Against Sewage, Surfrider Foundation, etc.).

The findings and discussion presented in this chapter suggest that there are both opportunities and limitations to employing social media to facilitate a climate change advocacy campaign. The opportunities that emerged for Sustainable Surf when using a social media platform during the DBL campaign included reaching an active community that shares the values of the organisation. Furthermore, the characteristics of the DBL allowed Sustainable Surf to effectively reach 'key influencers' within the online surfing community.

Limitations of using a social media platform were that, although there were benefits to a smaller scale campaign reaching a wider audience, the opportunity to reach a more global audience was suppressed. It also emerged that there was a disparity between the DBL goals and the posts; consequently, this research suggested this may distract an audience from the core goals, although it was found that the inclusion of the induced categories added value to the campaign. Additionally, it was found that news events offer the potential to disseminate climate change advocacy messages when aligned with a relevant post. Although there were attempts made by Sustainable Surf to use this strategy, the local nature of the news events used lacked global 'hype', so efforts to reach larger global audiences were bypassed.

It is suggested that an NGO's practice of social media within existing campaigning structures is "changing the mode of change" (Beck, 2016). Building on these observations, this chapter is able to critically complement the transitions literature by not only highlighting the impact of grass roots organisations but also addressing how digitalisation and hyper-engagement with organisational messaging can impact on the uptake of specific transition dynamics. Returning to the four dominant transition themes identified earlier in this chapter, the authors make the following observations. First, *co-evolution* is not only evident but exaggerated through the media reinforcement and direct engagement of a broader user community. With that in mind the boundaries of this co-evolution are significantly dissolved, complicating the identification of the transition mechanisms and actors. This observation impacts directly on the analytical frame of the *multi-level perspective.* The ability of social media to directly engage diverse yet connected user groups enables an overlaying of the informational governance dynamic on to this model, creating opportunities to develop a more robust and realistic model of transition processes. Climate change advocacy through Web 2.0 channels also calls for a rewriting of the multi-phase process of transition over time. Results presented in this chapter resonate with broader longitudinal research on sustainability transitions within the surfing industry that call for an expansion of the components of the multi-phase model with a closer alignment of the pre-development, take-off, acceleration and destabilisation phases (Borne, 2018). Finally, *co-design* in the context of this chapter is evident not in the context of other established societal stakeholders but instead through a diverse and diffuse engaged population, and this was particularly represented through the 'off-mission' Instagram posts of values and arts. These insights provide significant scope for future research on transitions not only within the coastal zone but also from a broader systemic perspective, especially as informational governance will arguably become more important as digital technologies advance and evolve.

Bibliography

Ackland, R., & O'Neil, M. (2011). Online collective identity: the case of the environmental movement. *Social Networks, 33*(3), 177–190.

Anderson, A., (1997) *Media, Culture and the Environment*. London: UCL Press.

Auer, M.R., Zhang, Y., & Lee, P. (2014). The potential of microblogs for the study of public perceptions of climate change. *Wiley Interdisciplinary Reviews: Climate Change, 5*(3), 291–296.

Beck, U. (2016). *The Metamorphosis of the World: How Climate Change Is Transforming Our Concept of the World*. John Wiley & Sons.

Borne, G. (2018). *Surfing and Sustainability*. London: Routledge.

Borne, G. (2017). Sustainability and surfing in a risk society. In Borne, G., & Ponting, J. (Eds.), *Sustainable Surfing*. London: Routledge.

Borne, G. (2015). Sustainable development and surfing. In Borne, G., & Ponting, J. (Eds.), *Sustainable Stoke: Transitions to Sustainability in the Surfing World*. Plymouth: University of Plymouth Press.

Borne, G. (2010). *Sustainable Global Development and the Effective Governance of Risk*. New York: Edwin Mellen Press.

Borne, G., & Ponting, J. (2017). *Sustainable Surfing*. London: Routledge.

Borne, G., & Ponting, J. (2015). *Sustainable Stoke: Transitions to Sustainability in the Surfing World*. Plymouth: Plymouth University Press.

Boykoff, M., & Goodman, M. (2009). Conspicuous redemption? Reflections on the promises and perils of the 'celebritization' of climate change. *Geoforum, 40*(3), 395–406.

Bruns, A., & Burgess, J. (2011). The use of Twitter hashtags in the formation of ad hoc publics. *Proceedings of the 6th European Consortium for Political Research (ECPR) General Conference 2011*.

Buttel, F. (2000). Ecological modernization as social theory. *Geoforum, 31*(1), 57–65.

Castells, M. (2009). *Communication Power*. New York: Oxford University Press.

Ehnert, F., Kern, F., Borgström, S., Gorissen, L., Maschmeyer, S., & Egermann, M. (2017). Urban sustainability transitions in a context of multi-level governance: a comparison of four European states. *Environmental Innovation and Societal Transitions, 26*(1), 101–116.

Geels, F., & Schot, J. (2010). The dynamics of transitions: a socio technical perspective. In Grin, J., Rotmans, J., & Schot, J. (Eds.) (2010). *Transitions to Sustainable Development: New Directions in the Study of Long Term Transformative Change*. London: Routledge.

Goldkind, L. (2015). Social media and social service: are nonprofits plugged in to the digital age? *Human Service Organizations: Management, Leadership & Governance, 39*(4), 380–396.

Gibson, C., & Warren, A. (2017). Surfboard making and environmental sustainability. Sustainable surfing, In Borne, G., & Ponting, J. (Eds.) *Sustainable Surfing.* London:Routledge, pp. 87.

Grin, J., Rotmans, J., & Schot, J. (2010). *Transitions to Sustainable Development: New Directions in the Study of Long Term Transformative Change*. London: Routledge.

Grin, J. (2010) Understanding transitions from a governance perspective, Grin, J., Rotmans, J., & Schot, J. (Eds.) *Transitions to Sustainable Development: New Directions in the Study of Long Term Transformative Change*, London, Routledge.

Hales, R., Ware, D., & Lazarow, N. (2017) Surfers and public sphere protest. In Borne, G., & Ponting, J. (Eds.), *Sustainable Surfing*. London, Routledge.

Hestres, L. (2015) Climate change advocacy online: theories of change, target audiences and online strategy. *Environmental Politics, 24*(2), 193–211.

Hestres, L. (2014) Preaching to the choir: internet mediated advocacy, issue public mobilisation, and climate change. *New Media and Society, 16*(2), 323–339.

Hoefnagel, E., de Vos, B., & Buisman, E. (2013). Marine informational governance, a conceptual framework. *Marine Policy, 42*, 150–156.

Instagram. (2015). Sustain Surf: Kelly Slater image. Retrieved 1 March 2016, from www.instagram.com/p/ 06l1Mdmp5B/?taken-by=sustainsurf

Lazarow, N., & Olive, R. (2017). Culture, meaning and sustainability in surfing. Sustainable surfing. In Borne, G., & Ponting, J. (Eds.), *Sustainable Surfing*, London, Routledge.

Lazarow, N., Miller, M.L., & Blackwell, B. (2008). The value of recreational surfing to society. *Tourism in Marine Environments*, *5*(2–3), 145–158.

Machado, R., & Toth, J. (2015). *Sustainability in the surf industry*. In Borne, G., & Ponting, J. (Eds.), *Sustainable Stoke: Transitions to Sustainability in the Surfing World*, Plymouth: University of Plymouth Press.

Martin, S.A., & O'Brien, D. (2017). Surf resource system boundaries. In Borne, G., & Ponting, J. (Eds.), *Sustainable Surfing*, London, Routledge.

Mol, A.P. (2006). Environmental governance in the information age: the emergence of informational governance. *Environment and Planning C: Government and Policy*, *24*(4), 497–514.

Mol, A.P. (2008). *Environmental Reform in the Information Age. The Contours of Informational Governance.* Cambridge University Press, Cambridge.

Mol, A.P. (2009). Environmental governance through information: China and Vietnam. Singapore. *Journal of Tropical Geography*, *30*(1), 114–129.

Moser, S.C., & Dilling L. (2011) Communicating climate change: closing the science-action gap. In Dryzek, J.S., Norgaard, R.B., & Schlosberg, D. (Eds.), *The Oxford Handbook of Climate Change and Society.* Oxford: Oxford University Press, 161–174.

Nulman, E., & Özkula, S.M. (2015). Environmental non-governmental organisations' digital media practices toward environmental sustainability and implications for informational governance. *Current Opinion in Environmental Sustainability*, *18*, 10–16.

O'Reilly, T. (2007). What is Web 2.0: design patterns and business models for the next generation of software. *Communications and Strategies*, *65*(1), 17–37.

Roose, M. (2012). Greenpeace, social media, and the possibility of global deliberation on the environment. *Indiana Journal of Global Legal Studies*, *19*(1), 347–364.

Rotmans, J., & Loorbach, D. (2010). Towards a better understanding of transitions and their governance. A systemic and reflexive approach. In: Grin, J., Rotmans, J.,

& Schot, J. (Eds.), *Transitions to Sustainable Development. New Directions in the Study of Long Term Transformative Change*. London: Routledge, pp. 105.

Saxton, G.D., Niyirora, J.N., Guo, C., & Waters, R.D. (2015). #AdvocatingForChange: the strategic use of hashtags in social media advocacy. *Advances in Social Work, 16*(1), 154–169.

Saxton, G.D., & Waters, R.D. (2014). What do stakeholders like on Facebook? Examining public reactions to nonprofit organizations' informational, promotional, and community- building messages. *Journal of Public Relations Research, 26*(3), 280–299.

Schäfer, M.S. (2012). Online communication on climate change and climate politics: a literature review. *WIRES Climate Change, 3*(6), 527–543. doi:10.1002/wcc.191

Shapiro, M., & Park, H. (2018). Climate change and YouTube: deliberation potential in post-video discussions. *Environmental Communication, 12*(1), 115–131.

Sheldon, P., & Bryant, K. (2016). Instagram: motives for its use and relationship to narcissism and contextual age. *Computers in Human Behavior, 58*, 89–97.

Soma, K., Termeer, C., & Opdam, P. (2016). Informational governance – a systematic review of governance from sustainability in the information age. *Environmental Science and Policy, 56*,89–99.

Stevens, T.M., Aarts, N., Termeer, C.J.A.M., & Dewulf, A. (2016). Social media as a new playing field for the governance of agro-food sustainability. *Current Opinion in Environmental Sustainability*, *18*, 99–106.

Surfing Magazine. (2016). Greg Long is the champion of the world. *Surfing Magazine*. Retrieved 10 March 2016, from /www.surfingmagazine.com/news/greg-long-is-the-champion-of- the-world/#4xtiWFS5wRIxWL4K.97

Sustainable Surf. (2015). About us. Sustainablesurf.org. Retrieved 12 May 2015, from http:// sustainablesurf.org/ about-us/

The Interia. (2013). ESPN names Kelly Slater most influential figure in action sports. The Inertia. Retrieved 12 March 2016, from www.theinertia.com/business-media/espn-names-kelly- slater-most- influential-figure-in-action-sports/

Thompson, S. (2015). Pro surfing and the art of inspiration. In Borne, G., & Ponting, J. (Eds.), *Sustainable Stoke: Transitions to Sustainability in the Surfing World*. University of Plymouth Press, pp. 76–79.

Toonen, H., & Mol, A. (2013). Putting sustainable fisheries on the map? Establishing no take zones for the North Sea plaice fisheries through MSC certification. *Marine Policy, 37*, 294–304.

Thorpe, H., & Rinehart, R. (2010). Alternative sport and affect: non-representational theory examined. *Sport in Society, 13*(7–8), 1268–1291.

Ware, D., Lazarow, N., & Hales, R. (2017) Surfing voices in coastal management – Gold Coast Surf Management Plan –a case study. In Borne, G., & Ponting, J. (Eds.), *Sustainable Surfing*, London, Routledge.

Wenger, E. (2009.) Communities of practice. *Communities, 22*, 5.

Whilden, K., & Stewart, M. (2015) *Transforming Surf Culture Towards Sustainability: A Deep Blue Life* In Borne, G., & Ponting, J. (Eds.), *Sustainable Stoke: Transitions to Sustainability in the Surfing World* . University of Plymouth Press, 130–139.

6 Cultivating diverse values by rethinking blue economy in New Zealand

Nicolas Lewis

1 Introduction

This chapter develops a conceptual, methodological and political framework for cultivating new forms of sustainable blue economy. Over the last decade, the term 'blue economy' has increasingly been used by commentators and institutional actors to capture a burgeoning interest in the coasts and oceans as sites for economic development (Silver *et al.*, 2015; Winder & Le Heron, 2017). Increasingly understood as resource frontiers and spaces for capital accumulation, the world's marine spaces are increasingly being opened up for investment, markets and new forms of governance. Much of this economisation of resources (Calışkan & Callon, 2009) and associated governmentalisation of economic subjects and spaces is being conducted within the discourse of the 'blue economy', a term used to corral, aggregate and measure the growth potential of diverse economic activities.

The term 'blue economy' is, however, used not only to encompass the economisation of marine resources, but also the associated concerns about the ecological and environmental consequences of exploiting these frontiers, and the impact they will have on coastal communities. Multiple actors are seeking to ensure that resource extraction is both sustainable and ordered; that is, that property rights are clarified and allocated to facilitate accumulation, that closer attention is paid to sensitive marine ecologies and local and global environments, and that, where these various concerns meet, regulation is clearly defined and transparent (Winder & Le Heron, 2017). Australia's National Marine Science Plan (NMSC, 2015), for example, places science at the centre of making a new and sustainable blue economy that is shaped by grand challenges such as energy security, marine sovereignty and climate change. The World Wildlife Fund (Hoegh-Guldberg, 2016), on the other hand, advocates a sustainable blue economy focused less on growth and more on enhanced livelihoods, revitalised local ecologies, social development, and vibrant and resilient community institutions.

In short, the oceans remain something of a frontier, in terms of our knowledge of their potential, our abilities to harness it, and our own potential to govern any exploitation. Internationally, governments, would-be resource exploiters and various voices of concern for community and environment are grappling with how to categorise, conceptualise and regulate the blue economy. This chapter examines how 'blue economy' thinking in New Zealand is being used as a platform to address these interests and concerns. It argues that such thinking directs attention in three novel directions: a re-categorisation of economy; new economic objects of concern; and to a potential politics of economic rent and social return. Taken together these moves promise to extend 'blue economy' thinking into more profound notions of sustainable development and possibilities for coastal transitions than those linked with socio-cultural and environmental values. In this way, the chapter complements work on socio-technical transitions. It calls for the economy to be internalised within debates over, and visions for, making transitions, and challenges the literature on sustainability transitions to stretch beyond values to accommodate the agency of business actors and address questions of social returns. An embodied and richly spatialised case study, the account points to the potential of co-designing futures with economic agents and to the importance of identifying transitions drivers beyond state institutions. The chapter develops two socio-technical devices of its own (a table and a schematic representation of the blue economy), which it offers up as artefacts that might operate performatively to shift investment practices and foster sustainable transitions.

2 Blue economy

The significance of coastal zones to community, regional and national economies is well rehearsed in the literature and alive in the politics of many places at different scales (Allison & Horemans, 2006; Béné *et al.*, 2010; Pinkerton, 2011; Smith & Chambers, 2015; Link & Scheffran, 2017). Asserting and maintaining control of the oceans remains as pivotal to the fates of empire, nations and peoples as it has for centuries. Property rights in coastal zones continue to be asserted and contested in relation to commons set within the wider jurisdictional frames of communities, regions and nations. The oceans and coasts have been mapped and laid out as zones of influence and control, especially in the years since the signing of the United Nations Convention on the Law of the Sea (UNCLOS) in 1982, which brought these resource fields and the activities, claims and concerns that animate them into a broad nation-state ordering framework. However, UNCLOS only partially domesticated the coasts and oceans. Property rights remain unclarified in many settings, especially coastal zones, while new aspirations, technologies and ecological concerns are driving demand for more ordered resource use.

Internationally, the term 'blue economy' has emerged as a discursive framework for this ordering work. It is used increasingly in policy circles and interpreted to refer to anything from highly capitalised investments designed to exploit marine resources, to radical green initiatives (Winder & Le Heron, 2017). The former definition uses 'blue economy' as a category that aggregates and frames economic activities for regulation, while the latter implies a moral economy centred on sustainable practices and ethics of low, slow, or no growth.

In New Zealand, both interpretations are being used to think about marine-based economic development. The term entered mainstream economic development discourse when the New Zealand Seafood Council named its 2014 industry conference 'Growing the Blue Economy'. Reference to 'blue economy' resurfaced in relation to international economic management efforts in the 2016 Statistics New Zealand effort to measure the national marine economy (StatsNZ, 2016). More recently, it has been reworked by researchers in the Sustainable Seas National Science Challenge (SSNSC) to refer not just to activities taking place in oceans and coastal zones, but to the processes by which they are valued, managed and developed to create economic and socio-cultural values.

The SSNSC is a 'grand challenge' framework for reassembling research trajectories and capabilities in marine science into a commitment to a new and more coherent national science programme that challenges business-as-usual. Whilst in effect charged with creating new economic objects in New Zealand's marine spaces, the SSNSC has in practice complemented its responsibilities to pose science and economic development questions with multiple research streams examining management and governance processes. These include those to do with indigenous values, resource rights and knowledge. Room has been created to examine alternative conceptions of blue economy and to animate a new form of blue economy that seeks to transcend limp references to sustainability in building blue economies. Recognising that this will require innovations in thinking, investment, production and regulation, SSNSC has adopted a definition of blue economy that highlights practices rather than sectors or firms. It defines 'blue economy' as economy that "works innovatively to enhance the capabilities of local people to work with the dynamics of marine ecosystems to generate livelihoods and healthy communities while maintaining ecological functioning" (SSNSC, 2018). In what follows, we elaborate on this reworking.

3 Reconceptualising economy

Post-structuralist political economists have argued that 'the economy' is a social construction, created in order to manage modern nation-states in the wake of the 1930s depression (Mitchell, 2008). The crucial point for this chapter is that it is unhelpful to consider the economy as a prior, ontologically stable object that awaits discovery, demystification and management. Rather, as Mitchell insists, *economy* is a set of practices and processes of stewarding resources to make livelihoods. By contrast, *an actualised economy* is what emerges in a series of concrete economic relations such as flows of investment capital, labour and goods, and the institutions necessary to secure those relations. *The* blue economy is in these terms a discursive project that seeks to assemble diverse economic activities taking place in coastal and ocean seascapes. Accounting for it in these terms is simply a question of aggregating similarly narrow understandings of actualised economies at the sectoral or regional scale.

I extend Mitchell's argument in four directions so as to realise the wider generative potential of the 'blue economy' metaphor. The first is to recognise that economic relations are never narrowly economic, or reducible to simple market exchanges, a well-rehearsed theme in critical economic thought. Updated to embrace socio-ecological politics, such exchanges are always simultaneously political, socio-cultural, socio-ecological and dependent on human-non-human relations and environmental valuation.

The second direction is to recognise that failing to adopt a broader conception of economy tends to externalise economy from socio-ecological decision-making. Instead, it is positioned as a self-actualising backdrop, or worse as a set of imperatives about how to think about socio-ecological futures. This comes with two significant sets of incapacitating consequences. First, it is seen to have its own, independent laws and logics, which are turned over to the science of economics or the demands of 'industry', themselves fundamentally different but often conflated. They tend then to be represented as constraints or imperatives in decision-making. Second, its significance, and its generative potential, is sidelined by socio-ecologists thinking about sustainable futures. Economy becomes the elephant (or whale) in rooms where a more radical blue economy is considered. Investment, the moment in which economy is actualised in space, is ignored. Instead, economic concerns are presented as stakeholder interests or inferred from the existence of activities assumed to have been a rational response to prefigured market imaginaries. Possibilities of doing economy differently are never considered.

The third new direction is to focus on economisation and market-making (Çalışkan & Callon, 2009). Economy in this reading is about stabilising and pacifying economic relations that are understood as being simultaneously political, socio-cultural and ecological. As Murray Li observes of a project to capitalise land on the frontiers in Sulawesi:

> To turn it to productive use requires regimes of exclusion that distinguish legitimate from illegitimate uses and users, and the inscribing of boundaries through devices such as fences, title deeds, laws, zones, regulations, landmarks and story-lines. Its very 'resourceness' is not an intrinsic or natural quality. It is an assemblage of materialities, relations, technologies and discourses that have to be pulled together and made to align. To render it investible, more work is needed.
>
> (Murray Li, 2014, p. 589)

Working from such a perspective, 'blue economy' might be reinterpreted as a generative project of economy-making that emphasises socio-ecological knowledge and local/regional livelihoods in making coasts and oceans resource-full. Actualised blue economies might be interpreted as entrepreneurial, investment and management initiatives that draw on marine environments to create economic and social values that sustain or enhance the resourcefulness of those environments and their communities. Here resourcefulness might be understood as 'life-making' or the fostering of a potential to create local/regional values and livelihoods through attention to more-than-human futures. Highlighted in Tsing's (2015) exploration of possibilities in the capitalist ruins of the matsutake economy in Oregon, being resourceful is about innovative and resilient economy(ising) of/with socio-natures. But what if we could be resourceful in advance of collapse?

In a 'resourcefulness' framework, blue economy becomes a process of economy-making that begins with socio-ecological knowledge and a commitment to fostering innovative economic practices that highlight possibilities. For the SSNSC, such an approach is argued to start with ecosystem management principles and protocols, participatory management processes, resourceful economic practices and radical socio-cultural and environmental values frameworks. As Murray Li (2014) foreshadows in her own work on 'Land's End', the radical thinking required in contemporary New Zealand is being spurred at least in part by thinking with indigenous knowledge and claims to coastal and oceanic resources.

Finally, I extend Mitchell's challenge to rethink economy to already actualised coastal and ocean economies. The question here is how to conceptualise, categorise and make diverse economic activities available for 'blue' economisation. In the New Zealand context, government agency Statistics New Zealand (SNZ) 'measured' New Zealand's marine economies in 2016 using standard economic methodologies and the formal Standard Industrial Codes (SIC). It produced sector-based measures that yielded a sense of the relative significance of different marine resource 'commodity' economies, but revealed little about blue economy. By SNZ's own admission, the measures are incomplete, based on multiple assumptions (many of them heroic), and contradict assumptions made by others using differing methods/assumptions. While useful for various lobbyists, without identifying connections, relationalities and dynamics such as investment trajectories, they offer little to those making either specific or strategic management decisions.

By contrast, it is possible to build knowledge of blue economy by creating new knowledge objects that highlight key investment priorities and trajectories and relationalities such as shared fates. One approach is to re-categorise actualised economies to build a new grid of intelligibility with which to think, act and generate understandings useful to decision-makers. Drawing on the academic literature, documents in the grey literature and informal conversations with blue economy actors, we imagine very different, 'lived' categories of actualised economies in the New Zealand context; economies whose defining actors and relations extend into and across coastal and ocean spaces. Table 6.1 offers a preliminary schematic of six such economies.

Table 6.1 Actualised economies in the New Zealand context

	Size	*Economic practice*	*Market orientation*	*Ethical co-ordinates*	*Labour relation*	*Investment structure*
Iwi/hapū/ whānau	Corporation to marae	Extractive to customary	Varied	Tikanga, Kaupapa Māori	Iwi-capitalist to customary	Hapū/whānau to iwi trust board
Community	Family, club, co-op	Cultural and use values	Non-market	Community environmental	Unpaid/ co-operative	Family, club/co-op, philanthropy
Commodity	Large	Extractive	Exports	Accumulation	Fully capitalist	(Global) corporation
Techno-science	High capital, low workers	Creative (techno)	Export/enabling	Knowledge economy	Relational/ immaterial	IP/angels/ government
Foundational	Varied (largely SME)	Service	Local; place-dependent	Survival/place accumulation	Family to owner-manager to firm	Family to corporation
Distinctiveness	Largely small (focused)	Creative (artisanal)	Value-added	Place, quality, distinction	Artisanship	Entrepreneur artisan (family)

The categories are heuristic, but point to definitive spheres of diverse economic activity in New Zealand's oceans and along its coasts. They include the two forms of non- or partially capitalist economy commonly identified in diverse economy accounts of New Zealand's economies: community economy and Māori economy. The latter are distinguished by collective ownership and commitments to distinctive ethical co-ordinates, but encompass significant diversity. Māori economy, for example, ranges from activities organised and carried out at the tribal scale to those organised at the scale of the hapū (descent group) or whānau (extended, multi-generational families). Iwi tribal development trusts manage assets and reinvest monies from Treaty of Waitangi settlements. They now have significant corporate-scale holdings in dairy industries, aquaculture and fishing, while customary economic practices such as gathering of foods from the sea and communal lands remain significant at the whānau and hapū levels, and whānau trusts may have small- or medium-sized investments in community-based ecotourism, cultural artisanship or other community framed activities.

Activities that are generally lumped together as capitalist business and then divided into industrial sectors by the SICs, might themselves be broken into different economies. Table 6.1 points to four possible categories. The first includes the major high volume, resource-based activities that dominate New Zealand's GDP and export statistics: dairying, horticulture, mass tourism, aquaculture, fisheries and mining. Techno-science economies are those reliant on higher levels of capital input, different sources of investment and knowledge economy discourses. They include public and private enterprises that produce knowledge and technological goods which enable other firms to operate, and techno-science-dependent producers of final goods such as mussel oils or nutraceuticals. Foundational economies (see Bentham *et al.*, 2013) are made up of heterogeneous locally focused enterprises that depend on relationships with local places and that meet everyday needs by providing taken-for-granted services and goods such as care, utilities, food and housing. Enterprises involved may take any size, but it is in this economy that SMEs tend to operate. Finally, the notion of distinctiveness economies highlights the shared attributes and dependence on quality, distinctiveness and value-adding that characterise many smaller, quality-focused enterprises that emphasise artisanship, commonly supported by claims about provenance.

Each sphere of activity points to different types of enterprise, economic relations and investment structures, market orientation, and organising principles or ethical co-ordinates (Gibson-Graham, 2006). Each also suggests different management requirements and possibilities, and potentially different strategic policy interests. The proposition is that this categorisation may be more meaningful in terms of understanding investment dynamics and connections within and between those 'economies', identifying collective interests and shared practice, and making management and policy decisions. The model begins with the recognition that the firm as the primary unit of measurement is neither a closed nor stable entity; rather, it generally exists in co-dependent and networked forms, and encompasses different types of economy. It also recognises that different economies are far from discrete. Any one enterprise may be operating in multiple economies, while there are significant overlaps, for example, between community and hapū/whānau-based Māori economies, commodity, techno-science and iwi economies, and provenance and whānau-based Māori economies. Instead, the categories are loosely drawn, emergent, incomplete and non-discrete. They endeavour to elucidate difference, possibility and policy problematics, rather than create measurable artefacts that define stable objects for forms of distanced management.

4 Highlighting resourceful practices in New Zealand blue economies

One potential use of the categories highlighted in Table 6.1 is to produce a model of the blue economy that allows us to highlight diverse activities across the alternative-mainstream divide that promise sustainable transitions to socio-ecologically sustainable economies. These extend from regulatory interventions to restocking the commons, the introduction of new, sustainable harvesting technologies, corporate social responsibility initiatives, experiments in collective ownership of infrastructure, the design and adoption of public and private environmental standards for harvesting and waste, and community seafood collection initiatives. In the New Zealand context they also include initiatives derived from indigenous knowledge, practice and resource ownership and stewardship. These include taiāpure and rāhui[1], commitments by iwi and whānau seafood and land-based organisations to quadruple bottom-line business models that commit to cultural and environmental values and 500-year investment horizons and eschew maximising approaches to extractive and commodity industries as well as community gathering.

Figure 6.1 maps out a model of blue economisation in New Zealand marine and coastal economies, using the economic categories developed in

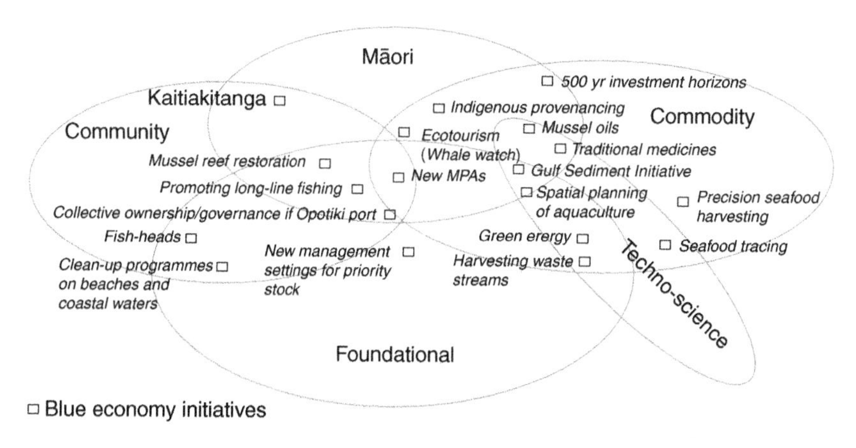

Figure 6.1 Promisingly blue initiatives in economising New Zealand coasts and oceans

Table 6.1. The schematic arranges various initiatives through which diverse actors are seeking to practise economy differently in the realms of regulation, production, community practice and ethical commitments. It points to a vital seascape of potential positive change, from multi-national fishing vessels developing new forms of net-less harvesting that avoid bottom damage and by-catch, to initiatives by community groups and small firms to seed mussel spat in the commons. They include the techno-science innovations that make it possible to imagine and produce new products, including those dedicated to developing green energy and others that make use of waste streams from aquaculture and fishing. They also involve ecotourism initiatives, both those developed by small-scale family enterprises as well as those catering for high-volume tourism. They include community clean-up exercises and the redistributive freefish-heads.co.nz initiative that donates unwanted fish heads to families who wish to make use of them, as well as the establishment in law of new marine protected areas and taiāpure. Initiatives in Māori economy energise most other economies, with their ethic of kaitiakitanga (environmental guardianship) underpinning community economy, and commitments to long-term investment horizons promising to inject multi-generation ethics of care for community and environment into commodity and foundational economies.

5 A new politics of rent

Borrowed from classical economists, the term 'economic rent' refers to a premium on returns to investments that derive from preferential access to key resources or other inputs into economic activities. These may arise from either the exercise of monopoly powers over pivotal economic resources or relations, or special access to the higher 'yielding' qualities of particular resources, natural or social. These qualities may inhere in particular natural resources, be produced and secured by protected technologies, or arise from historical investment in infrastructure, services or socio-cultural life. Both sources of rents are commonly produced and/or secured by regulation such as collective investment, state ownership, appellations, rules of access to commons, patents or import licences. In both cases they are likely to derive from place-specific factors.

Historically, rent discourses are associated with land and its differential qualities, scarcity and monopolisation by rentiers. For Vercellone (2008), contemporary cognitive capitalism is just as reliant on rents that arise from knowledge resources (education systems, public research, untraded interdependencies and productive co-operation). For contemporary critics of platform capitalism, the control of the internet and the big data increasingly generated and deployed by associated technologies represents a new rent platform (Birch, 2017). Rent discourses have also been deployed in relation to debates about the economy of wild fisheries, and especially to the effects of management systems that create and allocate special access to fisheries (Béné *et al.*, 2010; Torkington, 2016, Flaaten, 2010; Flaaten *et al.*, 2017).

Rent is generally understood and widely criticised as income that is not earned by those who are appropriating it (Sayer, 2015). Hudson (2011) invites us to distinguish between profit, earned returns on investment that mobilises capital to generate innovation, and rent, unearned income appropriated from natural or socially produced resources. It is regarded by neo-classical economists and Marxist political economists alike as the consequence of institutional failures. For the former it arises from and further produces market failures and inefficiencies, which are commonly held to arise from state intervention. For the latter, it arises from the uneven socio-political power that emerges from and reproduces capitalism. The place of rent in the history of economic thought, however, has been more complicated than the critique encapsulated in these two received positions (Hudson, 2011, 2012).

Hudson (2011, 2012) points to a classical political economy that interpreted economic rent generated by state monopolies as a justified return to support a national development programme. The regulatory and other monopoly rents arising from the mass public infrastructure projects necessary to roll out industrial capitalism not only generated economic rents but also the very possibility of private accumulation. As these projects rolled out in Europe, the USA and elsewhere, rent became a touchstone for debates about who appropriated, controlled and reinvested the returns. It became caught up and then lost in the great ideological debates of the 20th century, the rise of Keynesianism, and programmes of infrastructural and resource privatisation. Hudson suggests that this laid the ground for dispossessing labourers of their right to share in the natural resources of their lands. The neo-liberal reforms of the 1970s and 1980s locked in 'rentier rights', which became "legalised tollbooths to extract revenue that rightly should belong in the public sector". The private ownership of natural monopolies and calls to deregulate and privatise economies became increasingly entangled. Hudson adds that the erasure of rent in economic debates has in turn contributed to sustaining the myth that public enterprise is less able than private capital to make economies more productive and competitive. He directs us to challenge the reduction of rents to profits by the social groups that benefit.

A renewal of interest in rent offers up an economic rationale and political mandate for extracting a social return from investments in the foundational economy, which Bentham *et al.* (2013) interpret as investment in social franchises. It also revitalises a politics of regional development that might invest in places in order to stimulate activity and create jobs in those specific places. A concern with rent opens up at least four platforms for a progressive politics that extends beyond criticism of appropriation of rents by social elites.

The first of these political platforms is to unleash a progressive assault on rent as unearned income in institutional politics. This might, as *The Washington Post* (30 March 2015) suggests, "bring liberals and conservatives together" to challenge the primacy of the market as a just and efficient allocation strategy or one that can deliver the interests of nation, region or community. In relation to common pool natural resources such as fresh water or wild fish, a concern with 'rent' in any national setting might be used to pull together multiple lines of values-based politics and political economy to do with environmental justice, ecological damage, scarcity concerns, and privatisation or foreign ownership.

Second, as Harvey (2002) insists, there is a fundamental contradiction between the pursuit of rents and globalising capitalist modernity. He observes (2002, p. 106) the existence of rents "leads global capital to value distinctive local initiatives". The uniqueness, authenticity, particularity and originality that generate distinctiveness rents are all "inconsistent with the homogeneity presupposed by commodity production". Capital seeking to stimulate and exploit such rents must foster local cultural developments that are often transgressive and can be antagonistic to its own smooth functioning. Harvey goes on to add that not only is this a contradiction that capital must address in its own organisational strategies, but that it is within such commonly transgressive local cultural spaces that "all manner of oppositional movements can form" (Harvey, 2002, p. 108). That is, for example, in the context of New Zealand's blue economy, the added premium available to iwi corporations from indigenous branding will require government to be attentive to Māori demands with respect to resource governance and socio-cultural concerns, as well as requiring iwi trust boards to practise a capitalism that retains the support of the wider iwi. More generically, the premiums available from the application of green technologies in aquaculture derive in one way or another from green social movements, which in other settings may materialise as opposition to those very businesses. The contradiction opens up wider political possibilities for a progressive politics of rent.

Third, as Lewis *et al.* observe (2013), this is just one expression of a wider politics of possibility that emerges from seeing economic rent as geographical rent. That is, the rents that we have understood as economic, all derive to one extent or another from value that is added by producing goods/services 'here' rather than 'there'. It is *derived from place and investment in place*, from: the environment and/or non-human actors fixed or situated in place; place-based knowledge, trust and socio-cultural institutions; place-specific investments in education and research (institutions, organisations, networks, etc.); and provenance and credence values associated with place. In this sense, economic rents are demonstrably returns to place, and in any moral economy, framing ought to be returned to place. Here, foundational economy and arguments about social return have particular purchase.

Finally, each of these political possibilities can be tied to what have been termed enactive thought experiments with rent platforms (see Lewis *et al.*, 2016). Place-based thought experiments with various community, business and government co-producers of knowledge can be used to visibilise rent as a normal economic return, but one that is locally generated and demands a collective return. In this way rent can be rendered a productive force rather than collapsed into capital gains and income returns to capital and then either invisibilised as earned income or dismissed by critics as unearned. Situated thought experiments and the diverse politics that animates them can be directed away from the negative realms of unwarranted rents to imagining and legitimating interventions that return rents to localised ecologies and communities. In a capitalist political economy all concerns with, and investments in, local and regional development are, and have always been, about producing geographical rents and realising them in place. Recognising this might allow all participants in debates to turn their attention to co-producing collective institutions that stimulate a collective economy that creates rents and makes them 'stick' in place.

6 Conclusion

To what, then, does this experimental reading of 'blue economy' add up? The proliferation of blue economy discourse and the enabling and performative scientific and market-making investments assembled around it reveal a project of economisation that seeks to open up new frontiers for capitalist accumulation in the oceans. This chapter makes three disruptive yet generative interventions to the project. It reinterprets the discourse itself, points to progressive diversity in practice that might be encouraged as a foundation for a counter or deeper blue economy, and revitalises a politics of economic rent at a moment when the struggle over rent in a new frontier of economisation is pivotal. These interventions are grounded in an enactive politics of knowledge production that favours generative critique, rather than taking aim at easy targets from a comfortable distance.

This is not an account that begins by challenging capitalists, dismissing the language of economy, or seeking to displace economic with environmental or socio-cultural values. Instead it begins by troubling what we have come to accept as stable representations of each of those things by focusing on practices and highlighting what is normally hidden. It embraces contestation and transformation rather than seeking to stabilise knowledge and economic relations in relation to pre-set economic goals such as GDP or exports, or resistance to them. It argues that speaking economy more coherently and directly to economic actors (investors, entrepreneurs, scientists, bankers, policy agencies, consumers and publics) establishes a first and necessary platform for the critical work required by contemporary interest in the blue economy. It outlines a set of new knowledge objects that shift decision-making and in themselves call forth new thinking and invite political engagement at a time when discourses of the blue economy speak openly about transitions and making new economies. These promise to encourage New Zealanders to begin to imagine, pose and answer new and situated questions about economy, individually and collectively; and to normalise such questions:

- What economy do/will communities practise? What economy will allow environments to thrive? What activities *do we want* in *our* oceans/coasts?

- How are innovative New Zealanders working with marine resources to create diverse cultural uses, and exchange values and make livelihoods?
- How do we stimulate a new set of investment rationalities and practices that mobilise social, political, ecological and environmental capabilities within altered ethical co-ordinates so as to enhance resourcefulness?

The chapter also offers up a new politics of geographical rent that foreshadows alternative socio-ecological futures and pathways to more penetrating coastal transitions than those that direct attention to socio-cultural and environmental values. Echoing Callon, it offers up a framework for an alternative project of economisation "'in which new emerging forces are offered the possibility of becoming stronger [and] to limit the grip of established forces" (Callon, 2005, p. 18).

Acknowledgements

This project is funded as part of the Sustainable Seas Challenge in the New Zealand National Science Challenge programme. It is funded by Ministry of Business, Innovation and Employment contract C01X1515 (Sustainable Seas). I would like to thank Richard Le Heron and the other members of the Blue Economy research team for their insights.

Note

1 Rāhui is a temporary prohibition on access to an area or a resource. The practice derives from indigenous environmental stewardship and is provided for in the Fisheries Act 1996. Taiāpure designate areas of coastal and estuarine waters of special value to iwi or hapū as sources of food or cultural and spiritual values and give them certain regulatory authority with respect to fishing and related activities within the taiāpure (Fisheries Act 1996: s. 174).

Bibliography

Allison, E., & Horemans, B. (2006). Putting the principles of the sustainable livelihoods approach into fisheries development policy and practice. *Marine Policy, 30*(6), 757–766.

Béné, C., Hersoug, B., & Allison, E., (2010). Not by rent alone: analysing the pro-poor functions of small-scale fisheries in developing countries. *Development Policy Review, 28*(3), 325–358.

Bentham, J., Bowman, A., de la Cuesta, M., Engelen, E., Ertürk, I., Folkman, P., Froud, J., Johal, S., Law, J., Leaver, A., & Moran, M. (2013). *Manifesto for the Foundational Economy*. Centre for Research on Socio-Cultural Change, Manchester. Available at: http://hummedia.manchester.ac.uk/institutes/cresc/workingpapers/wp131.pdf

Birch, K. (2017). Rethinking value in the bio-economy: finance, assetization, and the management of value. *Science, Technology, & Human Values, 42*(3), 460–490.

Çalışkan, K., & Callon, M. (2009). Economization, part 1: shifting attention from the economy towards processes of economization. *Economy and Society, 38*(3), 369–398.

Callon, M. (2005). Why virtualism paves the way to political impotence: a reply to Daniel Miller's critique of "The laws of the market". *Economic Sociology: European Electronic Newsletter, 6*(2), 3–20.

European Commission (EC). (2017). Report on the blue growth strategy: towards more sustainable growth and jobs in the blue economy. Staff Working Papers, Brussels. Available at: https://ec.europa.eu/maritimeaffairs/sites/maritimeaffairs/files/swd-2017-128_en.pdf

Flaaten, O. (2010). Fisheries rent creation and distribution – the imaginary case of Codland. *Marine Policy, 34*(6), 1268–1272.

Flaaten O., Heen K., & Matthíasson T. (2017). Profit and resource rent in fisheries. *Marine Resource Economics, 32*(3), 311–328.

Gibson-Graham, J.K. (2006). *A Postcapitalist Politics.* Minneapolis: University Of Minnesota Press.

Harvey, D. (2002). The art of rent: globalization, monopoly and the commodification of culture. *Socialist Register, 38*, 93–110.

Hoegh-Guldberg, O. & Ridgway, T. (2016). Reviving Melanesia's Ocean Economy: The Case for Action – 2016. WWF International, Gland, Switzerland. Available at: http://ocean.panda.org.s3.amazonaws.com/media/Reviving_Melanesia's_Ocean_Economy_Full_Report-WWF-LowRes.pdf

Hudson, M. (2011). Simon Patten on public infrastructure and economic rent capture. *American Journal of Economics and Sociology, 70*(4), 874–903.

Hudson, M. (2012). Veblen's Institutionalist Elaboration of Rent Theory. Working Paper, Levy Economics Institute of Bard College, No. 729.

Lewis, N., Le Heron, R., Campbell, H., Henry, M., Le Heron, E., Pawson, E., Perkins, H., Roche, M., & Rosin, C. (2013). Assembling biological economies: region-shaping initiatives in making and retaining value. *New Zealand Geographer, 69*(3), 180–196.

Lewis, N., Le Heron, R., Carolan, M., Campbell, H., & Marsden, T. (2016). Assembling generative approaches in agri-food research. In Le Heron, R., Campbell, H., Lewis, N., & Carolan, M. (Eds.), *Biological Economies: Experimentation and the Politics of Agrifood Frontiers*. Routledge, 1–20.

Link, P.M., & Scheffran, J. (2017). Impacts of the German energy transition on coastal communities in Schleswig-Holstein, Germany. *Regions Magazine, 307*(1), 9–12.

Mitchell, T. (2008). Rethinking economy. *Geoforum, 39*(3), 1116–1121.

Murray Li, T. (2014). What is land? Assembling a resource for global investment. *Transactions of the Institute of British Geographers, 39*(4), 589–602.

National Marine Science Committee (NMSC) (2015). *National Marine Science Plan 2015–2025: Driving the development of Australia's blue economy*. Canberra. Available at: www.aims.gov.au/documents/30301/2094401/NMSP_FINAL_Aug2015.pdf/22bff822-097f-4d6b-8c68-c53f791cd892

Pinkerton, E. (Ed.). (2011). *Co-operative Management of Local Fisheries: New Directions For Improved Management and Community Development*. UBC Press.

Sayer, A. (2015). *Why We Can't Afford the Rich.* Bristol: Policy Press.

Silver, J., Gray, N., Campbell, L., Fairbanks, L., & Gruby, R.L. (2015). Blue economy and competing discourses in international oceans governance. *The Journal of Environment & Development, 24*(2), 135–160.

Smith, J., & Chambers, C. (2015). Where are all the fish?: Local fish networks in the Westfjords of Iceland. *Environment, Space, Place*, *7*(2), 15–40.

SNZ (Statistics New Zealand). (2016). New Zealand's marine economy 2007–13. Wellington: Statistics New Zealand. Available at: http://archive.stats.govt.nz/browse_for_stats/environment/environmental-economic-accounts/nz-marine-economy-2007-13.aspx

SSNSC (Sustainable Seas National Science Challenge). (2018). Strategy for Phase II 2019-2024: Draft for Consultation. Sustainable Seas National Science Challenge, Wellington. Retrieved 13 May 2018, from http://sustainableseaschallenge.co.nz/sites/default/files/2018-04/2018.04.06_DRAFT_Strategy%20for%20Phase%20II%20for%20Consultation_REVISEDdiagrams.pdf

Torkington, B. (2016). New Zealand's quota management system – incoherent and conflicted. *Marine Policy*, *63*, 180–183.

Tsing, A. (2015). *The Mushroom at the End of the World: On the Possibility of Life in Capitalist Ruins*. Princeton and Oxford: Princeton University Press.

UNDESA (United Nations Department of Economic and Social Affairs). (2014). *Blue Economy Concept Paper.* United Nations, New York.

Vercellone, C. (2008). *The New Articulation of Wages, Rent and Profit in Cognitive Capitalism. The Art of Rent*. Queen Mary University School of Business and Management, London.

Winder, G., & Le Heron, R. (2017). Assembling a blue economy moment? Geographic engagement with globalizing biological-economic relations in multi-use marine environments. *Dialogues in Human Geography*, *7*(1), 3–26.

Part II

Empirical approaches

7 Participatory processes for implementation in Aotearoa New Zealand's multi-use/user marine spaces?

Unacknowledged and unaddressed issues

Richard Le Heron, Paula Blackett, June Logie, Dan Hikuroa, Erena Le Heron, Alison Greenaway, Bruce Glavovic, Kate Davies, Will Allen and Carolyn Lundquist

1 Introduction

Highly visible participatory processes (PPs) directed at multi-use/user issues, as distinct from single or sectoral issues, in marine spaces are generally regarded internationally and in Aotearoa New Zealand (ANZ) as a phenomenon of the 21st century. In the ANZ setting, the governmental institutions embracing marine spaces are four-fold: those of central government, regional and local authorities, Māori entities and a series of recent initiatives focusing on resolving issues in multi-use/user marine spaces. The chapter explores multi-use/user marine spaces as objects of growing socio-scientific inquiry and as an historically resonating concept. Taking a broad view, PPs signal the presence and collision of competing actors whose investment and institutional activities, values and modes of operation problematise efforts to govern such spaces towards collectively acceptable sustainable outcomes. In the case of ANZ, marine spaces have been stewarded over the centuries by Māori. Their traditions of social organisation and collaboration over the use of marine resources have ensured that Māori have retained connections, through oral genealogies, with specific portions of land and sea. The contemporary ANZ marine scene is quite unique with its co-existence of Māori and European worldviews, knowledge and modes of governing. It provides an unprecedented opportunity to explore the wider social and collective work of PPs in complex historically and geographically trajectories of power relations. It is also a highly politicised field of emerging interactions. Our interest in the chapter is how to enable all sectors of society in ANZ to participate in efficient and effective decision-making and so move to co-producing and co-governing the marine environment and making new kinds of marine futures.

This social objective puts a premium on exploring the collective work and achievements of PPs in marine spaces.

In ANZ, single and multi-use/user PPs have originated in a context of regulatory and governance reforms featuring retreating governmental guidance. The PPs have faced over three decades of contingent mixes of influential structural trends and dynamics, and evolving individual and collectivised behaviours on the part of many participating actors. Individual PPs are viewed in the chapter as social entities of particular contexts and conditions, seeking to articulate and implement multiple goals. The relaxing regulatory conditions typical of neo-liberal reform in recent decades effectively mean the tasks of socially identifying and organising the futures of marine spaces has been passed over by central government, to those directly and indirectly involved in marine spaces, to sort out. This is a situation of new freedoms and unknown challenges. We argue that participation and PPs in particular are social technologies suited to do bridging work so necessary to advancing decision-making capacities and identifying, choosing and implementing named and negotiated sustainability outcomes.

The chapter has three chief interests. First, what did the retreat from command-and-control bureaucracy-centred governance mean for actors who were trying to grapple with marine issues and tensions and were evolving in their own right in marine settings whilst trying to reposition themselves? Second, how have the trajectories of Māori governmental aspirations impacted in different settings? The new governmental conditions are products of an unbalancing of the power relations and procedures of interventionist governance, with much scope as a result to put in place historically new governmental connections and relations, albeit working with residual and still influential structures of past intervention. Third, little is known about the actual developments that PPs have undergone, as they have sought to survive and influence the conditions and prevailing outcomes that energised their genesis. The chapter is thus an attempt to outline a stocktake of key PPs and their features, analyse how they have emerged both individually and as a participatory movement appearing as part of a transforming regulatory setting, and explore the plurality of collective implementation achievements that are part and parcel of newly formed PPs.

2 Altering regulatory-governance power relations

Our historically situated actor-centric investigation, drawing on socio-theoretic ideas from geography and the wider social sciences (Gelcich *et al.*, 2011; Hibbard *et al.*, 2008; Higgins & Larner, 2017, Nicolini, 2016; Rabe, 2014; Woodward & McTaggart, 2016), recognises the broad contextual influences impacting on the worlds of marine actors and how these actors seek to turn influences into generative rather than constraining factors to give focus, direction and tangible expression to their implementation hopes. It also recognises that the institutionalised approaches to marine governance and regulation in

ANZ are deeply embedded in the ideas and structures of colonialism, science and the Westminster parliamentary system (Taylor, 2013). The switch in regulatory and governmental model from top-down state policy and procedures positioning compliant actors, to actors having to take on responsibility to position and champion themselves horizontally through networking and negotiation in new ways with respect to other interests, is a profoundly significant development in this field of power relations. Brown (2017) has shown the last quarter century to be dominated by failures in compliance, monitoring and enforcement of environmental law. This major resetting of the regulatory context and conditions intersects with the renaissance of the Māori economy and knowledge, and a corresponding realignment of institutions and investors as well as communities and the public.

What were the main consequences of regulatory adjustment for actors at large? First the state, by virtue of inaction, moved away from an historical strategy of making actors exogenous to state processes. Instead, in any marine space an endogenous and unique complement of actors are now increasingly open to different kinds of association potentially unified by shared interests and aspirations to go beyond the status quo. This has been most prominent in the acknowledgement and material impact of the Treaty of Waitangi[1] and Crown partnership and Treaty settlement processes which have reinstated iwi and hapū[2] attachments to ancestral resources. For the most part, however, Māori have found it difficult to co-develop with new partners and the residual pākehā (non-Māori) systems of command-and-control governance and management, where conventional and colonial-inflected worldviews, knowledge and forms of social organisation and accepted behaviours continue to be major obstacles. Second, the historical contributions of major actors have become more visible and their positive and negative potencies more traceable, as the dominance and cloaking effects of state-shaping influences recede. This has fed through to a clearer appreciation of probable causal mechanisms of influential dynamics on marine spaces. In this regard, at least the following major dynamics are at work: 1) a huge expansion of the land-based economy, led by extensifying and intensifying dairying in traditional and new farming locations, a primary impact of which has been that more catchments, rivers and streams around the country are subjected to greater run-off of effluent and other toxic wastes (Foote *et al.*, 2015; Parliamentary Commissioner for Environment,2015); 2) the increasing range and scale of extractive industries in ocean waters, notable examples being oil and gas exploration and infrastructure, and deep sea mining proposals including phosphate from the Chatham Rise (Le Heron *et al.*, 2016); 3) record numbers of tourists, visiting coastal bays, ports and fjords, propelled by burgeoning container and cruise ship industries (Lewis, 2016; Pawson, 2018; PECC, 2017); 4) anxieties and conflicts over traditional and new uses and presence of additional numbers of users in marine spaces; 5) a firm commitment by central government to export-led growth (Ministry of Business Innovation and Employment, 2013), something that has forced its way into the thinking of those associated with the marine estate; 6) continuation sedimentation into rivers and coastal waters from urban and rural land development; 7) impacts from exotic forestry expansion and clear felling and replanting; and 8) a mixture of pressures from the scientific community about climate change impacts and loss of land and marine biodiversity (Lundquist *et al.*, 2016). These all involve issues and claims over rights, entitlements and ownership relating to activities co-inhabiting marine spaces.

Third, an accompanying trend has been that the standing, status and nature of the now more visible actors have been crucially altered, this being especially the case for Māori roles in marine governance and the gradual acceptance by other interests to Māori's changing influence. Governmental processes in the interventionist era generally failed to recognise either the significance of Māori collective organisation or the equivalent knowledge systems of Māori approaches relating to ecology and social processes (Maxwell *et al.*, 2017). Waitangi Tribunal research shows that, from the outset of their occupancy of ANZ, Māori culture was extensively connected with marine resources. Each iwi and hapū around the ANZ coast developed complex rules for controlling the waters adjacent to their land area. They also had considerable knowledge and understanding of what happened in mountains, catchments and rivers, and the flow-on effects upon habitation at the coast and the health of the ocean (Davies *et al.*, 2017).

European colonisers imposed a blanket of institutional and legal homogeneity over the pre-existing differentiated and localised iwi and hapū relations of ecology and economy. The subsequent land settlement process obliterated the customary assertion of iwi and hapū connections to the land and sea, and degraded and transformed the resource landscape upon which Māori depended (Le Heron & Roche, 2018). European governance yardsticks directed in general terms what land and sea activities and techniques of production were permissible and, most disruptively, rule-making anywhere was transferred to parliament, out of the hands of iwi and hapū.

The fundamental changes associated with the Treaty-Crown (TC) partnership that now characterises the ANZ scene bring into focus two key aspects of societal organisation – the power relations in which choices and decision-making possibilities are created and embedded and the decision-making processes over pre-given choices. The former is ontological, that is, it defines what can prevail. The latter is epistemological and methodological, detailing how decision-making is laid out and might occur. Thus, we need to be cautious about the basis on which the partnership is able to take effect and evolve. In the legal context of ANZ, any power vested in Māori to make decisions or enforce decisions comes from agreements with the Crown and these are endorsed in parliament. In the 20th century, Māori had advisory roles to the Crown, without power to make laws or to make much progress on their aspirations of co-governance, while pākehā institutions were aided by relations of dominance power (Hikuroa, 2017; Le Heron & Roche, 2018;, Simmonds, 2016) and could constantly remarginalise Māori entities, rights and entitlements (Came, 2013). The distinctions drawn expose the difficulties and dilemmas of building 'partnership' co-governance frameworks. Māori are no longer merely stakeholders, lesser stakeholders, providers of inferior ecological or social knowledge, or less well-equipped in terms of participatory expertise, hallmarks of past power relations. In today's governance world, such operational assumptions would be a retreat to the hierarchies Māori have fought against. Were there to be widespread acceptance that Māori culture, knowledge and social organisation have been ontologically displaced in the past, and that the TC redresses much of this prescription of how to engage, then we would expect quite different protocols being developed and embraced from the outset, in the PPs under scrutiny.

What might the seascape of participation arrangements look like around the coast of ANZ? Our empirically grounded research encourages us to hypothesise that every PP will be confronting many implementation challenges which will be politically mediated by both iwi and hapū strategies and the strategies of other parties. This is likely to include attempting to co-design a PP format, setting up the PP, clarifying the rules of engagement and the content of agenda, operationalising PP procedures, negotiating emergent differences and tensions amongst those involved, consolidating recommendations, championing adoption of proposed changes, and getting closure on changes in procedures expected to lead to new behaviours by actors.

3 Identifying and locating key and contemporary PPs

PPs have political agency in so far as they can showcase and attempt to make the most of competing views relating to the governance of marine spaces. ANZ has an enormously long coastline in relation to its land area. Very few if any portions of the country's coast and proximate waters are not the object of contemporary governance interventions. At present, coastal and ocean management in New Zealand is covered under 25 statutes governing 14 agencies across seven spatial jurisdictions (Bremer & Glavovic, 2013). To this official list can be added the 500 or so iwi and hapū marae-based collaborations (MaoriTelevision, 2017) that have endured despite being functionally marginalised by past government legislation. The Resource Management Act 1991 (RMA) established an effects-based planning system in which responsibility for the sustainable management of natural and physical resources is devolved to regional and district councils, which are expected to monitor and guide investor behaviours, making use of consent mechanisms. Regional Councils are the lead agency for coastal management and, as such, are responsible for managing marine ecosystems to 12 nautical miles (territorial sea). Beyond the territorial sea, activities related to the oil, gas and mineral industries, among others, occurring in the exclusive economic zone (from 12 to 200 nautical miles) are regulated by the Exclusive Economic Zone and Continental Shelf (Environmental Effects) Act (EEZ Act) 2012 and subject to appeals via the Environmental Protection Authority. Little is known, however, about the breadth, ramifications and implications of the legislative web encompassing marine activities (see Bingham *et al.*, 2018 for a first report on how ecosystem-based management is embedded in ANZ resource-management legislation).

A great many participatory initiatives are found in New Zealand's marine waters, but most are single-issue orientated. The most classic example is the Coast Care movement which has generated approximately 100 Coast Care projects (Coastal Restoration Trust (2017). Another popular movement is around Citizen Science (Peters, 2016) where the education dimension is paramount. Neither of these examples has explicitly addressed multi-use/users marine spaces as a focus. Harmsworth *et al.* (2014) have documented the rise of Treaty Settlement participation initiatives, some tied to existing institutions such as those of regional councils and others formed as independent entities, with distinguishing rights in institutional settings. The plethora of participatory initiatives is suggestive of a turn, for multiple reasons, towards participation and collaboration. We focus on the nature and scale of PPs in multi-use/user marine spaces in this century.

Initially the research team identified and mapped by regional council areas the nation's main PPs and compiled background information on their origins, evolution, purpose, procedures and so on (Blackett *et al.*, 2016). This scoping gave a basis for a preliminary assessment of material and discursive conditions associated with each initiative, an exposure to local and more general rationalities, and deliberating entities constituting the PP (Hobson, 2009; Warren & Visser, 2016). The emphasis is on evidence of higher-level normative goals being mobilised in strategic ways. We acknowledge that the methodology does not embrace operational planning. From a long list of 19 initiatives, 13 were investigated for their evolution and revealed practices. The investigation used websites, published works, submissions, government reports, newspapers and academic articles as principal sources, as they give tangible statements about what was being done and how. There was the added advantage that the sources required few resources to gather detail. A down side, however, is that information could be outdated or sketchy and even insufficient to reasonably identify moderate detail of key features, but positively, they are readily verifiable sources. On the few occasions where there was uncertainty about factual details, the nature of developments and the significance of any process changes and notable achievements, clarification was sought through phone calls or short interviews. Purely land-based initiatives were excluded in order to maintain the integrity of the marine initiatives' population.

4 Main features and collectivising efforts of PPs

Both Figure 7.1 and Table 7.1 are unique mappings and tabulations of ANZ's contemporary marine spaces. The figure and table are constructed to give initial visibility to the location of the PPs, what has been assembled and to the politics and power relations, at least as far as public sources allow, defining the conspicuous and publicised PPs. Significantly, the locations of the PPs

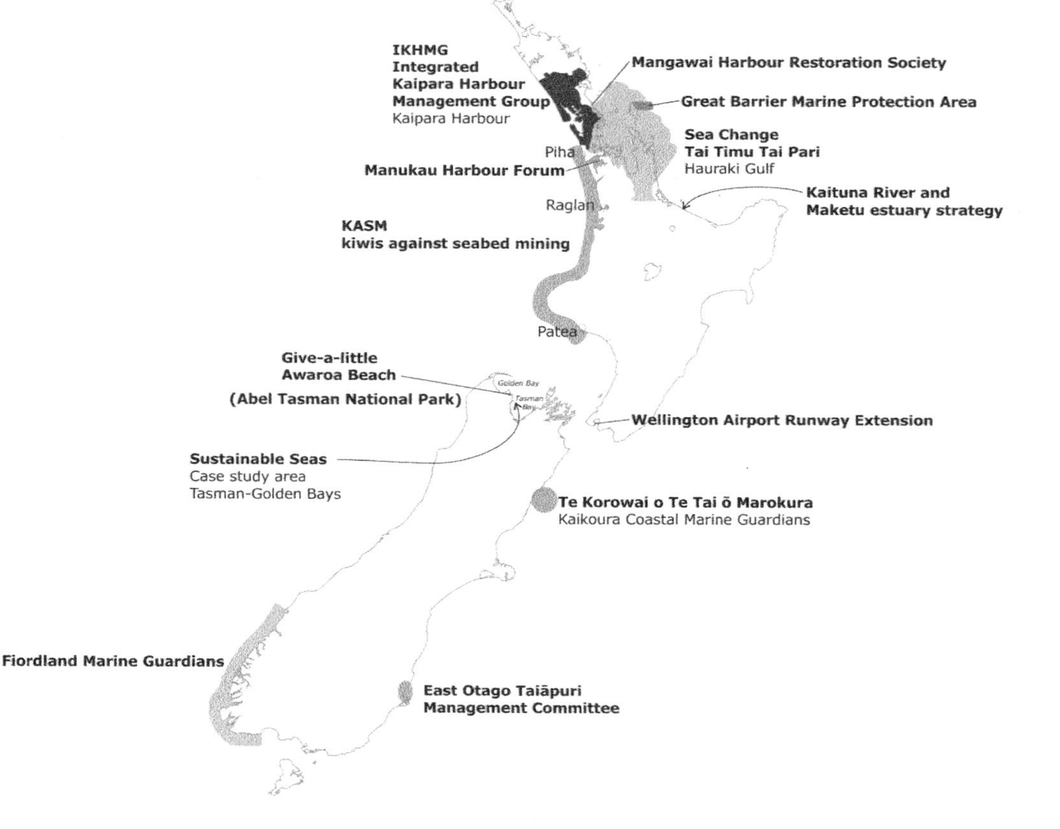

Figure 7.1 Aotearoa New Zealand Multi-Use Marine Participatory Processes Project Areas

Table 7.1 Features of main participatory processes in Aotearoa New Zealand's multi-use marine spaces

Participatory initiative	*Date created*	*Geographic scale/s*	*Formative environmental trigger*	*Main purpose/s, changing roles*	*Initiative significance as intervention in marine space*	*Main leadership features*	*Basis of political legitimacy*	*Key implementation achievements*
Kiwis Against Seabed Mining (KASM)	2005	National, also drawing in international constituency	Foreshore and seabed controversy in 2004 opened new issues around seabed mining	Prevent proposals to mine the foreshore and seabed in ANZ waters being approved	Public mobilisation, ongoing contestation over consent processes, direct opposition to extractive activities in ocean areas	Raglan community (2005 and iwi), widened over decade since inception, to become unofficial public portal into Environmental Protection Agency processes	Incorporated society (not for profit and in manner of charitable trust), government unable to withdraw charitable trust status	Two successful appeals and one defeat in consent process appeals
Integrated Kaipara Harbour Management Group	2005	Regional, catchment areas of Kaipara	Community and hapū, ongoing conflict with commercial fishing, landowners and land users	Promote integrated catchment management, inter-agency collaboration with kaitiaki of Kaipara Harbour and catchment Environs Holdings Ltd	Māori leadership in integrated catchment management in response to measurable cumulative effects, innovative flagship farms, inability to develop a restoration plan	Environs Holdings Ltd/Te Uri o Hau Settlement Trust through kaitiaki unit, Ngati Whatua Nga Rima o Kaipara, focus on scientific evidence and educational outcomes	Treaty settlement investment in collaborations around documented state of Kaipara	Effective catchment and estuary monitoring programme now in place, undertook gap analysis relating to Kaipara futures, co-ordinated successful ownership arrangements to fulfil goals for Kaipara, engagement of some interest groups (farmers) slow

Mangawhai Harbour Restoration Society	1994	Local	Accelerated erosion/ institutional inaction	Close second harbour, reinstate old harbour, in recent decade shift to contested focus on mangrove management	Social networks, re-addressed early processes, evolved into community and Māori led restoration	Community led by well-known locals	Local support, against planning procedures adopted, as these seen as basis of questionable harbour creation and later inaction over protests	Restoration work completed, unclear outcomes in relation to mangrove controversy
Sea Change Hauraki Gulf Marine Spatial Plan	2013	Regional, self-identified boundaries embracing gulf waters, islands and peninsulas	Many multi-use and user conflicts centred on commercial and recreational fishing, tourism, port activities, sedimentation from rural and urban sources, loss of shellfish beds	Secure a healthy, productive and sustainable resource shared by all associated with gulf	In marine policy and marine science circles seen as a 'trial' of regional marine spatial planning processes, variety of implementation impasses met as initiative evolved	Partnership combinations between manawhenua, central and local government (Auckland Council and Waikato Regional Council) and other stakeholders	Broad aspirations for better Hauraki Gulf futures springing from multiple sources	Sea Change Report with detailed recommendations by topics of concern but muted advocacy of implementation published late 2016, Auckland Council not enthusiastic about implementation, Waikato Regional Council developing implementation plan
Hauraki Gulf Forum	2000	Regional, boundaries mandated in Hauraki Forum Act	Environmental state of Hauraki Gulf reported in succession of widely distributed reports	Promote and facilitate integrated management, protection and enhancement of the Hauraki Gulf	Non-statutory body administered by Auckland Council – no powers to implement	Ministry for Environment, Ministry of Fisheries, Te Puni Kokiri, Councils, Tangata Whenua. Operational support from Auckland Council.	Pioneering Hauraki Gulf Maritime Park Act 2000, organised annual conferences on state of Hauraki Gulf, part of Sea Change processes	Exemplary monitoring framework in place early, four State of Hauraki Gulf reports published, no translation of science findings into implementation plans

(*continued*)

Table 7.1 (Cont.)

Participatory initiative	*Date created*	*Geographic scale/s*	*Formative environmental trigger*	*Main purpose/s, changing roles*	*Initiative significance as intervention in marine space*	*Main leadership features*	*Basis of political legitimacy*	*Key implementation achievements*
Manukau Harbour Forum	2010	Local	Restoration of declining ecology of Manukau Harbour	Ensure a rich and diverse marine and terrestrial environment for all	Collective of Auckland Council and local boards, with strong public input and general oversight	Local boards and community interests	Local public support and commentary on marine space	Local values delineated, monitoring proposals
Great Barrier Environmental Strategy Planning Committee	2016	Local	Addresses overfishing by advocating Marine Protected Area	Ensure abundance for future generations	Reveals phases of evolution in Marine Protected Area (MPA) processes	Department of Conservation, local board, Auckland Council, community, and Ngati Rehua Ngatiwai ki Aotea	Local agitation over state of marine environment, especially falling fish stocks	Advocacy for Marine Protected Area but no central government move to legislate for MPAs
Kaituna river and Maketu estuary strategy	2008	Local	Treaty Settlement opened opportunity for iwi to press for redress of encroachment of capital works for agriculture on ecosystem of river	Improve sustainable management of valuable river and estuary resources	Poised at implementation stage	Iwi/regional council supporting 'local people's stewardship (iwi not named in official documents)	Historical legacy over river diversion, wetland loss and degradation in Kaituna River and catchment	Mobilised interests into conversation, but hampered by legacy institutional politics and interests of mainly farmers

Wellington Airport runway extension	2016	Local	Restricted commercial decision, localised ecological impact	Infrastructure and runway for larger planes	Example of preference for consultation rather than collaborative processes	Greater Wellington Regional Council and Wellington City Council, iwi noticeably absent	Commercial imperative, seeking to justify economic over ecological concerns	Proposal stalled
Awaroa crowd-funded beach purchase Abel Tasman National Park	2015	National/ international	Threat of foreign privatisation of portion of land adjacent to Abel Tasman National Park	Buy a pristine beach for public park put up for sale by owners	Innovative funding mechanism, crowd funding Giveaittle.	Concerned citizens as champions, linked at outset with Department of Conservation and local iwi	Public empathy with acquisition of asset for public	Spectacular success as a funding model, sensitive engagement with local iwi and with government agencies
Kaikoura Coast Guardians (Te Korowai)	2009	Local	Need for fisheries protection for Ngati Kuri as the Tangata Moana of Te Tai o Marokura, whale and dolphin tourism and other recreation activity expansion	Achieve a flourishing, rich and healthy environment where opportunities abound to sustain the needs of present and future generations	Integrated land and water plan for the Kaikoura coast	Te Korowai (citizens and Nga Tahu), supported by Tai-o-Marokura Marine Management Act 2014, drew inspiration and practical experience from Guardians of Fiordland	Multi-user collaboration, informed by Guardians of Fiordland Fisheries Inc.	Pioneering co-governance implemented, facing huge challenges around restoration and revitalising marine economy post November 2016 Magnitude 7.8 earthquake

(*continued*)

Table 7.1 (Cont.)

Participatory initiative	*Date created*	*Geographic scale/s*	*Formative environmental trigger*	*Main purpose/s, changing roles*	*Initiative significance as intervention in marine space*	*Main leadership features*	*Basis of political legitimacy*	*Key implementation achievements*
Guardians of Fiordland Fisheries Inc/ Fiordland Marine Consortia	1995	Regional	Conflicts over fishing areas, later responded to influx of cruise ships in Fiordland waters	Ensure quality of Fiordland's marine environment and fisheries, including the wider fishery experience, maintained or improved for future generations	Fishers collaborating	Led by local Fiordland users and community representatives, became more institutionalised when focus widened to include tourism management	Willingness to collaborate and accept individual and collective commitments, style of collaboration altered as tourism grew	Moved from major conflicts amongst fishers to negotiated rules of operating in marine space, leading edge principles and practices in managing cruise ships and tourist numbers more generally in fiords
East Otago Taiapuri Management Committee	2001	Regional	Concern over depleting paua stocks, conflicts involving competing types of fishers	Management of the customary fisheries (using appropriate sustainable management measures such as rahui)	East Otago Taiapure Management Committee (EOTMC)	Members of the Kati Huirapa Runaka ki Puketeraki applied for a taiapure (local fishery)	Opportunity for Māori leadership in co-governance	Introduction of traditional rahui (ban on fishing) as management tool

(Figure 7.1) encompass marine spaces that have been subjected to dynamics sourced in land, coast and sea. This is also the case for PPs not investigated. In many instances the PPs have gained notoriety for the way they confront many unacknowledged or ignored dynamics that are part of the ANZ scene. They highlight how land-coast-sea relations are intertwined (Winder & Le Heron, 2017). The pākehā administrative severance of this holistic connectivity is an extraordinary legacy of ANZ's knowledge and institutional compartmentalisation.

Table 7.1's special significance is that it puts into one simple framework attributes that distinguish the complement of PPs, indicates notable feature/s of the origins and emergence of the selected PPs, provides simplified contextualisation of the PPs, links the PPs into contextual dynamics, and gives scope to gain some understanding of emergent PPs. The overview is thus a rare insight into the multiplicity of PP situations and therefore governmental approaches in ANZ's marine realm.

PPs are unambiguously sites of intervention in the marine space or with reference to issues of contention. Whether the interventions are localised or wider in their geographic reach, their responses and approaches to multi-use/user pressures is inevitably multi-scalar and rooted in multiple histories. The juxtaposition in the table of environmental triggers and main purpose elicited from website descriptions, for example, documents the degree of consistency underlying all of the PPs and reveals both a considerable variety and depth to the formative conditions and responses across the PPs. Indeed, it is reasonable to characterise the PPs as diverse sites of generative political action, the nature, extent and impulses of which are highly contingent, reflecting the stamps of different kinds of leadership and political legitimacy from the past and in new conditions.

Regardless of their origins and nature, PPs are unquestionably recent initiatives, formed in times of great contextual change, with in many cases minimum governmental guidance. The earliest, the Mangawhai Harbour Restoration Society (1994) and the Guardians of Fiordland (1995), struggled to achieve momentum, facing frustrations springing from the incompatibility of existing institutions with the collective vision of multiple interests (Mangawhai) and the reluctance of fishers to engage in efforts to improve their collective outcomes (Fiordland). In both cases it was a decade or more before learnings from participation were translated into tangible governance and management proposals and eventual implementation. The PPs formed in the 21st century have benefited from accumulating knowledge. The Guardians of Fiordland, for instance, were consulted by the Kaikoura Te Korowai initiative, as they fashioned their participatory processes. The ongoing development of these two PPs was materially assisted by specific legislation.

An important feature is that the PPs are largely independent of central government, but not independent of the historical relations enshrined in the TC partnership or the traditions of conventional models of governance still embraced by regional councils and, as suggested above, the mandating by parliament of mature PPs. This means the focus of analytical attention must swing towards the nature, directions and interactions of a plurality of agents in new relations. Further, each PP is a nexus of both Māori, iwi and hapū trajectories of participation meeting newly framed PP structures, and PPs engaging through their new structures with newly started Māori, iwi and hapū initiatives.

When viewed at the national level through a socio-theoretic lens, they are highly differentiated and could be seen as a series of independent and spontaneous experiments in social organisation for the sustainability of marine spaces. The PPs form a spectrum of governance modalities, from those guided by bureaucratic forms, competitive interactions amongst newly assembled players and efforts to promulgate network governance. Importantly, these modalities are co-constituted rather than standing in isolation. Yet they should not be regarded as some kind of system or whole. Rather, they are unique assemblages that when observed individually demonstrate mix after mix of different actors striving to find collective configurations to advance the health and wealth of their marine life space.

Are there aspects especially germane to their emergence in the regulatory conditions of their time? Foremost, the inventory as a whole is revelatory. It shows remarkable consistency amongst the PPs – all are dedicated to confronting the immediate and longer-term problems of multiple use and users – the new problematic. There is strong evidence that ANZ's coastal and near-shore realm is beset by a wide range of influences emanating from users attempting to retain or extend access and ownership to waters for their particular ends. What the conflicts look like is highly varied but each springs from issues over the attempted intrusion of ownership, property rights or comparable entitlements into marine spaces. This dimension is largely absent from public discourse though it lies behind the nature of environmental triggers. Interestingly and importantly, the statements of purpose are uniformly statements of collective hopes that signal the content of alternative marine space futures. The umbrella-like quality of guiding statements of each PP attests to the desire to bring interests, issues and insights together, instead of treating them in governmental isolation. Both the reasons driving the formation of the PPs and the ethos encapsulated in statements of purpose indicate a governance complexity that exceeds the administrative scope and capacities of existing institutions. In particular the economic, ecological and cultural impacts in the same marine spaces are centre-stage. This is a multiverse of influences and power possibilities and sought-after futures which are internalised in each PP. Le Heron & Logie (2017) report on first evidence about the ontological diversity and its implications of the Sea Change marine spatial-planning process relating to the Hauraki Gulf.

A conspicuous feature of the PPs is the high visibility of Māori leadership. As might be expected in the new regulatory climate, 12 of the PPs examined are distinguished by clear lines of Māori, iwi and hapū interactions in their formation and subsequent emergence. One PP, the Wellington Runway

Extension, followed a path of minimum consultation with Māori, a situation that contrasts with the diversity and complexity of the other PPs. Available documentation suggests, however, that this may only be a partial step towards embracing the TC ethos. Nonetheless, the ability to create a participatory platform having iwi and hapū centre-stage in co-developing a PP is a monumental step towards a style of co-governance, compared to the exclusionary and prescriptive politics and policy-making of the past. The acknowledgement of Māori in the names of some PPs is not a trivial or gratuitous move. The documents perused suggest the incorporation of Māori concepts and terminology, with the idea of integrated management, used in quite a few PPs, stands as code for both catchment-wide and the even more complete 'mountains to the seas' visioning.

Finally, what of implementation achievements? The key achievements column in Table 7.1 provides unambiguous evidence that the PPs are accomplishing much in spite of the nature of the governmental conditions they are working under. The examples included do not make up a comprehensive list. The case of KASM, a nation-wide marine citizenship movement, inspired by localised concerns (Figure 7.1), reveals difficulties imposed by working within the existing institutional setting to try to attain procedural improvements. The ledger for KASM indicates two successful appeals in the consenting process of the Environmental Protection Authority, though these were then followed by a declined appeal. At the other extreme, the activist crowdfunding initiative (Awaroa) concluded its mission very rapidly. We surmise as a tentative conclusion that it is likely that in the absence of the PPs in their marine spaces, progress would have been very limited or slow, since the PPs have provided an institutional framework to focus efforts on the wider gains relating to the health of the marine space.

5 Participation for implementation?

The chapter's title poses a very contemporary question with an international flavour: 'Have PPs in ANZ advanced the prospects for efficient, effective and power-cognisant implementation of sustainable outcomes in multi-use/user marine spaces?' The question goes to the heart of what should be governed and managed by whom in what sorts of collective arrangements and the kinds of approaches that might be pursued to delineate and achieve goals. The option of participatory processes in contrast to consultative or prescriptive governmental structures heralds fresh possibilities for engagement and new sorts of knowledge and learnings from being part of differently structured processes. The implicit assumption is that involving multiple and often competing interests into sharing and learning processes can foster transformations. In many respects the question troubles the orthodoxy that participatory processes should be viewed as somehow on trial. Instead, the potency of participation and participatory processes has been laid out. The chapter's empirical study of the ANZ scene and a selection of PPs has been informed by theoretical ideas that highlight the need to place and outline participatory initiatives in wider and changing structural influences, be cognisant of the collective strengthening contributions of PPs and to regard PPs as emergent phenomena that are co-shaped as much by their capacities and capabilities as structural pressures. Interpreted as social or collective technologies, PPs should be regarded as inherently emergent, replete with possibilities that are discoverable should there be a culture of exploration drawing on the interests associated with a PP. This view diverges from much of the literature where participation is seen as a final clip-on to established, usually hierarchical decision-making processes, and so remains poorly understood.

The chapter has made several breakthroughs relating to how implementation might be understood and used to convey achievements resulting from participatory processes. First, we now have a view of implementation that is a product of our methodology. By this we hold that our emphasis on structure and agency, and the fundamental importance of grasping new kinds of regulatory associations, sees us treating implementation as a malleable concept embedded in political and socially constructed realities, serving to identify a number of different sorts of developments. While implementation as a word is popularly seen as the end-point of a series of decision-making steps, it is actually as much context-created, exceedingly complex and context-creative in so far as it is unavoidably describing multiple change possibilities. The new regulatory environment identified in ANZ has meant the placement of implementation as 'the end-point' in a command-and-control governing regime no longer holds, or at least is matched by developments that suggest implementation is a synonym for multiple achievements. Which achievements are highlighted and what significance is attached to them depends on whose perspective or perspectives are being brought to bear in any evaluative and interpretive exercise. The idea of linear steps towards bureaucratic authorisation of implementation is altering into multiple expressions about the value and impacts of implementation claims. Thus, implementation remains an abstract idea which continues to be useful, but it is simultaneously about truth claims and such claims may be made in many settings in PPs and on behalf of PPs for all sorts of differing ends. This shift from outcome per se to a plurality of traceable detail from differing worldviews, and starting-points, runs counter to conventional interpretations.

Second, we need to be sensitive to the idea that PPs are assembling devices for making choices, and choices are made in explicit and informal decision-making processes. This interpretation of ongoing or emergent qualities directs attention to the potentialities of each PP (as indicated in the first instance by identifiable achievements) to move deliberation towards bolstering goals about the marine space's futures. This could be regarded as a kind of discovery process in which implementation aspirations are always under review.

Third, to what extent are PPs able to engender whole system redesign? Anecdotal and circumstantial evidence gathered when we were investigating each PP would suggest that each PP had come up against barriers in the existing system that translated into an unwillingness to take seriously proposals for new directions. In recent years, especially at Environmental Defence Society conferences and workshops, two simple questions have been raised time and time again: What next, and how? Were we to unpack the questions, we would find the elephant in the room is the incredibly important role that central government still plays, in either nurturing or holding back PP proposals for implementation, by direct moves to dismantle or re-assemble the background institutional and legislative infrastructure in which PPs find themselves. This situation amounts to an implementation impasse, a situation that we can make some sense of given the chapter's methodology and evidence in Table 7.1.

Yet any legislative and institutional fabric is created by people and potentially can be reassembled by people. This easy-to-say proposition was recently given unprecedented weight at the 2017 New Zealand Coastal Society conference, where the extent of change central government embarked upon to create conditions for rapid and fair implementation of disaster recovery after the Kaikoura earthquake were outlined. Keynote speakers from the North Canterbury Transport Infrastructure Recovery (NCTIR) unit (Bull & Sweeny, 2017) showed how the process of making things happen started with an empowering act and then moved through orders in council that delegated powers, the formation of NCTIR to govern and project manage subsequent steps, changes in consenting processes relating to slips and roads north and south of Kaikoura, approvals for the Department of Conservation to deviate from usual protocols relating to intervention in animal and plant communities, new authorities relating to archaeological undertakings and provisions to make rapid changes in property boundaries. These facilitative and supportive actions have enabled lower-order innovative and transformational activities to be conceived, promoted and carried out. The speakers fully acknowledged a central tension of the process: involving rebuilding quickly or rebuilding deliberatively. Clearly the urgencies of a disaster-recovery programme were enough to create an obligation for central government to act. The resultant collaborative processes amongst often competing interests depended on high levels of negotiation and trust. Whether the learnings from the Kaikoura experiences can be melded into those from participatory experiences elsewhere in ANZ is now a highly relevant and timely question.

Acknowledgements

This project was funded by a Ministry of Business, Innovation and Employment research contract CO1X1515 (Sustainable Seas National Science Challenge) through the NIWA (National Institute of Water and Atmosphere), Hamilton. The chapter draws on investigative, analytical and interviewing research undertaken as part of Project 1.2 'Testing EBM-supportive participatory processes for application in multi-use marine environments' co-led by Paula Blackett and Richard Le Heron, in the Our Seas programme of Sustainable Seas. The chapter reports on selected findings from the first phase of the project. We are indebted to the many people who have supported our research with time and insight or have been directly or indirectly involved in the project's research processes.

Notes

1 The Treaty of Waitangi is an agreement made in 1840 between representatives of the British Crown and more than 500 chiefs. Ostensibly the purpose of the Treaty was to enable British settlers and the Maori people to live in New Zealand under a common set of laws or agreements, but the reality was that it mandated the British Crown to oversee European settlement. The Treaty settlement process introduced late in the 20th century was designed to redress historical grievances arising from breaches of the Treaty almost from the time it was signed.

2 The largest political grouping in pre-European Maori society was the iwi (tribe). This usually consisted of several related hapū (descent groups).

Bibliography

Barney, D., Coleman, G., Ross, C., Steine, J., & Tembeck, T. (Eds.) (2015) *The Participatory Condition*. University of Minnesota Press, Minneapolis.

Bingham, K., Blackett, P., Greenaway, A., Glavovic, B., James, G., Gillespie, A., Peart, R., Ioins Magellans, C., Inglis, G., Le Heron, R., Logie, J., Lundquist, C., Whetu, J., Faulkner, L., Taylor, L., Reynolds, F., Scott, D., Te Whenua, T., Margaret, T., & McMahon, E. (2018). *How Current Legislative Frameworks Enable Ecosystem Based Management (EBM) in Aotearoa New Zealand.* Landcare Research, Auckland.

Blackett, P., Le Heron, R., Logie, J., Allen, W., Davies, K., Glavovic, B., Greenaway, A., Hikuroa, D., Lundquist, C., Simmonds, N., & Taei, S. (2016). Participatory processes for Sustainable Seas: a preliminary review of initiatives in New Zealand's ocean domain, Presentation to New Resource Geographies conference, Massey University, Palmerston North, 22 November.

Bremer, S., & Glavovic, B. (2013). Exploring the science–policy interface for integrated coastal management in New Zealand. *Ocean & Coastal Management, 84*, 107–118.

Brown, M. (2017). *Last line of Defence. Compliance, Monitoring and Enforcement of New Zealand's Environmental Law*. Environmental Defence Society, Auckland.

Bull, L., & Sweeney, M. (2017). North Canterbury earthquake – Environmental challenges and opportunities of the coastal reinstatement scheme, Keynote address, Tauranga Moana 2017, New Zealand Coastal Society conference, 14 November.

Came, H. (2013). Doing research in Aotearoa: a pakeha exemplar of applying Te Ara Tika ethical framework, Kotuitui. *New Zealand Journal of Social Sciences,. 8*(1–2), 64–73.

Coastal Restoration Trust. (2017). Available at: www.coastalrestorationtrust.org.nz/coastcare-groups/coastcare-groups-map/

Davies, K., Ratana, K., Lundquist, C., Fisher, K., Le Heron, R., Spiers, R., Foley, M., Greenaway, A., & Mikaere, H. (2017). From mountains to the seas: developing a shared vision for addressing cumulative effects in Aotearoa New Zealand. *Regions Magazine, 308*(4), 15–18.

Foote, K., Joy, M., & Death, R. (2015) New Zealand dairy farming: milking our environment for all its worth. *Environment and Management, 56*(3), 709–720.

Gelcich, S., Hughes, T., Olsson, P., Folke, C., Defeo, O., Fernandez, M., Foale, S., Gundersen, L., Rodriguez-Sickert, C., Scheffer, M., Stewart, R., & Castilla, J. (2011). *Navigating Transformations in Governance of Chilean Marine Resources.* Available at: www.pnas.org/cgi/doi/10.1073/pnas.1012021107

Harmsworth, G., Awatere, S., & Robb, M. (2014) Māori values and perspectives to inform collaborative processes and planning for freshwater management. Landcare Research Policy Brief, Lincoln.

Hibbard, M., Lane, M.B., & Rasmussen, K. (2008) The split personality of planning: indigenous peoples and planning for land and resource management. *Journal of Planning Literature, 23*(2), 136–151. doi:10.1177/0885412208322922

Hikuroa, D. (2017). Mātauranga Māori – the ūkaipō of knowledge in New Zealand. *Journal of the Royal Society of New Zealand, 47*(1), 5–10. doi:10.1080/03036758.2016.1252407

Hobson, K. (2009). On a governmentality analytics of the 'deliberative turn': material conditions, relationalities and the deliberating subject. *Space and Polity, 13*(3), 175–191.

Higgins, V., & Larner, W. (Eds.). (2017). *Calculating the Social: Standards and the Reconfiguring Of Governing.* Palgrave Macmillan, Basingstoke.

Hurlbert, M., & Gupta, J. (2015). The split ladder of participation: a diagnostic, strategic, and evaluation tool to assess when participation is necessary. *Science Direct, 50*, 100–113.

Innes, J., & Booher, D. (2004). Reframing public participation: strategies for the C21. *Planning Theory and Practice, 5*(4), 419–443.

Le Heron, E, & Logie, J. (2016). Insider -outsider perspectives about the emergence of the Hauraki Gulf Sea Change (Tai Timu Pai Pari) marine spatial plan participatory processes. Paper presented in special session on Making the most of participation for sustainable seas, Sustaining our Seas conference, University of Sydney, Sydney, 13 December.

Le Heron, R., & Roche, M. (2018). The taniwha economy. In Pawson, E. (Ed.) *New Biological Economies.* Auckland University Press, Auckland, New Zealand, in press.

Le Heron, R., Lewis, N., Fisher, K., Thrush, S., Lundquist, C., Hewitt, J., & Ellis, J. (2016). Non-sectarian scenario experiments in socio-ecological knowledge building for multi-use marine environments. Insights from New Zealand's Marine Futures project. *Marine Policy, 67*, 10–21.

Lewis, N. (2016). Making a 'blue economy' in New Zealand. Presentation to Agri-food XXIII conference, University of Adelaide, Adelaide, Australia.

Lundquist, C., Fisher, K., Le Heron, R., Lewis, N., Ellis, J., Hewitt, J., Greenaway, A., Cartner, K., Burgess-Jones, T., Schiel, D., & Thrush, S. (2016). Prioritising marine science needs in New Zealand: a focus on multi-user management and policy to address cumulative effects. *Frontiers in Marine Science*, on Ocean research priorities and prioritising ocean research. doi.org/10.3389/fmars.2016.00002

MaoriTelevision. (2017). Must-have marae map at your fingertips. Available at www.maoritelevision.com/news/education/must-have-marae-map-your-fingertips

Maxwell, K., Taiapa, C., Ratana, K., Davies, K., & Awatere, S., (2017). Mauri Moana, Mauri Tangata, Mauri Ora. Paper presented at Tauranga Moana 2017, New Zealand Coastal Society conference, 14 November.

Ministry of Business Innovation and Employment. (2013). *Statement of Intent 2013–2016* . Wellington, New Zealand.

Nicolini, D. (2016). *Knowing in Organizations. A Practice Based Approach*. Routledge.

Pacific Economic Cooperation Council (PECC). (2017). Available at: /www.pecc.org/event-calendar/past-events/event/429-managing-the-blue-economy-future-of-port-management-and-shipping-in-the-asia-pacific

Parliamentary Commissioner for the Environment. (2015). *Update Report. Water Quality in New Zealand: Land Use and Nutrient Pollution.* Wellington.

Pawson, E. (2018). Introduction. In Pawson, E. (Ed.) *New Biological Economies*. Auckland University Press, Auckland, New Zealand, in press.

Peters, M. (2016). *An Inventory of Citizen Science Programmes, Projects, Resources and Learning Opportunities in New Zealand.* NZ Landcare Trust, Hamilton.

Rabe, L. (2014). Participation and justice in marine governance. Abstract. Paper presented at the XVIII ISA World Congress of Sociology, Yokohama, Japan.

Simmonds, N. (2016). Thoughts on kaupapa for interviewing, 14 August email to Participatory Projects team.

Taylor, G. (2013). Case Note: Environmental policy-making in New Zealand 1978–2013. *Policy Quarterly, 9*(3), 18–27.

Warren, C., & Visser, L. (2016). The local turn: an introductory essay revisiting leadership, elite capture and good governance in Indonesian conservation and development programmes. *Human Ecology, 44*, 277–286.

Winder, G, & Le Heron, R. (2017). Further assembly work: a mountains to seas blue economy imaginary. *Dialogues in Human Geography, 7*(1).

Woodward, E., & McTaggart, T. (2016) .Transforming cross-cultural water research through trust, participation and place. *Geographical Research, 54*(2), 129–142.

8 Transition management in coastal agriculture

Evidence from the German dairy industry

Oliver Klein and Christine Tamásy

1 Introduction

This paper explores the burgeoning issue of transition management focusing on sectoral change in the coastal regions of north Germany. We use the example of the dairy sector, even though this industry is not typically associated with coasts. In the German context, however, milk production and processing are most prominently located in the wider coastal zones of the North and Baltic Seas, thus playing an important economic role in these otherwise disadvantaged regions that have been facing profound structural changes in light of the declining port industries. Therefore, the importance of other sectors such as agribusiness has tended to increase gradually, which is in line with the German Ministry of Food and Agriculture (BMEL) stating that livestock production in Germany has developed very successfully in terms of economic progress over decades (BMEL, 2015). Hence, the coastal regions, especially in northwest Germany, could to some extent benefit from the dynamic development of the domestic dairy industry. Nevertheless, the BMEL also sees an urgent need for change to raise sustainability levels in animal-based food production in Germany. This requirement is related to the persistent trajectory of intensification and its inherent deficits with respect to animal welfare and environmental protection, accompanied by a decreasing social acceptance of intensive livestock production.

However, the initiation of a sustainability transition does not happen by itself. This argument is underpinned by Markard *et al.* (2012) who regard socio-technical transitions as multiple sets of interwoven processes that lead to a fundamental shift in socio-technical systems including far-reaching changes along different dimensions: technological, material, organisational, institutional, political, economic and socio-cultural. In the course of such a transition, new products, business models and organisations emerge, while technological and institutional structures as well as consumer perceptions change more or less fundamentally (Markard *et al.*, 2012). If these changes explicitly aim at more sustainable modes of production and consumption, one can speak of a sustainability transition, which is often induced by mechanisms of guidance and governance (Smith *et al.*, 2005). With regard to agriculture and food production, some argue that such a sustainability transition requires innovative approaches considering the principles of multifunctionality and/or extensification instead of keeping prices low at any cost (Tamásy, 2013).

The purpose of this chapter is therefore to start by describing the recent trajectory of the German dairy industry with a specific focus on the coastal regions in the northwest and, based on this, to map out urgent needs and possible directions of transitions towards more sustainable production systems. This is followed by reflections on sustainability strategies in practice exemplified by means of a medium-sized dairy firm in Lower Saxony, northwest Germany (Molkerei Ammerland). In doing so, we aim to contribute to the multi-faceted "geography of sustainability transitions" (Truffer & Coenen, 2012, p. 3), while focusing on coastal regions in general and coastal agriculture in particular. More precisely, we will unveil how a firm-based strategy relates to a specific socio-cultural context and, simultaneously, involves 'visible' elements of this 'coastal (agri)culture' in order to construct a self-image as 'a pioneer in sustainability', even though the real effects of this strategy are quite remarkable.

The analysis uses secondary data including official statistics on the German dairy sector, which are prepared annually by the Association of the German Dairy Industry. These data give clear insights on current structures and long-term developments of milk production and processing, allowing us to trace the ongoing pathway of intensification of the dairy industry in Germany. In addition, we use further relevant documents (e.g., market studies, press articles, newsletters, expert commentaries) as well as scientific literature on the dairy sector. The firm case study provides first-hand information on the respective milk processing company (e.g., statistics, annual reports, advertising material and sustainability reports) in addition to information received by extensive expert interviews. Through these means we outline the company-specific philosophy regarding sustainability in practice, innovation activity, market positioning, as well as transition management in holistic terms.

The paper is organised in five sections. After the introduction, we outline the conceptual framework based on the idea of sustainability transitions and its underlying approaches with a particular focus on food systems. This chapter also includes a brief summary of the state of research on that topic. This is followed by a description of the German dairy industry and its persistent trajectory of intensification, which has resulted in sectoral as well as regional concentration processes. Thereafter, we point out a number of problems and challenges indicating the need for a sustainability transition in the German dairy sector. After that, we present an industry case study (Molkerei Ammerland) in order to explore and discuss specific strategies aimed at achieving more sustainable modes of dairy production (and consumption). In the conclusion of this paper, the implications and significance of our results are discussed.

2 Conceptual background

A sustainability transition requires a fundamental shift in the socio-technical system of food production. Transitions unfold through the reconfiguration of networks of actors (e.g., entrepreneurs, firms, non-governmental organisations, states and consumers), institutions (e.g., norms, regulations and standards of good practice) and knowledge as well as material artefacts (Markard *et al.*, 2012). Innovation plays an important role as a key driver for the development of new solutions along the value chain from 'farm to fork'. Previous analyses of transitions in food systems have focused on urban food policies (Cohen & Ilieva, 2015), eco-innovations in supply chains (Mylan *et al.*, 2015) and pig husbandry (Geels, 2009; Elzen *et al.*, 2011).

Different frameworks have been used to study sustainability transitions in various socio-technical systems. Markard *et al.* (2012) differentiate between the socio-technical regime, strategic niche management, the multi-level perspective (MLP) and technological innovation systems (TIS). According to Truffer & Coenen (2012), MLP and TIS are the two dominant approaches in sustainability transitions research, even though the relationship between them is a matter of scientific debate in the fast-broadening research community. Early work on the multi-level perspective stresses the importance of landscapes, regimes and niches, which can be understood as nested hierarchy (Geels, 2002). Over the years, the MLP has been used to analyse the dynamics of various systems such as the automotive industry (Geels & Penna, 2015; Geels *et al.*, 2012), water supply and personal hygiene (Geels, 2005), factory production (Geels, 2006b), aviation (Geels, 2006c), biogas (Raven & Geels, 2010), sewers (Geels, 2006a), highways (Geels, 2007), electricity (Geels *et al.* 2016), nuclear and renewable energy (Geels & Verhees, 2011; Verbong *et al.*, 2008), coal (Turnheim & Geels, 2013), and horticulture (Berkers & Geels, 2011). Significant criticism on the lack of a consideration of agency in the multi-level perspective, the operationalisation of regimes, a bias towards bottom-up change models, epistemology and explanatory style, the treatment of socio-technical landscape as residual category, and flat ontologies (versus hierarchical level) are responded to by Geels (2011).

Just recently, Geels (2014) has developed a 'Triple Embeddedness Framework' (TEF), which conceptualises firms (in industries) and their economic environments. A TEF distinguishes between the mechanisms through which actors exert pressure on firms. On the other hand, firms can use various strategies in response, which are also influenced by diverse regime elements enabling or constraining perceptions. The TEF was explicitly developed with the agrifood sector in mind (Geels, 2014). Livestock agriculture is strongly associated with huge societal problems such as climate change, and incumbent firms are often seen to be reluctant to address these properly. In addition, parts of the agrifood sector have a high public visibility and a strong political influence so that normative, cultural and political processes matter (Geels,

2014). Markard *et al.* (2016), for example, belong to an emerging strand of research on the politics of transitions.

Many analyses in relation to technological innovation systems were intended to inform policy and, therefore, much attention was paid to drivers and barriers of innovation (Markard, Raven, & Truffer, 2012). In general, the TIS concept investigates the emergence of new technologies and related institutional and organisational changes by using a chain-based business perspective (Tukker *et al.*, 2008). Most recently, Bergek *et al.* (2015) discuss interactions between technological innovation systems and wider 'context structures' – technological, sectorial, geographical, and political – in response to profound criticism (Markard *et al.*, 2016). This extended perspective supports the arguments of Hansen & Coenen (2015) who posit that sustainability transitions "are constituted spatially and unpacking this configuration will allow us to better understand the underlying processes that give rise to these patterns" (p. 6). However, most empirical analyses of sustainability transitions have focused on the national scale (e.g., Coenen *et al.*, 2012) or a particular city (e.g., Cohen & Ilieva, 2015), while neglecting specific spaces such as coastal zones or rural areas, which are dominated by conventional agrifood production.

Therefore, we see a need to increase our knowledge on how sustainability transitions within these types of space evolve and unfold. Particularly, coastal zones which are characterised not only by their relatively peripheral location, but also by the decline of formerly important industries such as shipbuilding or fisheries, show unfavourable conditions for entrepreneurial innovations at first glance. Against this background, and based on the premise that innovation activities of firms are crucial for the initiation of sustainability transitions, it seems useful to adopt an 'inside-out' perspective, as demanded by Geels (2014). This perspective puts the firm into focus by assuming "that firms-in-industries not only *adapt* to external pressures but also strategically attempt to *shape* their environments, which are thus partly enacted" (Geels, 2014, p. 268). In this regard, Dosi (2000), among others, has highlighted the relevance of innovation strategies with firms competing by offering improved products or doing things better and faster than competitors do. We adopt these arguments for our analysis of alternative pathways towards more sustainable food systems in coastal areas. The food system, in the sense we use it here, includes all those activities involving the production, processing, transport and consumption of food. It comprises a set of elements (e.g., actors, institutions, and technologies), which are bound together by interdependent network relations and governance modes. In the following chapter, we describe the specific trajectory of the German dairy industry as an example of a food system in transition whose elements are primarily located in the coastal areas of northwest Germany.

3 The dairy industry in Germany

The dairy industry is one of the largest sectors of the German food industry, with a production volume of 32.7 million tons of milk and an annual turnover of €21.9 billion in 2016. Currently, there are 67,319 dairy farmers and 152 milk-processing companies still active on the market, which has changed significantly during the last decades (Association of the German Dairy Industry, 2017). This becomes particularly evident in the number of dairy farms, which has dropped rapidly by more than 70% between 1991 and 2016 (ZMP, 1995; Association of the German Dairy Industry, 2017). In the same period, the total number of milk cows declined from 5.6 million to 4.2 million animals, thus indicating a significant growth of herd sizes per dairy farm. Today, the average herd size in Germany amounts to approximately 60 milk cows (compared to 22 milk cows in 1991). Despite the reduction of dairy farms and livestock numbers, the total amount of milk production in Germany has continuously risen to 32.7 million tons in 2016, compared to a production volume of only 29.1 million tons in 1991. This development is a result of the strong increase of the average milk yield per cow achieved by means of optimised feeding (e.g., larger proportions of feed concentrate). The average milk yield per cow and year is currently about 7,700 kg, representing an increase of almost 60% as against 1991, when the average annual output was around 4,800 kg per cow (ZMP, 1995; Association of the German Dairy Industry, 2017). These data provide the first evidence for the specific trajectory of the German dairy industry based on and shaped by an enormous intensification (see Figure 8.1).

This also applies to the milk-processing level characterised by ongoing sectoral concentration due to mergers and acquisitions. Referring to this, the merger of Nordmilch and Humana Milchunion in 2011 was likely the most remarkable amalgamation, resulting in today's DMK Deutsches Milchkontor. With respect to production volume and turnover, the two merging companies

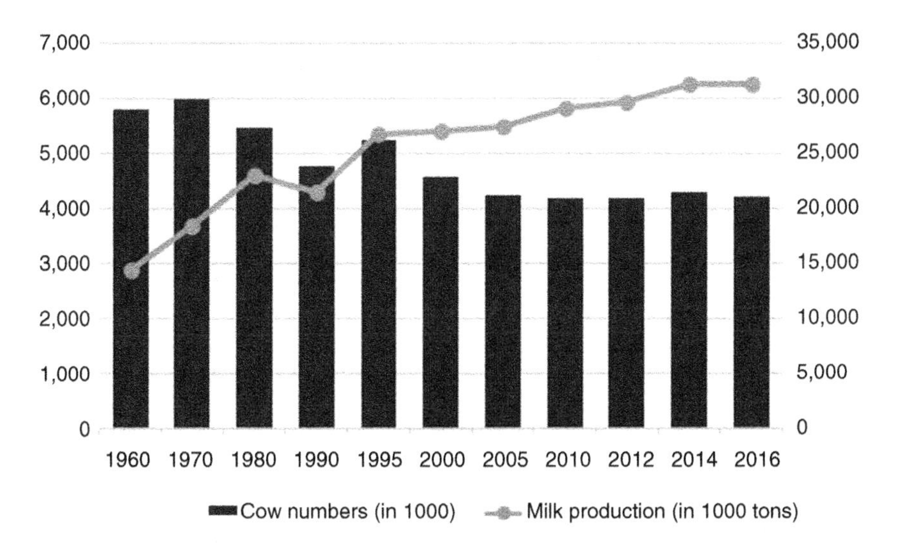

Figure 8.1 Development of dairy farming in Germany (based on AMI, 2016; Association of the German Dairy Industry, 2017); after 1990: statistical effects due to the reunion of Germany

Table 8.1 Top 10 dairy companies in Germany in 2016

Rank	*Company*	*Turnover (in € million)*
1	DMK Deutsches Milchkontor	4,600
2	Müller	1,800
3	Hochwald Foods	1,440
4	Arla Foods (DK)	1,400
5	FrieslandCampina (NL)	1,100
6	Bayernland	1,000
7	Zott	902
8	Ehrmann	755
9	Fude & Serrahn	647
10	Molkerei Ammerland	639

Source: based on Association of the German Dairy Industry (2017)

were ranked first and third in the German dairy market at that time. Thus, DMK is now by far the largest dairy company in Germany with a production volume of 6.7 billion kg per year and an annual turnover of €4.6 billion (DMK, 2016). The concentration process in the German dairy industry is further evidenced by the rapidly decreasing number of milk-processing companies. While there were still 379 dairy companies in 1991, this number dropped to only 152 in 2016 (see Figure 8.1). In particular, the smaller dairies were being pushed out of the market due to a lack of competitiveness within an increasingly contested economic environment. It is apparent from Table 8.1 that two of the top five dairy companies in Germany originate from abroad – the Danish Arla Foods and the Dutch FrieslandCampina – which is an indicator for the growing internationalisation of the dairy industry. Another specific feature of the German dairy market is the fact that about two-thirds of all German milk is processed by co-operatives (Schlecht & Spiller, 2009). Co-operatives are jointly owned by dairy farmers based on the principles of self-help, self-administration, identity, personal responsibility, democracy and solidarity (Rhodes, 1983; Theuvsen, 2006).

In Germany, the dairy sector is one of the most strictly controlled food industries, particularly in terms of quality management and product safety. As the quality management in dairy firms is based on the principles of self-responsibility and self-monitoring for risk analysis, the practical implementation includes a number of different quality management systems or the HACCP[1] control system for threat and risk assessment (Association of the German Dairy Industry, 2017). Moreover, the German dairy industry has established diverse quality programmes on a voluntary and individual basis. These programmes are flanked by a series of laws and regulations both at national and EU level which ensure quality assurance in dairy farms and milk-processing firms. In this regard, EU regulations (e.g., the EU dairy regulation) create uniform legal frameworks for all European member states, while being supplemented through national legal acts such as the "Milchgüte-Verordnung" in Germany. Other policy instruments with a strong impact on the European dairy industry focus on market regulation, particularly the introduction of the milk quota in 1984. The aim behind this intervention measure was to prevent the oversupply of milk production and the consequent drop in prices at EU level. However, as the milk quota regime was due to expire in April 2015, the European dairy market has become more liberalised again, even though policy instruments will continue to be implemented in times of crisis and price distortion (Weber & Hansen, 2014).

With respect to the geographical structures of the dairy sector in Germany, two main production areas have emerged over time: the Alpine foreland in Bavaria, southern Germany, along the border to Austria; and the coastal regions of Lower Saxony and Schleswig-Holstein in northern Germany. These regions are characterised by the fact that arable land use is only possible under aggravated conditions. In the Alpine region as well as in the low mountain range, the cultivation of crop plants, for example, is impeded by specific morphological and topographical features (e.g., sloping sites) and/or marginal soils. By contrast, the coastal zones in the north are characterised by flood basin soils and marshlands which, though being fertile and nutritious, are difficult to machine due to moisture and dampness. The use of agricultural land within these types of regions is therefore mostly restricted to the grazing of animals, particularly milk cows, or the production of staple feed for livestock farming (Nier & Tamásy, 2013; Nier *et al.*, 2013).

In recent years, milk production in Germany has increasingly moved to the north, where most counties could raise the number of milk cows from the year 2000 onwards (Nier *et al.*, 2013). The most significant increases were recorded in the counties of Cuxhaven, Wesermarsch and Stade, located near the North Sea coast in Lower Saxony. Over the same time period, the regions in the south have lost a large number of milk cows, especially the counties of Ravensburg, Oberallgäu and Unterallgäu. The main reason for this trend lies in the smaller dairy farm structures in the Alpine foreland. As these structures are disadvantageous for competition, particularly vis-à-vis the much larger dairy farms of the northern coastal regions, the observed spatial shift is thus a logical consequence. The strong presence of the market leader DMK in northwest Germany is another driver of this relocation process from the south to the north.

Disregarding the economic importance of the dairy industry above all in the coastal areas of Lower Saxony and Schleswig-Holstein, several authors have recently criticised the current trajectory of intensification of livestock production while emphasising potential ecological and socio-economic risks. In ecological terms, the deterioration of soils and groundwater due to enormous overflows of livestock excreta is probably the most vigorously debated problem. Further challenges of major relevance are seen in a high animal-disease risk, the overuse of antibiotics in factory-style livestock farming, ethical issues related to animal welfare, and increasing land-use competition which is reflected not least in the 'food or fuel' dilemma (Deimel *et al.*, 2011; FAO, 2006; Mose & Schaal, 2012; Mose *et al.*, 2007; Tamásy, 2013). At the economic level, the preservation and further development of dairy farms is currently one of the most challenging tasks, particularly in times of overproduction and low payout prices. With respect to animal welfare, the short lives of dairy cows and breeding methods selectively focused on a very high production of milk in a short period of time is receiving increasing attention. Intensive discussions on calves as 'waste products' of the dairy industry and the slaughtering of pregnant cows 'worthless' for production have further underlined sustainability issues related to dairy production in Germany. Changes in technology and globalisation forces have significantly impacted on the breeding of livestock with a great focus on a price-competitive production system (Urbanik, 2012).

The accumulation of all these problems makes evident that the future development of the German dairy industry is exposed to substantial risks, particularly from a sustainability viewpoint. The need for a sustainability transition in agriculture, recently articulated by Sutherland *et al.* (2015) in a broader European context, could thus be a viable approach for the long-term development of those regions whose agricultural basis is strongly dependent on dairy farming, such as the coastal zones of northwest Germany. When it is argued, however, that such a transition requires novel approaches towards multifunctionality, diversity and resilience (Tamásy, 2013), the hampering effects of regional and value chain lock-ins must be considered, since incumbent dairy farmers and processors are to a certain extent committed to the dominant 'regime' of conventional dairy production based on intensification. The mechanisms of 'being committed' do not only refer to ontologies of 'rational choice', but also to what Geels (2014) described as 'cognitive capital' (i.e., technical knowledge and competencies), shared mindsets and a common industry identity. Therefore, an understanding of the 'context structures' which are related to the provision of specific system-level assets[2] is urgently necessary to explore the initial conditions of sustainability transitions (Bergek *et al.*, 2015). This is all the more important as conventional agribusiness has increasingly come under pressure through political and societal concerns regarding 'new' values like food safety, animal welfare or environmental protection. Against this background, some dairy firms pursue new pathways towards more sustainable dairy production. In the following section, we present the specific approach of Molkerei Ammerland, a medium-sized co-operative dairy firm located in the wider coastal area of Lower Saxony.

4 Case study: Molkerei Ammerland

The foundations for Molkerei Ammerland were laid in 1885, when seven farmers decided to establish a co-operative in the northwestern part of Germany (see, for the detailed history of the Molkerei Ammerland, Paetrow & Sander, 2010). Today, 2,200 dairy farmers are the basis of the co-operative, and two production sites exist in Dringenburg (district of Ammerland) and north of the city of Oldenburg, both located within short distance of the North Sea. The main production site is Dringenburg, where cheese, butter, concentrates and milk powder are the main commodities, while fresh foodstuff is produced in Oldenburg. The maximum distance between milk suppliers and production sites is 80 km, to keep transport costs low. The brand name "Ammerländer" was chosen to emphasise the strong regional commitment of the co-operative. Production volume was around 1.7 million kg of milk in 2016, resulting in a turnover of €657.6 million created by 408 employees. Milk payout price for dairy farmers was at a minimum of 27.54 ct/kg in 2016 due to difficult world market conditions (Molkerei Ammerland, 2017). Today, about 50% of production of the co-operative is sold in more than 55 countries worldwide. Therefore, global market volatility directly affects key business data.

Sustainability as a general principle of production was explicitly communicated when the first sustainability report was published in 2013 (see Molkerei Ammerland, 2013); a second report followed in 2015 entitled "responsibility with vision" (Molkerei Ammerland, 2015). Sustainability reporting is based on guidelines of the Global Reporting Initiative (GRI). These guidelines offer reporting principles, standard disclosures and an implementation manual for the preparation of reports (GRI, 2015). A first step within the sustainability strategy was to inform staff personally about sustainability as an overarching principle of milk production. Therefore, written letters were sent by mail to all employees a couple of times to introduce the topic. In addition, detailed information was provided on the website of the co-operative and was incorporated into the firm's regular newsletter targeting milk suppliers. A discussion forum on sustainability has been introduced to discuss sustainability issues between the management and about 20 interested staff members (alternating participation) on a regular basis.

A core element of Molkerei Ammerland's sustainability strategy has been the introduction of the "pasture milk program" in 2011. Today, every other dairy farmer of the co-operative is involved in the pasture initiative. Furthermore, the milk is labelled "without genetic engineering" to directly convince customers in Germany with respect to the quality of product inputs, including feed and fertilisers. This means that dairy cows graze on pasture (for at least six hours) for a minimum of 120 days per year. The raw milk is treated and produced separately on different production lines. All participating milk suppliers diarise pasture time and used agricultural land. Independent external auditors check the provided documentation once a year. The rapid growth in pasture milk is supported by supermarket chains in Germany, which provide shelf-space for these products that are not as expensive as organically produced milk. Finally, the start of organic milk production has been scheduled by Molkerei Ammerland for autumn 2017. The target is to produce 30 million kg of organic milk annually. The overall objective is to build a strong national brand that satisfies consumer demands in the alternative market segments to create value in both markets.

5 Conclusion

In the previous sections, we have initially described the idea of sustainability transitions and its relevance for the German dairy industry, which is marked by ongoing processes of intensification and concentration. Besides outlining the major structures and developments underpinning this specific trajectory, we have also addressed a number of problems and challenges indicating the need for more sustainable modes of production in the German dairy sector. These challenges refer, among others, to the degradation of soils and groundwater due to enormous overflows of livestock excreta, the compliance with animal welfare standards and the preservation of dairy farms in a tough, competitive environment. Finally, we have presented a corporate sustainability strategy using the example of Molkerei Ammerland, a medium-sized dairy firm in Lower Saxony, northwest Germany.

The findings of our analysis reveal that the intensification pathway of the German dairy industry can be characterised by long persistence and stabilisation, maybe even path-dependence, in terms of becoming more and more productive. As a consequence, dairy farmers are often faced with the question: "Growing or giving way"? Against this background, it must be acknowledged that various milk-processing firms such as the Molkerei Ammerland have begun to distinguish themselves from their 'mass-oriented' competitors and the predominant regime of productivism. But what are the real effects of this strategy? In fact, there is plenty of grassland in the coastal regions of northwest Germany with grazing milk cows belonging to the 'normal' landscape picture for decades. Molkerei Ammerland, however, only now aims at capitalising on this imaginary of 'happy cows at pasture', even though the majority of milk suppliers have long kept their livestock exactly this way (i.e., on grassland). The pasture milk programme is thus more a marketing concept based on the rewriting of a likeable narrative which is already in place and is actually well-known. This approach can be understood as the firm's creative response to significant societal pressures regarding higher sustainability standards in animal-based food production.

Nevertheless, the strategy of Molkerei Ammerland has undoubtedly succeeded so far in achieving added value, while at the same time ensuring higher payout prices for the participating dairy farmers (as co-operative members). This may contribute to consolidating the regional dairy farm sector, which is otherwise strongly threatened by adverse economic conditions shaping the dairy market in Germany (and in the EU) for many years. In addition, Molkerei Ammerland contributes to raise public awareness of sustainable milk production, considering not only economic viability but also ecological balance and social fairness. This is supported by the fact that several other dairy firms in Germany have recently followed Molkerei Ammerland by introducing their own pasture milk programmes (e.g., Arla Foods, Schwarzwaldmilch). While these firms – as ambassadors of the existing regime – have adapted an innovative (though not really new) approach in response to various pressures, some references to the concepts of MLP and TEF can be made, even though further research is necessary, for example, to understand the specificities of niche-regime interaction in more detail. On the other hand, the case study has revealed the constructivist character of the wider coastal zone in northwest Germany as an integral part of Molkerei Ammerland's sustainability (and marketing) strategy, thus addressing the concerns by Murphy (2015) on insufficient conceptualisations of space and place as a general shortcoming of the transitions literature.

Basically, while sustainability transitions require a fundamental shift in the socio-technical system based on reconfigurations of networks, institutions, knowledge and material artefacts (Markard *et al.*, 2012), we must concede that the efforts of individual firms are only a first step to foster the needed transformations towards sustainability. A successful sustainability transition rather requires long-term efforts of all actors along the value chain as well as specific forms of guidance and governance (Smith *et al.*, 2005). Therefore, the creation of enabling and supporting 'context structures' is regarded as a viable way to overcome the 'mental obstacles' for a successful food system transformation towards sustainability (Bergek *et al.*, 2015). These 'context structures' could be shaped, for example, by regional network organisations which have the capability to structurally couple different innovation systems including both public and private actors. The implementation of co-operation structures in the dairy industry is thus a possible strategy to achieve a higher sustainability level, particularly in the coastal regions of Lower Saxony and Schleswig-Holstein where institutionalised agrifood networks have not yet been sufficiently developed. In general, while the binary between 'conventional' and 'alternative' food systems becomes increasingly permeable, we need a much deeper understanding of how innovative approaches come to emerge within established food systems and under what conditions they may trigger a holistic transition towards sustainability.

Notes

1 HACCP stands for Hazard Analysis and Critical Control Points.
2 Think of political support for technology-specific policies, the need for trained personnel, or the provision of venture capital.

Bibliography

Agricultural Market Information Company (AMI). (2016). *Market Balance Sheet Dairy 2016*. Bonn: author's edition.

Association of the German Dairy Industry. (2016). Top ten dairies in Germany 2016. Berlin. Retrieved 2 January 2017, from http://milchindustrie.de/uploads/tx_news/TOP_Molkereien_DE_Homepage_01.pdf

Association of the German Dairy Industry. (2017). Fakten Milch: Milch und mehr – die deutsche Milchwirtschaft auf einen Blick. Berlin. Retrieved 8 November 2017, from https://milchindustrie.de/wp-content/uploads/2017/10/Fakten_Milch_September_2017_A4.pdf

Bäurle, H., & Tamásy, C. (2012). *Regionale Konzentrationen der Nutztierhaltung in Deutschland.* Vechta: ISPA-Mitteilungen 79.

Bergek, A., Hekkert, M.P., Jacobsson, S., Markard, J., Sanden, B.A., & Truffer, B. (2015). Technological innovation systems in contexts: Conceptualizing contextual structures and interaction dynamics. *Environmental Innovation and Societal Transitions, 16,* 51–64.

Berkers, E., & Geels, F.W. (2011). System innovation through stepwise reconfiguration: the case of technological transitions in Dutch greenhouse horticulture (1930–1980). *Technology Analysis and Strategic Management, 23*(3), 227–247.

Coenen, L., Benneworth, P., & Truffer, B. (2012). Toward a spatial perspective on sustainability transitions. *Research Policy, 41*(6), 968–979.

Cohen, N., & Ilieva, R. (2015). Transitioning the food system: a strategic practice management approach for cities. *Environmental Innovation and Societal Transitions, 17,* 199–217.

Deimel, M., Arens, L., & Theuvsen, L. (2011). The influence of clusters on the competitiveness of hog production: the example of Northwestern Germany. *International Journal on Food System Dynamics, 2*(2), 155–166.

DMK Deutsches Milchkontor. (2016). Facts and figures. Retrieved 3 January 2017, from https://www.dmk.de/en/company/facts-figures/

Dosi, G. (2000). *Innovation, Organization and Economic Dynamics.* Cheltenham: Edgar Elgar.

Elzen, B., Geels, F.W., Leeuwis, C., & van Mierlo, B. (2011). Normative contestation in transitions 'in the making': animal welfare concerns and system innovation in pig husbandry. *Research Policy, 40*(2), 263–275.

Food and Agriculture Organization of the United Nations (FAO). (2006). *Livestock's Long Shadow: Environmental Issues and Options.* Rome.

Geels, F.W. (2002). Technological transitions as evolutionary reconfiguration processes: a multi-level perspective and a case-study. *Research Policy, 31*(8–9), 1257–1274.

Geels, F.W. (2005). Co-evolution of technology and society: the transition in water supply and personal hygiene in the Netherlands (1850–1930): a case study in multi-level perspective. *Technology in Society*, 27(3), 363–397.

Geels, F.W. (2006a). The hygienic transition from cesspools to sewer systems (1840–1930): the dynamics of regime transformation. *Research Policy, 35*(7), 1069–1082.

Geels, F.W. (2006b). Major system change through stepwise reconfiguration: a multi-level analysis of the transformation of American factory production (1850–1930). *Technology in Society, 28*(4), 445–476.

Geels, F.W. (2006c). Co-evolutionary and multi-level dynamics in transitions: the transformation of aviation systems and the shift from propeller to turbojet (1930–1970). *Technovation, 26*(9), 999–1016.

Geels, F.W. (2007). Transformations of large technical systems: a multilevel analysis of the Dutch highway system (1950–2000). *Science Technology and Human Values, 32*(2), 123–149.

Geels, F.W. (2009). Foundational ontologies and multi-paradigm analysis, applied to the socio-technical transition from mixed farming to intensive pig husbandry (1930–1980). *Technology Analysis & Strategic Management, 21*(7), 805–832.

Geels, F.W. (2011). The multi-level perspective on sustainability transitions: responses to seven criticisms. *Environmental Innovation and Societal Transitions, 1*(1), 24–40.

Geels, F.W. (2014). Reconceptualising the co-evolution of firms-in-industries and their environments; developing an inter-disciplinary Triple Embeddedness Framework. *Research Policy, 43*(2), 261–277.

Geels, F.W., & Penna, C.C.R. (2015). Societal problems and industry reorientation: elaborating the Dialectic Issue LifeCycle (DILC) model and a case study of car safety in the USA (1900–1995). *Research Policy, 44*(1), 67–82.

Geels, F.W., & Verhees, B. (2011). Cultural legitimacy and framing struggles in innovation journeys: a cultural-performative perspective and a case study of Dutch nuclear energy (1945–1986). *Technological Forecasting and Social Change, 78*(6), 910–930.

Geels, F.W., Kern, F., & Mylan, J. (2016). Reformulating the transition pathways typology: a comparative multi-level analysis of the German and UK low-carbon electricity transitions (1990–2014). *Research Policy, 45*(4), 896–913.

Geels, F.W., Kemp, R., Dudley, G., & Lyons, G. (2012). *Automobility in Transition? A Socio-Technical Analysis of Sustainable Transport.* London & New York: Routledge.

Global Reporting Initiative. (2015). *Reporting Principles and Standard Disclosures.* Amsterdam.

Hansen, T., & Coenen, L. (2015). The geography of sustainability transitions: review, synthesis and reflections on an emergent research field. *Environmental Innovation and Societal Transitions, 17,* 92–109.

Markard, J., Raven, R., & Truffer, B. (2012). Sustainability transitions: an emerging field of research and its prospects. *Research Policy, 41*(6), 955–967.

Markard, J., Suter, M., & Ingold, K. (2016). Socio-technical transitions and policy change: advocacy coalitions in Swiss energy policy. *Environmental Innovation and Transitions Management, 18,* 215–237.

Molkerei Ammerland. (2013). *Nachhaltigkeitsbericht 2006–2011 der Molkerei Ammerland EG.* Wiefelstede-Dringenburg.

Molkerei Ammerland. (2015). *Verantwortung mit Weitblick. Nachhaltigkeitsbericht 2012–2013 der Molkerei Ammerland EG.* Wiefelstede-Dringenburg.

Molkerei Ammerland. (2017). *Geschäftsbericht 2016.* Wiefelstede-Dringenburg.

Mose, I., & Schaal, P. (2012). Probleme der Intensivtierhaltung im Oldenburger Münsterland. Lösungsstrategien im Widerstreit konkurrierender Interessen. *Neues Archiv für Niedersachsen, 2012, 2,* 50–69.

Mose, I., Peithmann, O., & Schaal, P. (2007). Probleme der Intensivlandwirtschaft im Oldenburger Münsterland – Lösungsstrategien im Widerstreit der Interessen. In Zepp, H. (Ed.), *Ökologische Problemräume Deutschlands* (pp. 133–156). Darmstadt: Wissenschaftliche Buchgesellschaft.

Murphy, J.T. (2015). Human geography and socio-technical transition studies: promising intersections. *Environmental Innovation and Societal Transitions, 17,* 73–91.

Mylan, J., Geels, F.W., Gee, S., McMeekin, A., & Foster, C. (2015). Eco-innovation and retailers in milk, beef and bread chains: enriching environmental supply chain management with insights from innovation studies. *Journal of Cleaner Production, 107,* 20–30.

Nier, S., & Tamásy, C. (2013). *Strukturen und Dynamik in der niedersächsischen Milchverarbeitung.* Vechta: Weiße Reihe 38.

Nier, S., Bäurle, H., & Tamásy, C. (2013). *Die deutsche Milchviehhaltung im Strukturwandel.* Vechta: ISPA-Mitteilungen 81.

Paetrow, S., & Sander, T. (2010): *Unwiderstehlich. Seit 1885. Die Geschichte der Molkerei Ammerland.* Hamburg.

Raven, R., & Geels, F.W. (2010). Socio-cognitive evolution in niche development: comparative ana-lysis of biogas development in Denmark and the Netherlands (1973–2004). *Technovation, 30*(2), 87–99.

Rhodes, V.J. (1983). The large agricultural cooperative as a competitor. *American Journal of Agricultural Economics, 65*(5), 1090–1095.

Schlecht, S., & Spiller, A. (2009). Procurement strategies of the German dairy sector: Empirical evidence on contract design between dairies and their agricultural suppliers. Paper presented at the 19th Annual World Forum and Symposium "Global Challenges, Local Solutions", IAMA Conference, 20–23 June 2009, Budapest. Retrieved 3 January 2017, from /www.ifama.org/resources/files/2009-Symposium/1150_paper.pdf

Smith, A., Stirling, A., & Berkhout, F. (2005). The governance of sustainable socio-technical transitions. *Research Policy, 34*(10), 1491–1510.

Sutherland, L.-A., Darnhofer, I., Wilson, G. A., & Zagata, L. (Eds.) (2015). *Transition Pathways Towards Sustainability in Agriculture. Case Studies from Europe.* Wallingford & Boston: CABI.

Tamásy, C. (2013). Areas of intensive livestock agriculture as emerging alternative economic spaces? *Applied Geography, 45,* 385–391.

Theuvsen, L. (2006). European cooperatives: are they prepared for food product innovations? In Strada, A., & Sikora, T. (Eds.), *The Food Industry in Europe: Tradition and Innovation* (pp. 65–87). Cracow: Cracow University of Economics.

Truffer, B., & Coenen, L. (2012). Environmental innovation and sustainability transitions in regional studies. *Regional Studies, 46*(1), 1–21.

Tukker, A., Emmerst, S., Charter, M., Vezzoli, C., Sto, E., Andersen, M.M., Geerken, T., Tischner, U., & Lahlou, S. (2008). Fostering change to sustainable consumption and production: an evidence based view. *Journal of Cleaner Production*, 16(11), 1218–1225.

Turnheim, B., & Geels, F.W. (2013). The destabilisation of existing regimes: confronting a multi-dimensional framework with a case study of the British coal industry (1913–1967). *Research Policy, 42*(10), 1749–1767.

Urbanik, J. (2012): *Placing Animals: An Introduction to the Geography of Human-Animal Relations.* Plymouth: Rowman & Littlefield Publishers.

Verbong, G., Geels, F.W., & Raven, R. (2008). Multi-niche analysis of dynamics and policies in Dutch renewable energy innovation journeys (1970–2006): hype-cycles, closed networks and technology-focused learning. *Technology Analysis and Strategic Management*, *20*(5), 555–573.

Weber, S.A., & Hansen, H. (2014). Can a monitoring agency efficiently regulate the EU milk market? Thünen Working Paper 34a. Braunschweig. Retrieved 3 January 2017, from http://literatur.ti.bund.de/digbib_extern/dn054618.pdf

Wissenschaftlicher Beirat für Agrarpolitik beim Bundesministerium für Ernährung und Landwirtschaft (BMEL). (2015). *Wege zu einer gesellschaftlich akzeptierten Nutztierhaltung. Gutachten.* Berlin.

Zentrale Markt- und Preisberichtsstelle (ZMP). (1995). *ZMP Bilanz: Milch 1995. Deutschland, EU, Weltmarkt.* Bonn: In-house publication.

9 They sow the wind and reap bioenergy

Implications of the German energy transition on coastal communities in Schleswig-Holstein, Germany

P. Michael Link, Jürgen Scheffran and Kesheng Shu

1 Introduction

The catastrophic earthquake in Japan in March 2011 which triggered the nuclear meltdown of the plant in Fukushima led to a cascade of consequences that included a drastic change in German energy policy, aiming at the complete phase-out of nuclear energy by the mid-2020s (Kominek & Scheffran, 2012). In order to achieve this ambitious goal, there has to be a substantial substitution of energy production using renewable sources such as wind energy, solar energy or bioenergy. Even before the 2011 decision to initiate the energy transition in Germany ("Energiewende"), there had already been a considerable development of renewable energy sources in Germany (Wüstenhagen & Bilharz, 2006). The low-lying flat coastal areas of northern Germany are particularly suited for energy production from wind, but there is considerable intra-state heterogeneity, with wind-energy production being most prominent in the coastal counties along the North Sea coast and in Ostholstein near the Baltic Sea (Goetzke & Rave, 2016). Furthermore, large parts of the coastal areas are also used by agriculture, which makes it relatively easy to engage in biomass production for bioenergy plants. Both kinds of energy sources have become increasingly important in the German energy mix since the beginning of the 21st century. Additionally, the potential for energy production from agricultural leftovers such as straw is substantial in Schleswig-Holstein and could become an important alternative to the growth of energy plants in the northernmost German state (Weiser *et al.*, 2014).

In contrast to most fossil fuel sources, the renewable sources that are used for energy production in northern Germany require substantial space: land that is used to grow energy crops cannot be used for any other purpose simultaneously. While an individual wind turbine does not require a lot of space by itself, there are defined minimum distances between generators that make large wind parks also quite land-intensive endeavours (McKenna *et al.*, 2014). This may lead to land use-conflicts, as the coastal zones in northern Germany are usually utilised in multiple ways. Agriculturally used land is generally used to grow food crops or for grazing cattle or sheep on grassland. Furthermore, tourism is an important economic sector in the coastal zones of Schleswig-Holstein (Homp & Schmücker, 2015; Statistikamt Nord, 2015), particularly as the majority of the coastal waters around the German coasts are protected areas or national parks (Goeldner, 1999; Schiewer, 2008).

Assessments of the possible consequences of changes in land use to fulfil constantly growing bioenergy production have been conducted for the Eiderstedt peninsula on the west coast of Schleswig-Holstein. Despite the fact that land use in that area has changed over the course of the past centuries (Link & Schleupner, 2007), a quick shift of agricultural land use from grassland to arable farm land to grow large amounts of corn for bioenergy production would have substantial implications for key breeding populations of wading bird species (Schleupner & Link, 2008). Also, tourism is affected to some extent, albeit the impacts are much less pronounced (Link & Schleupner, 2011). In addition, it has to be noted that there are substantial disagreements between local farmers and environmental conservationists about land-use issues in conjunction with bioenergy production that have not yet been resolved (Schleupner & Link, 2009).

This assessment considers the two most important renewable energy sources in Schleswig-Holstein: wind energy and bioenergy. Applying a newly developed agent-based model, which extends a corresponding version covering the Jiangsu Province in China (Shu *et al.*, 2015), the development pathways of renewable energy in the northernmost German state are assessed for different scenarios of demand growth. The analysis of consequences for the individual counties focuses on the coastal regions of the state to determine whether there are significant differences in impacts for coastal communities in comparison with the landlocked parts of Schleswig-Holstein. Because of the limited number of land uses in the coastal regions of northern Germany – the prime focus being agricultural production – it is important to identify and follow sustainable development pathways that successfully increase the adaptive capacity of coastal zones, particularly in times of climate change. In the rural parts of northern Germany, renewable energy can be readily produced due to the given geomorphological setting, but its large-scale implementation requires a socio-technological transition to devise land-use strategies that simultaneously satisfy the needs for food and energy provision (Geels, 2002) in order to avoid or minimise possible local or regional conflict.

Recent assessments have highlighted the potentials for renewable energy production in Schleswig-Holstein but have also pointed to some conflict risks that need to be considered (Link & Scheffran, 2017; Scheffran *et al.*, 2017; Link *et al.*, 2018). The aim of this study is to assess possible development pathways of renewable energy production in Schleswig-Holstein, Germany, utilising an agent-based model that allows agents to allocate their resources among food, wind energy or bioenergy production. Simulations are conducted for various levels of subsidies for renewable energy production and results are evaluated for three sub-regions of Germany's northernmost federal state. The following section outlines the current state of renewable energy production in Schleswig-Holstein. Subsequently, the agent-based model used in this assessment is described in detail. Section 4 provides the results of the model simulations for different scenarios of future renewable energy production in northern Germany. Section 5 discusses the model results and concludes.

2 Renewable energy production in northern Germany

Northern Germany has already experienced a substantial development of renewable energy sources during the past three decades. In this case study, we focus on the state of Schleswig-Holstein, which lies north of the metropolitan region of Hamburg. Schleswig-Holstein is the northernmost federal state of Germany, with an area of almost 16,000 km^2 and a population of slightly more than 2.8 million people. It consists of three fundamentally different geomorphological zones: the west of Schleswig-Holstein is very flat marshland with fertile soils, which makes this the prime agricultural area of the state. The central part of Schleswig-Holstein consists of geest with sandy soils, while the east of the state consists of rolling hills and fjords, as this part of Schleswig-Holstein has been subject to glaciation during the last ice age (Liedtke & Marcinek, 2002). The overall terrain in Schleswig-Holstein is very flat, the highest elevation being the Bungsberg, which is a mere 168 m above sea level.

Because of these characteristics, Schleswig-Holstein is the state with one of the best potentials for wind-energy production throughout Germany (Ender, 2015). This particularly applies to the western marshlands and the central geest as well as the island of Fehmarn in the Baltic Sea. Outside Schleswig-Holstein, similarly well-suited areas can only be found in the coastal regions of Lower Saxony and Mecklenburg-West Pomerania. Consequently, wind turbines are primarily located in these preferred zones. Some parts of these areas are already saturated with wind turbines or are already close to being saturated. In other parts of Germany wind-energy production is somewhat less pronounced in comparison with other renewable energy sources.

This is reflected by the development of wind-energy production in Germany and particularly in Schleswig-Holstein in the past decades. There was a first sharp increase in wind-energy production in the 1990s, with the installed capacity increasing along with the number of installed wind turbines (Ender, 2015). By the turn of the millennium, the installed capacity had exceeded 5000 MW (Bundesministerium für Wirtschaft und Energie, 2017). However, some areas in Germany had become saturated with installations since the early 2000s, so a further expansion of wind-energy production could only occur by increasing the size of the individual installations. As space for onshore wind-energy installations is limited, other options have to be considered in order to maintain the expanding trend of wind-energy production in Germany. Consequently, offshore solutions have been sought, mainly in the North Sea area. After a long planning period, the first German offshore wind park started to produce energy in 2009. At the beginning of 2018, there were nine offshore wind parks in Schleswig-Holstein's part of the North Sea that were in operation or under construction, totalling 560 individual turbines (Bundesamt für Seeschifffahrt und Hydrographie, 2018). Further offshore wind parks are currently being planned.

Economic benefits from wind-energy production have consistently grown since the beginning of the millennium, reaching approximately €1.7 billion in 2015 (Bundesministerium für Wirtschaft und Energie, 2017). Investments have experienced a slight decline as a consequence of the economic crisis in the second half of the 2000s, but have soared in recent years to a new high of almost €7 billion for onshore wind turbines and more than €3 billion for offshore wind-energy production.

Among the German federal states, Schleswig-Holstein is the leader with regard to wind-energy production (Ender, 2015). More than 450 wind turbines were newly installed in Schleswig-Holstein in 2014, which amounts to almost one-quarter of all new German wind-energy installations. Lower Saxony is a distant second with half as many wind turbines as Schleswig-Holstein. In that year, the installed capacity in Schleswig-Holstein grew by more than 1300 MW, totalling more than 20% of the entire additionally installed capacity of almost 6200 MW. Numerous small installations have been decommissioned recently, making room for larger wind turbines. Repowering has become an important option to replace out-of-date equipment with state-of-the-art technology to secure wind- energy production in the upcoming decades. This emphasises the substantial dynamics in the wind-energy sector in Germany.

The geographic distribution of wind turbines in Schleswig-Holstein corresponds well with the wind zones. Wind-energy production mainly occurs along the North Sea coast in the counties of Northern Frisia and Dithmarschen (Landesamt für Landwirtschaft Umwelt und ländliche Räume Schleswig-Holstein, 2017) and at the Baltic Sea in the county of Ostholstein, mainly because of the vast amount of turbines on the island of Fehmarn. Some wind parks are also in operation in the western part of Schleswig-Flensburg and in Steinburg near the Elbe River. It has to be noted that wind turbines are generally not evenly distributed throughout the countryside but have been erected in large clusters in areas with particularly favourable conditions. Along the North Sea coast of Schleswig-Holstein, this corresponds to practically all coastal areas that are not protected by the regulations of the Wadden Sea National Park.

The considerable growth of wind-energy production in Schleswig-Holstein in the past decades has initially occurred by mainly installing wind turbines on agricultural land so that farmers could harvest energy in addition to their regular crops, as individual turbines do not require large amounts of land. This has led to the construction of numerous turbines that have redefined the appearance of the landscape in parts of western Schleswig-Holstein. There is vast potential for wind-energy production in the German northernmost state, but there are also substantial uncertainties associated with the land-use changes occurring in the direct vicinity of the turbines and with local effects of public acceptance (McKenna *et al.*, 2014).

However, despite the advantages of wind-energy production with regard to emissions reduction, opponents to the continued expansion of renewable energy sources point to the adverse effects of wind turbines: these include noise in the vicinity of the generators (Dai *et al.*, 2015), reflections of sunlight from the turbines, killed animals from the rotating hands of the turbines, such as birds (Dürr & Langgemach, 2006; Hüppop *et al.*, 2006) and bats (Rydell *et al.*, 2010), as well as aesthetic reasons (Gee & Burkhard, 2010). Considering that the Wadden Sea and vast parts of the adjacent coastal areas in Schleswig-Holstein are national parks and of particular ecological value to wildlife, land-use conflicts between environmental conservationists and proponents of wind-energy production have occurred locally, leading to limitations on the number of wind turbines in the vicinity of protected areas (Gatzert & Kosub, 2016).

To minimise such conflicts, there has been a considerable push to shift the wind-energy production in Schleswig-Holstein to offshore sites. However, the construction of offshore sites is associated with substantially higher costs, which also extends to the maintenance of installations and the transport of the generated energy to the consumers, who are usually located in considerable distances from the suppliers (Maubach, 2014). Furthermore, there has to be a considerable upgrade in the German power grid to transport the energy from the production sites to southern Germany and metropolitan areas where demand is greatest (Erlich *et al.*, 2006). The future success of wind-energy production will rely to a large extent on the development of production costs. Investment costs have to be reduced, particularly as feed-in tariffs are going to be discontinued and wind energy has to compete with other means of energy production on the German market as of 2020 (Nordensvärd & Urban, 2015). In this context, the focus on offshore sites far from the coast may need to be adjusted to sites in closer proximity to the coast, as this would reduce construction and operation costs substantially (Maubach, 2014) and offshore production in areas close to the coasts has proven to be a successful alternative in the neighbouring countries in the North Sea region. To identify suitable sites, measurable market services of the environment have to be balanced against qualitative, often not clearly tangible, indicators, which are nonetheless equally important. In coastal zones and coastal communities, the application of coastal ecosystem service assessments thus seems a feasible approach (c.f. Link & Borchert, 2015).

When looking at renewable energy production in Schleswig-Holstein, wind power is not the only source that is prominent in this mainly rural state of Germany. The production of biogas by fermentation has also expanded substantially in the past decade. First initiatives to promote the production of bioenergy in Schleswig-Holstein were successful even before the turn of the millennium, and it took only a few years to install the first significant capacity (Bundesministerium für Wirtschaft und Energie, 2017). In 2001, an equivalent of slightly more than 100 MW was installed, which had increased 40-fold a decade later. Until then, there was a constant expansion in the production of biogas, although there has been stagnation in recent years, which may be attributed to changes in the legal boundary conditions regarding the production of renewable energy in Germany. The currently installed capacity of bioenergy production in Schleswig-Holstein is just above 5000 MW.

In contrast to wind-energy production, biogas plants are more evenly spaced throughout Schleswig-Holstein, as they are not dependent on any particular physical setting such as favourable wind speeds. Nonetheless, there are far more biogas plants in the north of Schleswig-Holstein – particularly in the counties of Northern Frisia and Schleswig-Flensburg – than in the metropolitan region of Hamburg. This has to do with the fact that biogas plants require a substantial amount of feed-crops that should ideally not have to be transported too far before being processed in the plants. The north of Schleswig-Holstein and the area close to the Danish border is almost entirely used for agricultural production, and in this region there is more space for the production of energy crops next to food crops than in the south of the state. Also, most energy crops in Schleswig-Holstein are grown in the geest area, which roughly corresponds to the central (landlocked) areas of the state. The soils in this part of Schleswig-Holstein are good enough for these crops while the very fertile marshlands are used more for food crop production.

Similar to wind energy, the increase in the number of biogas plants in Schleswig-Holstein has sparked some serious discussions, as there are numerous side-effects associated with the production of energy plants for fermentation. Due to its considerable spatial demand, substantial bioenergy production in a given region such as the west coast of Schleswig-Holstein is likely to shape this area as an 'energy landscape', which needs to be managed accordingly in order to avoid land use and public acceptance conflicts (Blaschke *et al.*, 2013). A research project has been conducted on the Eiderstedt peninsula on the west coast of Schleswig-Holstein to determine how the landscape has changed due to the continued transformation of grassland to arable farm land to grow corn for biogas production, and what the possible implications are for tourism and breeding bird species if the trend of transforming the agricultural land continued (Link & Schleupner, 2007, 2011; Schleupner & Link, 2008, 2009). Depending on the conversion pattern, ecological impacts can vary, despite the fact that the general trends remain similar.

Ecological impacts include adverse effects on breeding bird populations, which require sufficient grassland for their breeding success. A substantial conversion of grassland into arable farm land not only decreases the overall amount of land available to the bird populations, it also diminishes the quality of the remaining grassland (Schleupner & Link, 2008). This leads to a disproportionately strong decline in the carrying capacity for several key bird species that can only partly offset this effect by relocating to areas in the vicinity of the peninsula.

Economic impacts of land conversion for energy crop production are harder to quantify. These mainly consist of changes in tourist frequency in the areas of biogas production because of the altered landscape appearance (Link & Schleupner, 2011). Even though tourists are aware of such changes, hardly anyone would reconsider their destination choice because of increased energy crop production. Therefore, the economic implications in the region are only marginal. While these local impacts of land-use changes, which have been assessed for one particular region in Schleswig-Holstein, point to certain trends, they are by no means indicative of possible developments in other parts of the federal state.

Another critical aspect of renewable energy production in northern Germany is that a lot of the energy produced in Schleswig-Holstein cannot be consumed locally but needs to be transported into adjacent metropolitan regions, such as Hamburg, where the demand for energy is particularly high. There is also some production of renewable energy in Hamburg, but this is by no means sufficient to meet the ever-growing demand in the entire metropolitan region (Energymap, 2018; Groscurth, 2005). Therefore, it becomes necessary to transport the energy from the suppliers in the rural areas to the consumers in the urban area. Furthermore, this may cause land-use issues in the metropolitan regions as well (Gertz *et al.*, 2015).

Investments in the energy transportation infrastructure are a key aspect of the energy transition in Germany. However, it appears to lag behind the expansion of production sites of renewable energy (Erlich *et al.*, 2006) and efforts have to increase to make the energy transition a successful endeavour in the long term (Scholz *et al.*, 2014). Critics of the strong emphasis on renewable energy stress this point, as well as the fact that the public acceptance of large energy transportation projects is also limited (Bertsch *et al.*, 2016). Nonetheless, efforts to install adequate capacities to transport energy from its sources to the areas of demand are well under way and should be able to bridge this gap in upcoming years (Beveridge & Kern, 2013).

As Schleswig-Holstein is a key provider of renewable energy in Germany, it is important to determine what potentials can actually be realised for different political and economic boundary conditions, and how such developments affect land use in the rural parts of the state where the energy is mostly produced – particularly in the coastal regions. This is done using an agent-based model that simulates possible developments of renewable energy production in Schleswig-Holstein, with a particular focus on the two main renewable energy types: wind and bioenergy. The model output consists of possible development pathways of agricultural and energy production until the middle of the 21st century for various sets of constraints. The feasibility and resulting conflict potentials of these development trajectories can then be analysed with sudden substantial land-use changes pointing to possible sources of conflict.

3 The dynamic agent-based model

The agent-based model of renewable energy production in Schleswig-Holstein simulates the annual decisions of farmers with regard to intensity and type of crop production in order to meet the demands for both food crops and energy biomass. Furthermore, the farmers have the option to invest in the construction of wind turbines on their property to supplement income from agricultural production by producing electricity from wind.

3.1 Model framework

The model combines a dynamic agent-based model and a partial equilibrium model of agriculture in Schleswig-Holstein (Figure 9.1). The geographical characteristics of the individual agents are maintained in the model by using a spatial grid that positions the agents in the landscape. This way, it is possible to distinguish agents in coastal regions from those in landlocked parts of Schleswig-Holstein. The agents are aggregated into three regions, where the communities adjacent to the North Sea or Baltic Sea are divided into two coastal regions and the remaining landlocked part of the state is considered as a third agent. The distinction between the North Sea coast and the Baltic Sea coast is made because of the different geomorphological settings of the two coastal areas, which leads to different renewable energy production patterns in these regions.

In the agricultural sector model, all farmers of one of the regions distinguished in the model are considered as one agent. This part of the

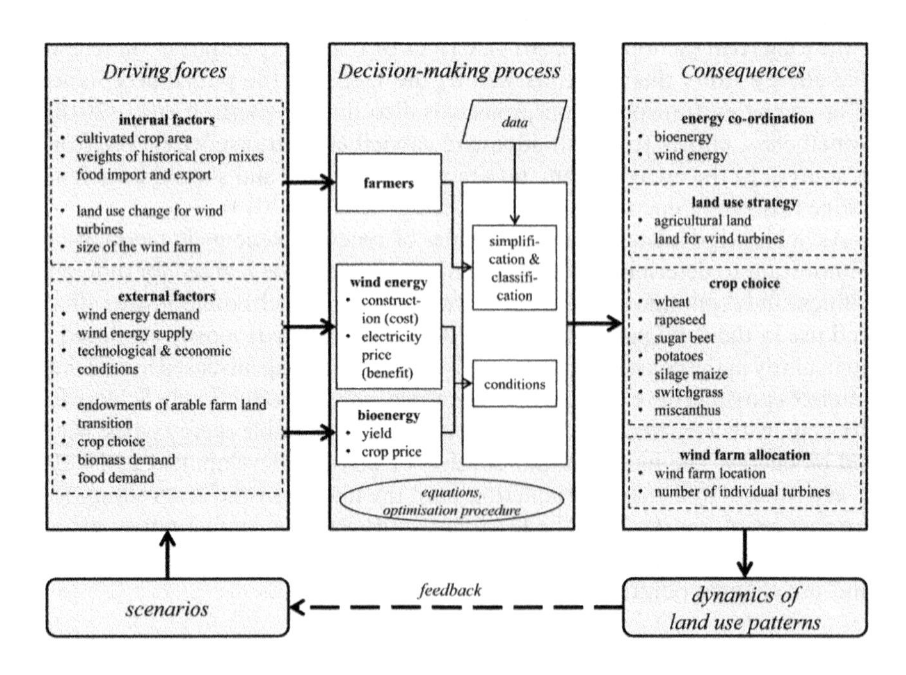

Figure 9.1 The framework of the dynamic agent-based model of optimised energy landscapes in Schleswig-Holstein

Source: adapted from Shu *et al.* (2015)

agent-based model is adapted from the Agricultural Sector and Mitigation of Greenhouse Gases (ASMGHG) model (Schneider & McCarl, 2003), which relates local cropping systems to agricultural commodity markets. The model also relates each agent's agricultural production scheme to the local bioenergy and wind energy markets. This way, the dynamic agent-based model can effectively consider the local geographic characteristics of each agent in the context of the given boundary conditions related to energy crop production. By relating biomass supply from all farmers to bioenergy production goals, the decisions of all agents in the market become interlinked. Similarly, decisions on wind-energy production are based on the respective development goals for the energy type. Note that all production goals are based on political decisions and are thus in reality subject to substantial possible change if an administration chooses to regionally change the course of the energy transition that was set by its predecessor.

3.2 Model structure

The model applied in this assessment determines the most cost-efficient land use pattern that not only meets the regional demand for food but also addresses the demand for biomass for bioenergy production. Production of wind energy is considered separately, as the unit area requirement of an individual wind turbine is rather limited in comparison to the space requirement of agricultural land for energy plants. The model is programmed in GAMS and consists of an objective function, a group of decision variables and a set of constraints.

Farmers can plant crops that will either be used as food or as input to bioenergy production (Figure 9.2). Furthermore, farmers can invest in wind turbines that will be constructed on their property to produce electricity from wind. Decisions regarding which crops to plant for which purpose and whether to invest in wind-energy production depend on the developments of the energy market and the market for agricultural goods. Therefore, the decisions of an individual agent, i.e., one region of the state, have an influence on the entire sector and subsequently on the other agents as well.

The objective function maximises the net present value of profits from agriculture and energy production over the 37-year time horizon of the simulation (2013–2050). Decisions are determined on an annual basis, adhering to the constraints that are based on environmental limits and economic demand for agricultural goods. Mathematically, these constraints define the convex feasible region[1] for all decision variables and are described in greater detail in the appendix.

In the simulations, an optimisation is performed for each time step to obtain the best possible set of values for the variables that not only are consistent with all the constraints but also maximise the objective function. In economic terms, this maximum of the objective function is referred to as competitive market equilibrium (McCarl & Spreen, 1980). In addition to the production allocation in each region, the energy and crop yields for each time

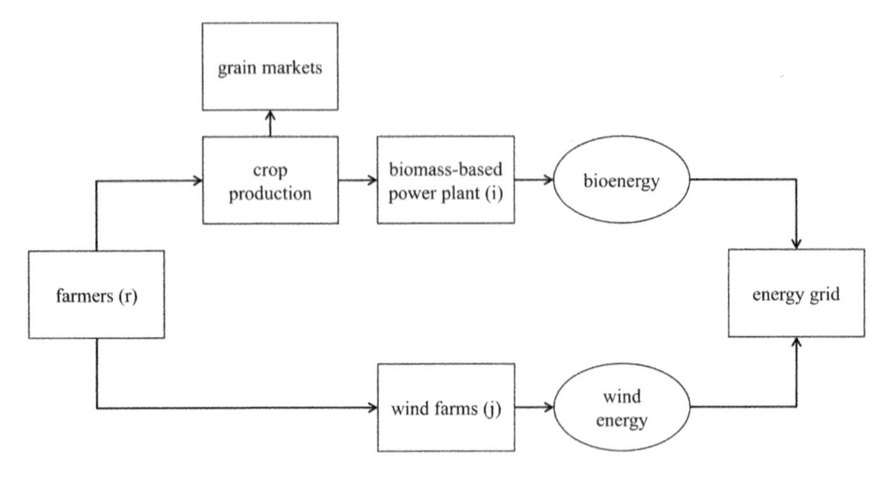

Figure 9.2 Implementation of the framework in the dynamic agent-based model

step as well as the associated market prices for the given set of political, economic, and technological boundary conditions can be determined.

3.3 Model parameterisation and scenarios

In this assessment, the model is applied to simulate the possible development of bioenergy and wind-energy production in Schleswig-Holstein from present until the middle of the 21st century. Regionally, the model distinguishes between the North Sea coastal area, landlocked Schleswig-Holstein and the Baltic Sea coastal area. These three areas have different characteristics with regard to suitability as locations for wind turbines, with the North Sea coastal area being more suitable for wind-energy production than the other regions. In contrast, the landlocked part of Schleswig-Holstein is slightly favoured for bioenergy production, as geest soils are somewhat disadvantaged in comparison with the rich marshland soils, for example, with regard to food crop production, leaving this region as a prime candidate area for energy crop production. These relative advantages are included in the model as increased energy output per unit area.

All scenarios analysed are based on the fundamental goal of reducing carbon dioxide emissions in Schleswig-Holstein by 40% by the year 2020 and by more than 80% by the middle of the 21st century in comparison to 1990 (Ministerium für Inneres und Bundesangelegenheiten des Landes Schleswig-Holstein, 2017). This corresponds to a necessary increase of renewable energy production to 37 TWh in 2025 for Schleswig-Holstein and a goal for 2050 that is accordingly higher. This drives the investments into additional wind turbines and more bioenergy production. A sensitivity analysis is conducted with regard to the level of subsidies granted to renewable energy production.

At the onset of the energy transition in Germany, investments into the construction of renewable energy infrastructure were highly subsidised by the government. In recent years, the desire to continue such subsidies has decreased considerably and it remains to be seen whether they might disappear entirely. At current levels of subsidies or without subsidies at all, investment decisions follow the same pattern. This is referred to as the reference case. To qualitatively alter development pathways, considerably higher subsidies would be necessary. Two scenarios are considered: a first fundamental shift in production patterns occurs if subsidies reach a level of approximately seven times the current subsidies (medium subsidy case); another occurs if subsidies exceed ten times the current level (high subsidy case). These possible development pathways of renewable energy production in Schleswig-Holstein are assessed in the subsequent sensitivity analysis (Figure 9.3). A special emphasis is placed on the coastal areas along the North Sea and the Baltic Sea to see whether some parts of the state are particularly useful for renewable energy production.

Simulations are conducted for the time period 2013–2050. The initial values of the parameters and decision variables (Table 9.2 and Table 9.3) are based on official statistical data provided by the Statistisches Amt für Hamburg und Schleswig-Holstein.

4 Simulation results

As an exemplary application of the model, a sensitivity analysis is conducted that shows how renewable energy production in Schleswig-Holstein is affected by different levels of subsidies. These vary from zero, which is a possibility if changes in state policy lead to altered energy feed-in regulations, to ten times the current level.

Subsidies granted for renewable energy production in Germany are currently at a level that does not have a profound impact on agricultural land use. If subsidies remain at present rates or vary only slightly, there is practically no effect on agricultural land use patterns (Figure 9.3a). The amount of land devoted to food crop production remains constant in total and there are only marginal shifts in the kind of crops produced. Right now, bioenergy plants predominantly use corn as fuel crops while energy crops such as switchgrass, different kinds of reed, and miscanthus are only of comparably low importance (Figure 9.3b).

However, this would change if subsidies for renewable energy production increased by a factor of more than five. In that case, agricultural production patterns shift substantially from food production to the growth of perennial energy plants. Approximately one-third of the arable farm land that is mainly used to produce wheat (Figure 9.3a) is reallocated to grow mostly reed. This shift affects agricultural land in all regions of Schleswig-Holstein, so the relative change is distributed fairly evenly between coastal and landlocked parts of the state (Figure 9.3b).

This shift in agricultural production patterns is reflected in the development of the number of bioenergy plants in Schleswig-Holstein. For current

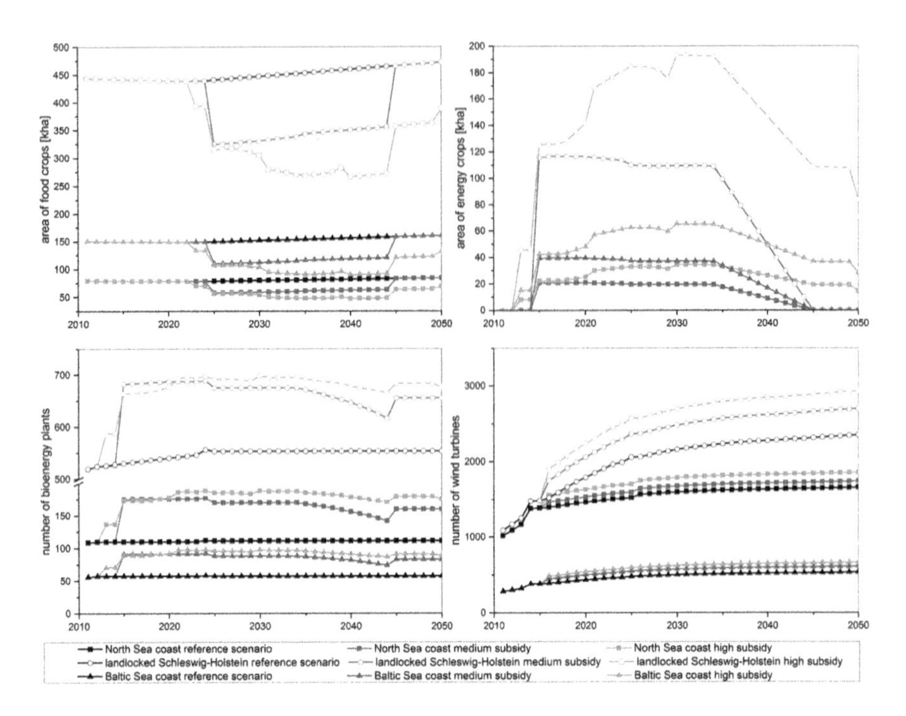

Figure 9.3 Selected results of the simulation model: a) agricultural land devoted to food crop production; b) agricultural land devoted to energy crop production; c) number of bioenergy plants; d) number of wind turbines

subsidies of renewable energy production, there is only a marginal increase in the number of bioenergy plants in the future (Figure 9.3c). If subsidies are discontinued altogether, the number of plants does not increase at all any more. In contrast, if the production of special energy plants is substantially intensified, this coincides with a pronounced increase in the number of plants as well. These are distributed fairly evenly geographically, so that there is no magnified impact on the coastal regions of Schleswig-Holstein. Based on the model, a substantial further expansion of bioenergy production is only possible with adequate financial incentives.

The situation is somewhat different for wind-energy production. There is a continued increase in the number of wind turbines in Schleswig-Holstein in all scenarios (Figure 9.3d). As expected, the growth in the wind energy sector is most pronounced if the financial incentives are higher, but there are some limits to the expansion due to the substantial investment costs involved. Also, the current concentration of wind-energy production along the North Sea coast has the effect that the future expansion is somewhat inhibited, which is most likely a consequence of the minimum distance constraint. This does not affect the other regions of Schleswig-Holstein, so their relative importance in wind-energy production increases despite the fact that these regions have a somewhat smaller wind potential.

5 Discussion and conclusion

The results of the simulation model indicate that renewable energy production will continue to play a fundamental role in Schleswig-Holstein and in particular in its coastal regions. In most of the scenarios the share of renewable energy production continues to grow, although it takes a considerable effort to achieve the goals set by the administration of Schleswig-Holstein.

A substantial increase in renewable energy production has implications on land use in the long run. In the scenarios with high subsidies for bioenergy, agricultural production shifts from conventional food crops to energy crops. That the shift is of similar magnitude in all parts of the state indicates that for bioenergy there is no significant difference between coastal areas and the landlocked parts of Schleswig-Holstein. Interestingly, there is practically no increase in energy crop growth over time if there are little or no subsidies for bioenergy production. This corresponds well with the recent development in Schleswig-Holstein that saw practically no new plants for bioenergy production being built during the past couple of years. New facilities require considerable investments and there appears to be little economic potential for further expansion of bioenergy production in Schleswig-Holstein should the current boundary conditions persist.

On the other hand, wind energy still has the potential to expand. However, there is a saturation effect with regard to wind energy in the coastal areas near the North Sea while there is continued expansion in the other parts of the state. Wind turbines are already quite prevalent in this part of Schleswig-Holstein and the model does not allow the number of wind turbines to exceed a predefined limit in the coastal region of the North Sea, which is based on current regulation and the fact that substantial areas of the North Sea coast are protected by the National Park 'Wadden Sea of Schleswig-Holstein'. Without this constraint, the optimal model solution would be to produce all the desired renewable energy from wind turbines in the North Sea coastal area, a result that is unrealistic under current regulation and because of the associated adverse environmental impacts on biodiversity as well as public opposition. This model version considers only onshore wind-energy production. An inclusion of possible offshore wind parks would avoid this effect and open up additional potential for wind-energy production.

An expansion of renewable energy production based on economic considerations will affect all parts of the state to some extent. While impacts from installations of new bioenergy plants are similar for both coastal and the landlocked regions of Schleswig-Holstein, implications of an increased number of wind turbines should be less pronounced along the North Sea coast as the expansion would mainly affect the other regions of the state. However, it has to be noted that the model does not consider local opposition to renewable energy projects, which could locally decrease the profitability of renewable energy production if increased costs to address concerns of local interest groups were to be included.

While the given model version already yields some useful insights into the possible development pathways of renewable energy in Schleswig-Holstein, further model development is advisable with respect to various aspects. First of all, the spatial resolution should be increased from three regions to county level as it would be then possible to capture the effects of varying regulation concerning renewable energy production between counties. Furthermore, agricultural production patterns (i.e., the decision on whether food or energy crops are planted) are based on a continuation of current trends in market prices for the different commodities. While this is a reasonable assumption to start with, sudden and unexpected developments on food and energy crop markets can have profound impacts that are also worth studying. Finally, energy demand and supply are currently considered in the model only for Schleswig-Holstein in isolation. However, the energy transition in Germany can only be successful if sufficient infrastructure is established to transport the vast amounts of energy generated in the mostly rural areas – such as Schleswig-Holstein – to the mostly urban areas, in which energy demand is greatest. In the case of Schleswig-Holstein, this would be the adjacent metropolitan area of Hamburg, but also connections to areas in Denmark appear reasonable. The nature and extent of such relationships should also be assessed in model simulations.

Nonetheless, it is already possible to analyse potential pathways of renewable energy production in Schleswig-Holstein for different economic and regulatory boundary conditions. The continued expansion of renewable energies in the German northernmost state is necessary if the desired production goals set by the state administration are to be reached. And whether it is wind or bioenergy that is reaped in Schleswig-Holstein, within the German context the overall success of the energy transition is considerably dependent on the amount of renewables produced in the state between the North Sea and the Baltic Sea. Furthermore, the energy transition provides a substantial opportunity for a socio-technological transition predominantly in the coastal areas, in which so far land use has been generally limited to agriculture, nature conservation and tourism. Renewable energy production may considerably increase the economic importance of these previously relatively disadvantaged regions. It is important to identify sustainable development pathways that minimise conflict potential between energy and food production and other associated land uses.

Acknowledgements

Research for this paper was supported through the Cluster of Excellence 'CliSAP' (EXC177), University of Hamburg, which is funded through the German Science Foundation (DFG).

Notes

1 In mathematical optimisation, a feasible region is the solution space of all possible solutions of an optimisation problem that satisfy the problem's constraints. It can be considered to be the initial set of candidate solutions to the problem prior to the start of the optimisation. If the feasible region is convex, i.e. line segments connecting any two feasible points only go through other feasible points, this helps the optimisation process as any local maximum that is determined simultaneously is a global maximum as well.
2 Terminal values are estimated for each crop. For energy crops, it is calculated as the net present value of future profits for the remaining productive life of the cultivation: $PV = \sum (P_t \cdot Y_t - PC_t) \cdot (1+r)^{-t}$, where P_t is the price of the crop in period t, Y_t is the yield, and PC_t is the production cost.

Bibliography

Bertsch, V., Hall, M., Weinhardt, C., & Fichtner, W. (2016). Public acceptance and preferences related to renewable energy and grid expansion policy: empirical insights for Germany. *Energy, 114*, 465–477.

Beveridge, R., & Kern, K. (2013). The Energiewende in Germany: background, developments and future challenges. *Renewable Energy L. & Pol'y Rev., 3*.

Blaschke, T., Biberacher, M., Gadocha, S., & Schardinger, I. (2013). 'Energy landscapes': meeting energy demands and human aspirations. *Biomass and Bioenergy, 55*, 3–16.

Bundesamt für Seeschifffahrt und Hydrographie. (2018). Genehmigte Windparkprojekte in der Nordsee. Retrieved 20 January 2018, from Bundesamt für Seeschifffahrt und Hydrographie, www.bsh.de/de/Meeresnutzung/Wirtschaft/Windparks/Windparks/PNS.jsp

Bundesministerium für Wirtschaft und Energie. (2017). *Zeitreihen zur Entwicklung der erneuerbaren Energien in Deutschland*. Dessau, Germany.

Dai, K., Bergot, A., Liang, C., Xiang, W.-N., & Huang, Z. (2015). Environmental issues associated with wind energy – a review. *Renewable Energy, 75*, 911–921.

Dürr, T., & Langgemach, T. (2006). Wind turbines as mortality factor for birds of prey. *Populationsökologie Greifvogel- und Eulenarten, 5*, 483–490.

Ender, C. (2015). Wind energy use in Germany – status 31.12.2014. *DEWI Magazin, 46*, 26–37.

Energymap. (2018). Energieregion Schleswig-Holstein. Retrieved 20 January 2018, from Deutsche Gesellschaft für Sonnenenergie e.V. (DGS), www.energymap.info/energieregionen/DE/105/119.html

Erlich, I., Winter, W., & Dittrich, A. (2006). Advanced grid requirements for the integration of wind turbines into the German transmission system. Paper presented at the Power Engineering Society General Meeting, 2006. IEEE.

Gatzert, N., & Kosub, T. (2016). Risks and risk management of renewable energy projects: the case of onshore and offshore wind parks. *Renewable and Sustainable Energy Reviews, 60*, 982–998.

Gee, K., & Burkhard, B. (2010). Cultural ecosystem services in the context of offshore wind farming: a case study from the west coast of Schleswig-Holstein. *Ecological Complexity, 7*(3), 349–358.

Geels, F.W. (2002). Technological transitions as evolutionary reconfiguration processes: a multi-level perspective and a case-study. *Research Policy, 31*(8), 1257–1274.

Gertz, I.C., Thöne, M., Siedentop, I.S., Albrecht, D.-I.M., Goris, D.-W.-I.A., Altenburg, D.-G.S., … Gerhards, D.-V.E. (2015). Auswirkungen von steigenden Energiepreisen auf die Mobilität und Landnutzung in der Metropolregion Hamburg – Ergebnisse des Projekts €LAN–Energiepreisentwicklung und Landnutzung. *Schriftenreihe des Instituts für Verkehrsplanung und Logistik, 13*, 223.

Goeldner, L. (1999). The German Wadden Sea coast: reclamation and environmental protection. *Journal of Coastal Conservation, 5*(1), 23–30.

Goetzke, F., & Rave, T. (2016). Exploring heterogeneous growth of wind energy across Germany. *Utilities Policy, 41*, 193–205.

Groscurth, H.-M. (2005). Grundlagenstudie „Erneuerbare Energien in Hamburg“. Endbericht–März.

Homp, C., & Schmücker, D. (2015). Im echten Norden geht Tourismus (fast) alle an. *Die Gemeinde, 67*(10), 256–258.

Hüppop, O., Dierschke, J., Exo, K.-M., Fredrich, E., & Hill, R. (2006). Bird migration studies and potential collision risk with offshore wind turbines. *Ibis, 148*(s1), 90–109.

Kominek, J., & Scheffran, J. (2012). Cascading processes and path dependency in social networks. *Transnationale Vergesellschaftungen:* (pp. 1288 auf CD-ROM): Springer VS.

Landesamt für Landwirtschaft Umwelt und ländliche Räume Schleswig-Holstein (Cartographer). (2017). Windkraftanlagen Übersichtsplan.

Liedtke, H., & Marcinek, J. (2002). *Physische Geographie Deutschlands*: Klett-Perthes Gotha.

Link, P.M., Böhner, J., Held, H., & Scheffran, J. (2018). Energy landscapes: modeling of renewable energy resources with an emphasis on Northern Germany. *Bulletin of the American Meteorological Society, 99*(4), ES71–ES74.

Link, P.M., & Borchert, L. (2015). Ecosystem services in coastal and marine areas – scientific state of the art and research needs. *Coastline Reports, 24*, 67–85.

Link, P.M., & Scheffran, J. (2017). Impacts of the German energy transition on coastal communities in Schleswig-Holstein, Germany. *Regions Magazine, 307*(1), 9–12.

Link, P.M., & Schleupner, C. (2007). Agricultural land use changes in Eiderstedt: historic developments and future plans. *Coastline Reports, 9*, 197–206.

Link, P.M., & Schleupner, C. (2011). How do tourists perceive and accept changes in landscape characteristics on the Eiderstedt peninsula? *Coastline Reports, 17*, 133–146.

Maubach, K.-D. (2014). *Energiewende: Wege zu einer bezahlbaren Energieversorgung*. Springer-Verlag.

McCarl, B.A., & Spreen, T.H. (1980). Price endogenous mathematical programming as a tool for sector analysis. *American Journal of Agricultural Economics, 62*(1), 87–102.

McKenna, R., Hollnaicher, S., & Fichtner, W. (2014). Cost-potential curves for onshore wind energy: a high-resolution analysis for Germany. *Applied Energy, 115*, 103–115.

Ministerium für Inneres und Bundesangelegenheiten des Landes Schleswig-Holstein. (2017). *Gesetz- und Verordnungsblatt für Schleswig-Holstein*, Issue No.: C 3232 A, 30, 124–129.

Nordensvärd, J., & Urban, F. (2015). The stuttering energy transition in Germany: wind energy policy and feed-in tariff lock-in. *Energy Policy, 82*, 156–165.

Önal, H., & McCarl, B.A. (1991). Exact aggregation in mathematical programming sector models. *Canadian Journal of Agricultural Economics/Revue canadienne d'agroeconomie, 39*(2), 319–334.

Rydell, J., Bach, L., Dubourg-Savage, M.-J., Green, M., Rodrigues, L., & Hedenström, A. (2010). Bat mortality at wind turbines in northwestern Europe. *Acta Chiropterologica, 12*(2), 261–274.

Scheffran, J., Link, P.M., Shaaban, M., Süsser, D., & Yang, J. (2017). Technikfolgenabschätzung in Energielandschaften – Agentenbasierte Modellierung von Energiekonflikten. *TATuP, 26*(3), 44–50.

Schiewer, U. (2008). *The Baltic Coastal Zones Ecology of Baltic Coastal Waters* (pp. 23–33). Springer.

Schleupner, C., & Link, P.M. (2008). Potential impacts on important bird habitats in Eiderstedt (Schleswig-Holstein) caused by agricultural land use changes. *Applied Geography, 28*(4), 237–247.

Schleupner, C., & Link, P.M. (2009). Eiderstedt im Spannungsfeld zwischen Naturschutz- und Agrarpolitik–Entwicklung eines methodischen Ansatzes für ein nachhaltiges Ressourcenmanagement. *Marburger Geographische Schriften, 145*, 33–49.

Schneider, U.A., & McCarl, B.A. (2003). Greenhouse gas mitigation through energy crops in the US with implications for Asian pacific countries. *Global Warming and the Asian Pacific.* Edward Elgar Publishing Limited: Cheltenham, UK, 168–184.

Schneider, U.A., McCarl, B.A., & Schmid, E. (2007). Agricultural sector analysis on greenhouse gas mitigation in US agriculture and forestry. *Agricultural Systems, 94*(2), 128–140.

Scholz, R., Beckmann, M., Pieper, C., Muster, M., & Weber, R. (2014). Considerations on providing the energy needs using exclusively renewable sources: Energiewende in Germany. *Renewable and Sustainable Energy Reviews, 35*, 109–125.

Shu, K., Schneider, U. A., & Scheffran, J. (2015). Bioenergy and food supply: a spatial-agent dynamic model of agricultural land use for Jiangsu Province in China. *Energies, 8*(11), 13284–13307.

Statistikamt Nord. (2015). Beherbergung im Reiseverkehr in Schleswig-Holstein 2015. Statistische Berichte Kennziffer: G IV 1 - j 14 SH, 2014 Herausgegeben am: 5. März 2015.

Weiser, C., Zeller, V., Reinicke, F., Wagner, B., Majer, S., Vetter, A., & Thraen, D. (2014). Integrated assessment of sustainable cereal straw potential and different straw-based energy applications in Germany. *Applied Energy, 114*, 749–762.

Wüstenhagen, R., & Bilharz, M. (2006). Green energy market development in Germany: effective public policy and emerging customer demand. *Energy Policy, 34*(13), 1681–1696.

6 Appendix: detailed dynamic agent-based model specifications

The general formulation of the regional dynamic agent-based model, which is based on the model of renewable energy production in a Chinese province (Shu et al., 2015) maximises the present value of the total profits across the entire simulation period of a system that considers both wind energy and bioenergy production. With regard to bioenergy, it covers the cultivation of both conventional crops and energy crops, subject to resource endowment constraints, energy crop transition constraints, cultivation selection constraints, and product demand constraints. With respect to wind energy, it determines how many wind turbines should be installed for the given constraints on allowed wind turbine density and wind potential.

6.1 Indices

Table 9.1 Model indices

Index	*Description*
r	regions
allcrop	all crops
fc	conventional food crops
ec	energy crops
pr	agricultural products
grains	grains
biomass	bioenergy feedstock
time	time horizon
ht	historical year
t	simulation year
s	policy scenario
a	crop age
l	biomass-based power plant locations
j	wind farm locations
number	the order of construction of wind turbines at a given location
bio-device	type of biomass-based power generation turbine
wind-device	type of wind power generation turbine

6.2 Parameters

Table 9.2 Model parameters

Parameter name	*Description*
$y_{t,r,fc,pr}^{conventional\ crop}$	yield of conventional crop (t/ ha)
$y_{ec,biomass,a}^{energy\ crop}$	yield of perennial crop (t/ ha)
$ps_{t,pr,s}$	price subsidy (EUR/ t)
$v_{pt,biomass}$	price of biomass in year t (EUR/t)
$sub_{t,r,fc,s}^{conventional\ crop}$	land subsidy for conventional crops (EUR/ha)
$sub_{t,r,ec,s}^{energy\ crop}$	land subsidy for perennial crops (EUR/ha)
$b_{t,r}^{land}$	total arable land area (ha)
$h_{ht,r,fc}$	historical cultivation area (ha)
k_{ec}	expected lifespan of energy crops (year)
$dema_{t,grains}^{grains}$	demand of grains (t)
discount	discount rate

Table 9.2 (Cont.)

Parameter name	*Description*
$\eta_t^{biomass}$	share of straw used for energy production in relation to its total amount (%)
$\alpha^{biomass}$	ratio of straw from main conventional crops to local biomass potential (%)
$c_{t,r,fc}^{conventional\ crop}$	plantation cost of conventional crops *fc* in region *r* year *t* (EUR/ha)
$c_{t,r,ec,a}^{energy\ crop}$	plantation cost of energy crops *pc* at age of *a* in region *r* year *t* (EUR/ha)
$c_{biomass,t}$	transportation costs per unit amount and distance in year *t* (EUR/km)
k_{device}	life span of fixed equipment in power plant (years)
dem_t^{power}	demand of electricity in year *t* (kwh)
$d_{r,j}$	distance between one location *r* and another location *j* (km)
$b_t^{biopower}$	annual fixed investment cost for biomass-based power plant in year *t* (EUR/MW)
$o_t^{biopower}$	other operational costs for biomass-based power plant in year *t* (EUR/MW)
$v_t^{biopower}$	fuel cost for electricity generation from biomass-based power plant in year *t* (EUR/MW)
$i^{\max,biopower}$	maximum capacity of biomass-based power plant in practice (MW)
$\alpha^{biopower}$	conversion factor from biomass to bioelectricity (kwh/t)
$\eta^{biopower}$	efficiency of bioelectricity generation (no unit)
$\sigma^{biopower,max}$	maximum annual utilisation hours of biomass-based power plant (hours)
$\widehat{i}_{l,t}$	newly added capacity of the existing biomass-based power plant built at location *l*, year *t* (MW)
$cap_{device}^{biopower}$	unit capacity of each type of biomass-based power generation turbine (MW)
$b_t^{windpower}$	annual fixed investment cost for wind farms in year *t* (EUR/MW)
$o_t^{windpower}$	other operational costs for wind farms in year *t* (EUR/MW)
$i^{\max,windpower}$	maximum capacity of wind farms in practice (MW)
$\eta^{windpower}$	efficiency of wind power generation (unitless)
$\sigma^{windpower,max}$	maximum annual utilisation of wind farms (hours)

(*continued*)

Table 9.2 (Cont.)

Parameter name	*Description*
$\widehat{i}_{j,t}$	newly added capacity of the existing wind farms built at location j, year t (MW)
$cap_{device}^{windpower}$	unit capacity of each type of wind turbine (MW)
$\rho_{j,t}$	pre-defined density of wind farms in region j year t

6.3 Decision variables

Table 9.3 Model variables

Variable name	*Description*	*Variable type*
$PRICE_t^{grains}$	price of grains (EUR/t)	non-negative
$PRICE_t^{power}$	price of electricity (EUR/kwh)	non-negative
$LAND_{t,r,fc}^{conventional\ crop}$	cultivated area for food crops (10^3 ha)	non-negative
$LAND_{t,r,ec,a}^{energy\ crop}$	cultivated area for perennial crops on arable farm land (10^3 ha)	non-negative
$CMIX_{t,r,ht}$	weighting coefficient of historical data	non-negative
$Y_{r,l,t}^{biomass}$	amount of biomass shipped from biomass production location r to biomass-based power plant at location l in year t (t)	non-negative
$Y_{l,number,t}^{biopower}$	amount of electricity generated by the *number*th biomass-based power plant located at l in year t (kwh)	non-negative
$\widehat{Y_{l,t}^{biopower}}$	amount of electricity generated by the biomass-based power plant located at l in year t (kwh)	non-negative
$Y_{j,number,t}^{windpower}$	amount of electricity generated from the *number*th wind farm located at l in year t (kwh)	non-negative
$\widehat{Y_{j,t}^{windpower}}$	amount of electricity generated by the wind farm located at j in year t (kwh)	non-negative
$I_{i,number,bio-device,t}$	number of new turbines of capacity *device* in the *number*th biomass-based power plant built at location l in year t	integer
$I_{j,number,wind-device,t}$	number of new turbines of capacity *device* in the *number*th wind farm built at location j in year t	integer

6.4 Objective function

Max WELF =

$$\sum_t (1+discount)^t \cdot \left\{ \begin{array}{l} \sum_{r,fc,grains} \left[y_{t,r,fc,grains}^{conventional\,crop} \cdot LAND_{t,r,fc}^{conventional\,crop} \cdot \left(PRICE_t^{grains} + ps_{t,grains,s} \right) \right] \\ + \sum_{r,fc} \left[LAND_{t,r,fc}^{conventional\,crop} \cdot sub_{t,r,fc,s}^{conventional\,crop} \right] \\ + \sum_{r,ec,a} \left[LAND_{t,r,ec,a}^{energy\,crop} \cdot sub_{t,r,ec,s}^{energy\,crop} \right] \\ + \sum_t \left[\left(PRICE_t^{power} + ps_{t,s}^{biopower} \right) \cdot \left(\sum_{l,number} Y_{l,number,t}^{biopower} + \sum_l \widehat{Y_{l,t}^{biopower}} \right) \right] \\ + \sum_t \left[\left(PRICE_t^{power} + ps_{t,s}^{windpower} \right) \cdot \left(\sum_{j,number} Y_{j,number,t}^{windpower} + \sum_j \widehat{Y_{j,t}^{windpower}} \right) \right] \end{array} \right\}$$

$$- \sum_t (1+discount)^t \cdot \left\{ \begin{array}{l} \sum_{r,fc} \left[c_{t,r,fc}^{conventional\,crop} \cdot LAND_{t,r,fc}^{conventional\,crop} \right] \\ + \sum_{r,ec,a} \left[c_{t,r,ec,a}^{energy\,crop} \cdot LAND_{t,r,ec,a}^{energy\,crop} \right] \\ + \sum_{r,l} \left(c_{biomass,t} \cdot d_{r,l} \cdot Y_{r,l,t}^{biomass} \right) \\ + \sum_{l,number} \left[\left(b_t^{biopower} + o_t^{biopower} \right) \cdot \sum_{device,t1 \in t-14 \to t} \left(I_{l,number,device,t1} \cdot cap_{device}^{biopower} \right) + v_t^{biopower} \cdot Y_{l,number,t}^{biopower} \right] \\ + \sum_l \left(o_t^{biopower} \cdot \sum_{t1 \in allt-14 \to t} \widehat{i_{l,t1}} + v_t^{biopower} \cdot \widehat{Y_{l,t}^{biopower}} \right) \\ + \sum_{j,number} \left[\begin{array}{l} \left(b_t^{windpower} + o_t^{windpower} \right) \cdot \sum_{device,t1 \in t-14 \to t} \left(I_{j,number,device,t1} \cdot cap_{device}^{windpower} \right) \\ + v_t^{windpower} \cdot Y_{j,number,t}^{windpower} \end{array} \right] \\ + \sum_j \left(o_t^{windpower} \cdot \sum_{t1 \in allt-14 \to t} \widehat{i_{j,t1}} + v_t^{windpower} \cdot \widehat{Y_{j,t}^{windpower}} \right) \end{array} \right\}$$

$$\forall s \qquad (1)$$

The objective function (1) of the model maximises the net present value of cash flows of the renewable energy sector in Schleswig-Holstein over the entire simulation period. The revenues consist of the sale of agricultural products, energy, governmental subsidies, and terminal values[2]. The costs are mainly related to land resources, labour resources, fertilisers, pesticides, and other auxiliary inputs.

The revenue terms account for:

- the sales revenue from conventional food crops
- the subsidy for conventional crops
- the subsidy for energy crops
- the sales revenue from electricity generated from energy crops
- the sales revenue from electricity generated from wind turbines

The cost items are:

- the cost of production inputs for conventional crops
- the cost of production inputs for energy crops
- investments in the construction of bioenergy plants
- investments in the construction of wind turbines

6.5 Constraints

6.5.1 Land endowment constraint

The most fundamental physical constraint on crop cultivation arises from the use of scarce and immobile resources. Particularly, the use of agricultural land is limited by given regional endowments of arable land. In equation (2), b denotes total arable land area in region r in year t.

$$\sum_{fc} LAND_{t,r,fc}^{conventional\,crop} + \sum_{ec,a} LAND_{t,r,ec,a}^{energy\,crop} \leq b_{t,r}^{land} \qquad \forall t,r \tag{2}$$

Equation (2) requires the sum of the arable farm land allocated to particular types of crops, which includes conventional and energy crops, to be smaller than the amount of locally accessible arable farm land during a given season; no matter which kind of field management has been adopted. This, to some extent, reflects the possible land use conflict between food crops and energy crop production.

6.5.2 Energy crop consistency constraint

Equation (3) focuses on energy crop consistency. Considering its natural death or the farmers' active eradication, the plantation area of perennial energy crops can never be larger but only smaller than or equal to their area in the prior year.

$$-LAND_{t-1,r,ec,a-1}^{energy\,crop} + LAND_{t,r,ec,a}^{energy\,crop} \leq 0 \Big|_{1<a\leq k_{ec}} \qquad \forall t,r,ec,a \tag{3}$$

6.5.3 Crop mix constraint

This constraint addresses aggregation-related aspects of farmers' decisions. Equation (4) forces farmers' cropping activities $LAND$ either in summer or in autumn to fall within a convex combination of historically observed seasonal choices h. Based on decomposition and economic duality theory, Önal & McCarl (1991) show that historical crop mixes represent rational choices embodying numerous farm resource constraints, crop rotation considerations, perceived risk reactions and a variety of natural conditions. In equation (4), the h coefficient contains the observed crop mix levels for the historical years. $CMIX$ are positive, endogenous variables indexed by historical year and region whose level is during the optimisation process.

$$-\sum_{ht}\left(h_{ht,r,fc}\cdot CMIX_{t,r,ht}\right)+LAND_{t,r,fc}^{conventional\,crop}=0 \qquad \forall t,r,fc \tag{4}$$

However, crop mix constraints are not applied to the crops, which under certain policy scenarios can be expected to expand far beyond the upper bound of historical relative shares (Schneider *et al.*, 2007). As the cultivation area of energy crops is expected to increase in the future, these crops are naturally excluded from this equation.

6.5.4 *Food security constraint*

This constraint defines the satisfaction of the requirement of food security in the context of expanding bioenergy production. The first term of equation (5) denotes the demand of a given food crop, the following terms denote the produced food from conventional crops.

$$dema_{t,grains}^{grains}-\sum_{r,fc}\left(y_{t,r,fc,grains}^{conventional\,crop}\cdot LAND_{t,r,fc}^{conventional\,crop}\right)\leq 0 \qquad \forall t,grains \tag{5}$$

6.5.5 *Biomass demand-supply constraint*

This constraint deals with the amount of biomass that is available for energy production. Equation (6) provides that the biomass amount used for energy production cannot exceed the total biomass produced by the agricultural sector in a given time period.

$$\begin{aligned}\sum_{l}Y_{r,l,t}^{biomass}&-\sum_{fc,biomass}\eta_t\cdot\frac{y_{t,r,fc,biomass}^{conventional\,crop}\cdot LAND_{t,r,fc}^{conventional\,crop}}{\alpha}\\&-\sum_{ec,biomass,a}y_{t,r,ec,biomass,a}^{energy\,crop}\cdot LAND_{t,r,ec,a}^{energy\,crop}\leq 0 \qquad \forall r,t\end{aligned} \tag{6}$$

6.5.6 Electricity demand constraint

This constraint addresses the demand for renewable energy. Equation (7) requires the produced electricity from wind turbines and from bioenergy plants to at least match the associated energy demand in a given time period.

$$\sum_{l,number} Y_{l,number,t}^{biopower} + \sum_{l} \widehat{Y_{l,t}^{biopower}} + \sum_{j,number} Y_{j,number,t}^{windpower} + \sum_{j} \widehat{Y_{j,t}^{windpower}} \geq dem_t^{power} \quad \forall t \tag{7}$$

6.5.7 Total investment constraint

This constraint limits the amount of investments that can be made in a given time period into the expansion of renewable energy production. Equation (8) not only considers the capital costs for the installation of new energy infrastructure but also includes the regularly occurring operation costs.

$$(1+discount)^t \cdot \left\{ \begin{array}{l} \sum_{r,fc} \left[c_{t,r,fc}^{conventional\ crop} \cdot LAND_{t,r,fc}^{conventional\ crop} \right] \\ + \sum_{r,ec,a} \left[c_{t,r,ec,a}^{energy\ crop} \cdot LAND_{t,r,ec,a}^{energy\ crop} \right] \\ + \sum_{r,l} \left(c_{biomass,t} \cdot d_{r,l} \cdot Y_{r,l,t}^{feedstock} \right) \\ + \sum_{l,number} \left[\begin{array}{l} (b_t^{biopower} + o_t^{biopower}) \cdot \sum_{device, t1 \in t-14 \rightarrow t} \left(I_{l,number,device,t1} \cdot cap_{device}^{biopower} \right) \\ + v_t^{biopower} \cdot Y_{l,number,t}^{biopower} \end{array} \right] \\ + \sum_{l} \left(o_t^{biopower} \cdot \sum_{t1 \in allt-14 \rightarrow t} \widehat{i_{l,t1}} + v_t^{biopower} \cdot \widehat{Y_{l,t}^{biopower}} \right) \\ + \sum_{j,number} \left[\begin{array}{l} (b_t^{windpower} + o_t^{windpower}) \cdot \sum_{device, t1 \in t-14 \rightarrow t} \left(I_{j,number,device,t1} \cdot cap_{device}^{windpower} \right) \\ + v_t^{windpower} \cdot Y_{j,number,t}^{windpower} \end{array} \right] \\ + \sum_{j} \left(o_t^{windpower} \cdot \sum_{t1 \in allt-14 \rightarrow t} \widehat{i_{j,t1}} + v_t^{windpower} \cdot \widehat{Y_{j,t}^{windpower}} \right) \end{array} \right\} \leq Invest_t \quad \forall t \tag{8}$$

6.5.8 Density of wind farms constraint

According to German regulation, there must be a defined minimum distance between two wind turbines. This distance is set to avoid adverse effects of one wind turbine on adjacent turbines via effects on local wind patterns. Equation (9) determines whether the installation of a wind turbine is possible in a given location depending on the location of already existing wind turbines.

$$I_{j,number,t1} = \begin{Bmatrix} 0, & when \sum_{wind-device} I_{j,number,wind-device,t1} = 0 \\ 1, & else \end{Bmatrix}$$
$$\sum_{number,t1 \in t-14} I_{j,number,t1} + \sum_{t1 \in allt-14 \to t} \widehat{i_{j,t1}} \leq \rho_{j,t1} \qquad \forall j,t \tag{9}$$

6.5.9 Set of constraints concerning biomass-based power energy generation

These constraints govern the transformation process from biomass to bioelectricity. Equation (10) denotes how much bioenergy can be generated from a given amount of biomass. The conversion rate depends on the kind of biomass that is used for energy generation. Equation (11) limits the amount of bioenergy that can be generated in a particular power plant in any given period of time. Equation (12) recognises that a given power plant cannot be operated all year round. The amount of energy generated depends on a plant's capacity and the maximum time of operation.

6.5.9.1 Bioenergy conversion constraint

$$\sum_{number} Y_{l,number,t}^{biopower} + \widehat{Y_{l,t}^{biopower}} \leq \sum_{r} Y_{r,l,t}^{biomass} \cdot \alpha^{biopower} \cdot \eta^{biopower} \qquad \forall l,t \tag{10}$$

6.5.9.2 Biomass-based power plant capacity limitation

$$\sum_{device,t1 \in t-14 \to t} I_{l,number,device,t1} \cdot cap_{device}^{biopower} \leq i^{\max,biopower} \quad \forall l, number, t \tag{11}$$

6.5.9.3 Bioenergy generation constraint

$$Y_{l,number,t}^{biopower} \leq \sigma^{biopower,max} \cdot \sum_{t1 \in device,t-14 \to t} \left(I_{l,number,device,t1} \cdot cap_{device}^{biopower} \right)$$
$$\forall l, number, t \tag{12}$$

6.5.10 Set of constraints concerning wind power generation

Similar to bioenergy production, these constraints govern the energy generation process from wind. Equation (13) denotes how much wind energy can be generated from a given wind potential. The conversion rate also depends on the efficiency of the given wind turbine. Equation (14) limits the amount of wind energy that can be generated by a particular wind turbine in any given period of time. Equation (15) recognises that a given wind turbine cannot be operated all year round. The amount of energy generated depends on the turbine's capacity and the maximum time of operation.

6.5.10.1 Wind energy conversion constraint

$$\sum_{number} Y_{j,number,t}^{windpower} + \widehat{Y_{l,t}^{windpower}} \leq Y_{j,t}^{windpotential} \cdot \alpha^{windpower} \cdot \eta^{windpower} \qquad (13)$$

$$\forall j,t$$

6.5.10.2 Wind farm capacity constraint

$$\sum_{device, t1 \in t-14 \to t} I_{j,number,device,t1} \cdot cap_{device}^{windpower} \leq i^{\max,windpower} \qquad (14)$$

$$\forall j, number, t$$

6.5.10.3 Wind energy generation constraint

$$Y_{j,number,t}^{windpower} \leq \sigma^{windpower,max} \cdot \sum_{t1 \in device, t-14 \to t} \left(I_{j,number,device,t1} \cdot cap_{device}^{windpower} \right) \qquad (15)$$

$$\forall j, number, t$$

10 Considering social carrying capacity in the context of sustainable ecological aquaculture

Teresa R. Johnson and Samuel P. Hanes

1 Introduction

Globally, coastal regions face growing threats, including those from declining natural resources, environmental change, pollution, urban growth and rural restructuring. A key sustainability challenge in coastal regions involves new uses and users that have consequences for local and regional spatial planning. A notable socio-technical transition (Geels, 2002) in these regions is under way, arising from an increase in seafood production away from wild fisheries towards aquaculture, a trend expected to continue (FAO, 2016). New innovations such as aquaculture in the coastal zone, while presenting opportunity in the economic sense, can threaten traditional uses and ways of life historically focused on wild fisheries. Opposition to aquaculture is widespread and significantly impacts project development (Gibbs, 2009), although growth of this sector in some places occurs without conflict (Hanes, 2018). As such, there is urgent need to consider ways to promote and integrate new directions like aquaculture in coastal regions that follow both socio-culturally and environmentally responsible paths needed for sustainability.

Sustainable ecological aquaculture is viewed as an alternative model of aquaculture development that embraces ecosystem principles while considering the concerns of the wider social, economic and environmental contexts of aquaculture, and thus calls for an interdisciplinary perspective to aquaculture research and planning (Costa-Pierce, 2010). Inclusion of sustainable ecological aquaculture offers opportunities to enhance livelihoods and resilience, especially as it serves as a diversification opportunity for those living in coastal economies (Bunting, 2013). However, the narrative around moving this "new" industry forward is coloured to some extent by its history of environmental impacts, particularly from experiences associated with industrial salmonid aquaculture. While more sustainable practices have been adopted in many places, perceptions from the industry's past shape the present discourse and often manifest in conflicts over new aquaculture development proposals.

Literature on aquaculture conflicts stresses the trade-offs coastal residents observe between economic opportunities and environmental quality (e.g., Mazur & Curtis, 2008; Whitmarsh & Palmieri, 2009). Many places feature these "jobs versus environment" debates, especially locations with finfish (as opposed to shellfish and sea vegetable) aquaculture, which is seen as more environmentally damaging because of disease transmittal to wild fish, water quality impacts and the quantity of food needed to produce farmed fish (Ridler *et al.,* 2007). There are also conflicts between fishermen and aquaculture farmers because of concerns over privatisation and enclosure by large firms limiting access to fishing grounds (Walters, 2007; Wiber *et al.,* 2010; Gibbs, 2004). However, the jobs versus environment dichotomy can oversimplify local debates (Suryanata & Umemoto, 2005; Murray & D'Anna, 2015). Perceptions of impacts to view scape or community character, while linked to environmental impacts, are in and of themselves important drivers of aquaculture conflicts (D'Anna and Murray 2015).

As the American state with the oldest and most rural population, Maine (USA) faces a unique social-ecological sustainability challenge and a "post-productive transition", which is a term for myriad changes taking place in many rural parts of the global North. These include the decline in resource extraction and commodity production, the growth of amenity migration and the rise in multifunctionality (e.g., remaining commodity production that needs to also meet conservation or other goals). Although historically diverse, Maine's fishing communities are now overly dependent upon a single resource, the American lobster fishery; in 2016, lobsters accounted for about 74% of the total value generated from commercial fisheries in Maine (ME DMR, 2017). Entry into the lobster fishery is limited at the local level as part of a co-management governance system, with some individuals waiting decades to access the fishery. Although abundance and catch levels of lobsters remain at historically high levels, scientists warn the industry to prepare for an uncertain future as water temperatures warm at an alarming rate (Mills *et al.,* 2013). Many participants and observers of the fishery are concerned about its ability to respond to future social and ecological changes, and about its social resilience (Henry & Johnson, 2015). At the same time, Maine's coastal communities are also vulnerable to rural restructuring, driven by amenity migration and youth outmigration (Thompson *et al.,* 2016). Previous research has found that Maine fishing families are being displaced from coastal regions due to increasing property values arising from amenity migrants (Thompson *et al.,* 2016), which has been countered with state and NGO efforts to protect the working waterfront (Donahue, 2014). Given these trends, which for simplicity we refer to as gentrification[1], Maine is looking to diversify its coastal economy. The expansion of aquaculture, or farming the sea, is arguably the most visible diversification strategy and is gaining considerable momentum. This is seen, for example, in programmes like the Aquaculture in Shared Waters project that provides aquaculture training for commercial fishermen (Pianka, 2016).

A transition towards sustainable ecological aquaculture in Maine makes sense not only because of a long commitment to working waterfronts, but particularly as the state is already a leader in aquaculture in the

USA. Historically important species grown in the aquaculture industry are finfish, oysters and mussels, while interest in new resources like sea vegetables, clams and scallops is emerging rapidly. A recent economic impact assessment estimated that aquaculture contributes $128 million to the state's economy (Gabe & McConnon, 2016). Another recent report recommended that the state invests heavily in shellfish aquaculture as a way to capitalise on this emerging economic opportunity, suggesting that Maine's shellfish aquaculture industry should expand by 35–40 new acres annually if Maine is to meet its full potential (Hale Group, 2016). However, where and how this growth should take place and if there should be any limits to it, remain unclear.

The limits of an area's aquaculture production are often described in terms of its physical, production, ecological and social carrying capacities (McKindsey *et al.*, 2006). Physical carrying capacity refers to the total area potentially available in an ecosystem for aquaculture; production carrying capacity is the maximum yield that an area can produce; and ecological carrying capacity is the maximum level of production possible without having an unacceptable ecological impact. Social carrying capacity, the focus of this study, has been called the "level of farm development that causes unacceptable social impacts" (McKindsey *et al.*, 2006, p. 452) or alternatively the "space of culture that the community is willing to allow (Gibbs, 2009, p. 85). Most aquaculture proposals are rejected on the basis of social reasons and thus understanding how to assess and understand causes of social carrying capacity is critical.

Techniques for inferring social carrying capacity remain a research frontier (Gibbs, 2009; Byron *et al.*, 2011). Indicators typically recommended include social acceptance and conflict levels with these being derived from surveys, GIS/spatial analyses and economic analyses focused on market barriers and willingness to pay. One way to determine the amount of aquaculture that society is willing to accept is via models that predict potential impacts of aquaculture on the ecological system and then allowing stakeholders (representing society) to decide what level is acceptable (Byron *et al.*, 2011). However, this kind of modelling-based approach begs the question of which stakeholders' preferences should weigh more and if the final assessments are representative of the values and interests of the stakeholders impacted (real or perceived). As well, such a process is time and data intensive. Following McKindsey *et al.* (2006, p. 459), we share the view that "the social aspects are qualitatively different" compared to the biophysical dimensions of carrying capacity and thus "require their own treatment". We agree with others that stakeholder input is central to decision-making and that ultimately societal discussions about social carrying capacity is necessary for a sustainable transition for aquaculture. In this chapter we aim to advance understanding of the social carrying capacity of aquaculture in the context of Maine's post-productive transition through an analysis of government documents to develop an empirical metric of conflicts, as a proxy for social carrying capacity. We then seek to explain the patterns observed using these documents, community metrics and ethnographic and stakeholder interviews. After describing the context in which this research takes place, we describe our empirical approach, our findings and the future directions of this research.

2 Background

Our study is part of research conducted through Maine's Sustainable Ecological Aquaculture Network, known as SEANET. Leveraging a $20 million grant from the National Science Foundation to the Maine EPSCoR office, the SEANET project adopts a sustainability science approach aimed at informing decision-making related to aquaculture development that embraces interdisciplinary research, stakeholder engagement and a coupled social-ecological systems perspective. Biophysical, social science and engineering research on aquaculture is occurring across Maine's diverse social-ecological landscape with targeted, comparative research in six traditional or emerging aquaculture hotspots (Figure 10.1), described later. Our research specifically takes place at the intersection of SEANET's Human Dimensions (social science) research theme and the social-ecological systems framework developed to guide interdisciplinary research across the complex project. The framework applies Ostrom's SES framework (e.g., McGinnis & Ostrom, 2014) to research on aquaculture – and the project identified four overlapping sustainability questions from this framework that guide the research. These questions focus on social-ecological carrying capacity, location or siting, efficiency and social-ecological resilience in the context of aquaculture development. What factors influence the social and ecological carrying capacities of aquaculture? What social and ecological factors influence the location or siting of aquaculture? What social and ecological factors influence the technical, economic and social efficiency of aquaculture? What factors influence the resilience of social-ecological systems in the context of aquaculture? Here we report on research exploring social carrying capacity across diverse socio-cultural, economic and historical geographies.

Maine offers an exciting laboratory to study factors influencing the social carrying capacity of aquaculture. To farm seafood in Maine one must secure a lease or licence from the state government that provides exclusive access to a defined area for farming. Standard leases are for up to 40 acres and a maximum of ten years, with opportunity for renewal. Experimental leases are for up to four acres and a maximum of three years, with no opportunity for renewal except for the purposes of scientific research. Limited Purpose Aquaculture Licenses (LPAs) are for 400 square feet for one year, with opportunity for renewal. An individual is limited to four LPAs. Standard leases always require extensive public input, including public hearings in the communities where farms are proposed, whereas LPAs do not and experimental leases usually do not. Public hearings can be very contentious with the most

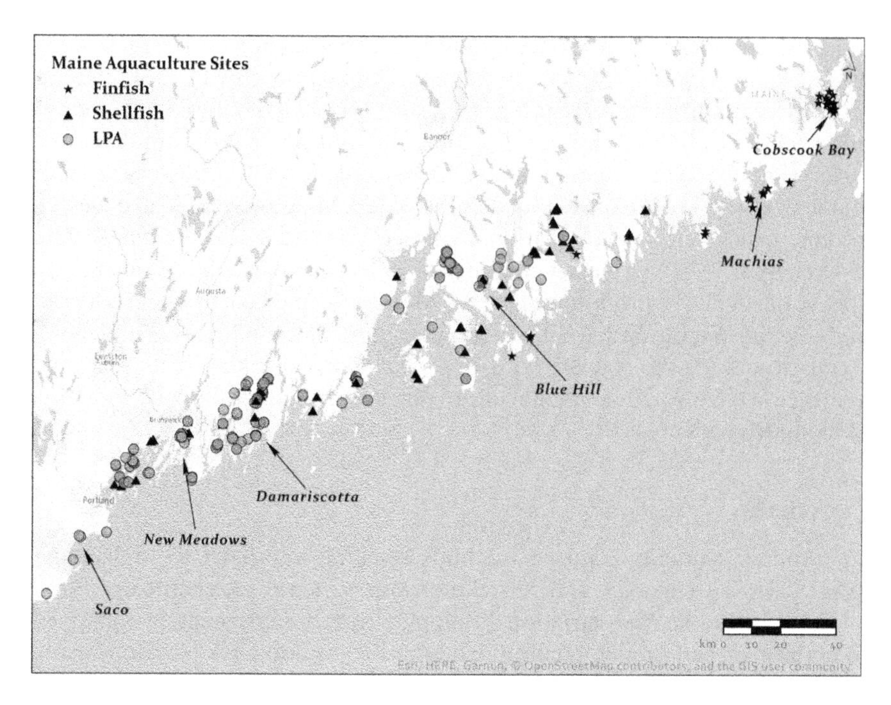

Figure 10.1 Map illustrating spatial variation in aquaculture and showing the six study sites in Maine (Cartographer: M. Kimble)

opposition from riparian landowners, while in some areas little conflict is seen (Hanes, 2018).

Aquaculture is not uniformly distributed across Maine's coast and for simplicity, one can think of three aquaculture regions (Figure 10.1). In the urbanised south region, aquaculture development is relatively nascent, with increasing interest in shellfish and seaweed farming. In the relatively more gentrified mid-coast areas, shellfish (primarily oyster) farming dominates and is expanding. In the rural, natural resource-dependent region to the east, known as Downeast Maine, finfish farming (primarily salmon) dominates and indeed is only found here. Historically important areas are the Damariscotta Bay in the mid-coast region, where oyster aquaculture began in the 1970s, and Cobscook Bay in Downeast Maine, where salmon aquaculture developed in the 1980s and has since consolidated under a large company with multiple sites in the region.

Maine is also an ideal location to study the impact of rural gentrification on aquaculture. The state began losing agricultural land in the late-1800s and other natural resource sectors followed after 1900. Population shrank from 1910 onward. Then, in the 1960s, population rebounded. Explanations include inexpensive land and housing, proximity to the Northeastern US metropolitan region, the interstate highways, the back-to-the-land movement and environmental amenities (Reilly & Renski, 2008). The newcomers avoided cities: between 1960–2000, 95% of Maine's population growth took place in rural areas. Natural resource job growth was not a major cause of the population rebound; Maine's farm population fell another 75% during this same period (Sherwood & Mageean, 2004). Young people continued to leave rural Maine and the in-migrants tended to be older. Maine's average age was 29 in 1960, which was near the national average, and today it is the oldest state at 44 (USCB, 2016). Amenity migration has transformed Maine since 1960. This is particularly pronounced along Maine's scenic coast, where tourism and amenity migration have driven up coastal real estate prices (Vail, 2004). Maine now leads the country with 16% of all coastal homes being seasonal residences (NOAA, 2013). Aquaculture farmers encounter amenity migrants all along Maine's coast.

3 Methods

We adopt a sequential, mixed methods research approach to understand social carrying capacity and its relationship to rural restructuring. First, to visualise the spatial variation in aquaculture development, we mapped all current or pending aquaculture standard lease and LPA locations using data collected by the Maine DMR and distributed by Maine Office of GIS and compiled by UMaine GIS[2]. We generated a map of all farm locations by licence type to show the spatial variation in aquaculture across the state and the location of the six study sites selected for comparative analysis: Saco Bay and the New Meadows within Casco Bay (in the urbanised south); Damariscotta River and Blue Hill Bay region (gentrified mid-coast); and Machias Bay and Cobscook Bay (rural Downeast) (Figure 10.1). These sites represent historical or emerging aquaculture hotspots in Maine. Aquaculture development in Saco Bay and New Meadows is very nascent and thus offers an interesting contrast to the other four sites where aquaculture is more established.

Social carrying capacity is often inferred from conflicts and social acceptance (McKindsey *et al.*, 2006) and we focused our analysis by measuring conflicts expressed at public aquaculture lease hearings. Data were obtained from transcripts sourced from the Maine's Department of Marine Resources (DMR). The DMR notifies the public of all lease applications, except 400 square foot Limited Purpose leases, by posting notices in local newspapers, alerting harbourmasters and sending notices to other parties of interest. If more than five persons express concerns or questions to DMR, it holds a public hearing. DMR creates transcripts of all lease hearings. Of the 263 lease hearings transcripts DMR possesses, 90 had been public hearings. Public hearings transcripts range from 15–75 pages and often include verbatim dialogue. DMR keeps these for all active leases and for all leases terminated in the past three years. Existing transcripts cover lease hearings between 1981 and 2015. We scanned all 263 hearing documents into PDF files and analysed them using NVivo 11. The research team coded transcripts for types of concerns members of the public expressed using inductive coding (Hanes, 2018). We quantified the number of times a person raised each specific type of concern in a hearing. We counted one person expressing the same concern multiple times as one concern, whereas we counted five persons raising the same concern as five concerns. This gave us a measure of how widespread each type of concern was. Most members of the public testifying at lease hearings were coastal landowners and transcripts provide an extensive record of their concerns with aquaculture through which we infer drivers of social carrying capacity.

To further explain patterns in conflicts seen, we characterised each aquaculture area using historical and current community-level socio-economic data obtained from the US Census (Table 10.1) and we conducted key informant interviews. We adopted common community metrics indicative of gentrification: % of population over 65 years old; % of population retired; % seasonal housing; median income; and median housing prices. We used % employment in agriculture, fishing and forestry as a proxy for natural resource dependence. We collected data at the town level and aggregated or averaged across all communities surrounding the site to produce a single metric for each aquaculture site (i.e., bay/river area). Our analysis of the context of conflicts and factors influencing the limits of aquaculture development, i.e., social carrying capacity, is further supported by 51 aquaculture key informant interviews with stakeholders across Maine and with farmers in each study site. These interviews covered a variety of topics including conflicts and limits on development. We entered interview transcripts into NVivo 11 for qualitative analysis via inductive coding. In this analysis, we drew on interviews as way to 'ground truth' findings and better understand patterns evident in the conflicts and socio-economic data.

4 Results

4.1 Inferring social carrying capacity via conflicts analysis

Before discussing regional variation it is important to note what did not change: the distribution of types of concerns. Coastal landowners expressed concerns about the same subjects in all six regions and they express these at broadly similar rates. The most common concerns were impacts on boating (30% of all concerns), environmental impacts (23%) and aesthetic impacts (23%). Impacts on commercial fishing came next at 17% of all concerns. These fell outside the scope of our study as we were only interested in conflicts between coastal landowners and aquaculture farmers. Conflicts with commercial fishermen can certainly limit social carrying capacity, but the factors affecting these conflicts are quite different from those affecting conflicts with coastal landowners and deserve separate treatment. The remaining 7% included a wide range of other coastal landowner concerns.

The number of leases varies greatly in these six regions so we measured conflict by calculating the number of concerns per lease in each region. Conflict was fairly uniform with a large jump upwards in the Blue Hill Bay region (Table 10.1). Clearly, Blue Hill Bay has the most conflict in Maine. It is worth noting that while Blue Hill Bay is a major centre for Maine aquaculture (Figure 10.1), it still has only 46 hearings in this dataset. Most of the state experiences much less conflict over aquaculture compared to this one area.

Another method for looking at regional variation is to draw an arbitrary cut-off to differentiate hearings with many versus few concerns. We did this at ten concerns, deeming that less than ten constituted a classification of 'few concerns' for a given area. All hearings with ten or more concerns featured speakers opposing the lease, whereas this was uncommon for hearings with nine or fewer. And, as noted below, hearings with higher concerns were markedly different in tone. Of the 90 public hearings, only 14 had ten or more concerns. Of these, 12 were in the Blue Hill Bay region and one was in the Damariscotta River. Another hearing fell into this category in Casco Bay,

Table 10.1 Conflict levels and socio-economic attributes of each study site

	Saco	*New Meadows*	*Damariscotta*	*Blue Hill*	*Machias*	*Cobscook*
Socio-economic attributes						
% change 2000–2010	6%	0%	2.4%	5%	-4%	-12%
% housing units – seasonal	11%	18%	35%	34%	16%	24%
Median housing prices	$238,475	$260,725	$251,571	$236,200	$127,475	$130,575
Median household income	$55,555	$59,444	$51,478	$48,911	$41,892	$37,035
% employment – from natural resources	0.63	3.38	5.09	6	15.73	7.58
Lease hearing analysis						
% of hearings – no concerns	60	44	50	32	45	61
% hearings – uncertainty	40	55	47	39	55	39
% of hearings – conflict	0	0	2	29	0	0

but this is outside our New Meadows River study site. This is another way of showing how different this region is: 12 out of 46 Blue Hill Bay hearings had high levels of conflict by this standard (26.1%), compared to two out of 217 for the rest of the state (0.9%).

Hearings with higher numbers of concerns also differed sharply in tone. In the coding process, we came to refer to the lower conflict hearings (under ten per hearing) as 'uncertainty' hearings and the high conflict ones as 'NIMBY' hearings. Although they express concern about the same subjects, the way they do this is quite different. In uncertainty hearings, members of the public tend to be unsure about aquaculture farms. They ask many questions about what it will look like, how it will operate, etc. They want to know how this will affect their view, sailing, dock, water quality or wildlife. NIMBY hearings are far more divisive. Members of the public are far more opposed to leases from the start. They almost always hire lawyers, which is rare in uncertainty hearings. The lawyers tend to question every criterion, apparently probing for any weakness that could block an applicant. When members of the public speak, they talk more about the harm they already assume the lease will inflict.

Recreational use provides an apt example. In one typical uncertainty hearing, a coastal landowner asked if the proposed farm would prohibit him from sailing and building a mooring which they had planned to construct. The aquaculture farmer explained that the layout of the farm, depicted in a map at the hearing, allowed plenty of room for sailing and mooring. The landowner seemed satisfied with this and had no further questions. In a NIMBY hearing, on the other hand, a landowner detailed recreational use of the space and then stated that the proposed farm would forever ruin the character of the place.

Another important facet of uncertainty hearings is the conditions DMR places on leases. Often aquaculture farmers agree to conditions that restrict their leases in highly specific ways. Some are clearly designed to make the lease comply with state law, but others are designed more to appease coastal landowners. For example, one farmer might agree to move their farm further away from the shore if a coastal landowner planned to build a dock on their property. Most leases with public hearings have these conditions (most without do not).

Uncertainty hearings educate landowners. They come with questions. Aquaculture farmers and sometimes the DMR lease site inspector educate them about what the farm will look like and how it will operate on a daily basis. If the landowners still have concerns, DMR often places legally binding conditions on the lease and the farmers generally agree to these without complaint at the hearings. The end result is that landowners come to understand what aquaculture is and how it can fit in with their own priorities. Crucially, hearings give landowners and aquaculture farmers a chance to meet face to face. Interviews indicate that face-to-face conversations are important for building trust (see below) and there are studies of rural restructuring elsewhere that bear this out (e.g., Sturtevant & Lange, 2003). Hearings provide a place to meet and have a facilitated conversation about how to mesh differing priorities.

Of course, face-to-face conversation carries risk too. When the two sides cannot agree, hearings can become confrontational and forge lasting personal animosity. Interviewees discuss this unfortunate fallout in the Blue Hill Bay region (see Section 4.3). This seems to create a path-dependency where past distrust prevents constructive relationships from developing, similar to "latent distrust" often observed between fishermen and fisheries scientists that can ultimately prevent co-operation among even well-intentioned parties (Johnson & McCay, 2012).

4.2 Socio-economic indicators

We then sought to understand our finding that the Blue Hill Bay region is likely approaching its social carrying capacity, if it has not already. Our starting hypothesis, based on our previous research in Maine's coastal communities (Thompson *et al.,* 2016), was that gentrified regions are likely to have a lower social carrying capacity (i.e., more conflict), compared to rural, resource-dependent or urban communities. As noted earlier, gentrification in Maine is driven primarily by amenity migration and these migrants often have different preferences regarding coastal landscapes from those found in natural resource-based communities. The US Census in each area shows that both Blue Hill Bay and Damariscotta regions have experienced rural restructuring, with relatively high seasonal housing, high median incomes and high home values (Table 10.1), all typical indicators of gentrification. In contrast, the Downeast region, including Machias and Cobscook Bay, is the most dependent on natural resources and is losing population, with the highest levels of employment based on agriculture, fishing and forestry (Table 10.1). These factors, along with lower incomes and housing values, suggest amenity migrants have not arrived in this region (Table 10.1). Saco Bay and New Meadows sites are located primarily within urban areas, characterised by slightly higher median incomes and housing prices.

4.3 Interviews

Given socio-economic indicators pointing to gentrification in both the Blue Hill and Damariscotta regions, our interviews indicate that what appears to affect social carrying capacity more is not gentrification, as both regions have experienced that, but rather personal relationships and trust-building. One important finding from the interviews is found in the responses to the question: "What are the main barriers the industry faces?" The most common responses were social/cultural (n=39), followed by environmental (n=27), economic (n=21) and technical (n=14) challenges. Stakeholders clearly view social carrying capacity as more significant than other limits at this point.

Our analysis of stakeholder interviews in Damariscotta provides many examples of farmers' efforts to engage with and inform the broader community. For example, one Damariscotta farmer described building relationships with local landowners, saying, "a lot of things benefit from proximity and the ability to talk things through or people actually seeing firsthand what's happening. But unless you work on relationships within the community, it just doesn't happen." Damariscotta farmers have worked to build better relationship in several ways. They founded the Damariscotta Oyster Festival in 1991, they supply many of the local restaurants, they offer oyster farm tours and they have conducted water-quality monitoring for a local NGO (the Damariscotta River Association). They even helped start and support the Damariscotta River Association's Gardening Program that helps coastal landowners set up hobby farms. The overall effect is that they have integrated oyster farming into coastal landowners' idealisation of rustic Maine.

Another farmer described coastal landowners' prevailing view toward their farms, saying:

> Damariscotta oysters are famous worldwide. So people take some pride in the industry. And it's harder to start a farm in some of the other estuaries because they don't have the history that we have. And we've proven that what we're doing is not causing any damage. It's actually net positive to the environment.
>
> (Informant SHF7)

Coastal landowners in this area have learned to trust that the industry is not harming the environment and landowners have come to see the industry in much the same way as they view Maine's lobster industry: as a valued (and delicious) part of Maine's heritage. Interviews stressed the importance of face-to-face contact and educating amenity migrants.

On the other hand, in Blue Hill Bay we heard more stories characterised by conflict and mistrust among farmers and riparian landowners. For example, one farmer described how at his first lease application hearing,

> I had three lawyers sitting at the table as interveners, hired by rich people to try to keep me from getting this lease. I shouldn't have to be a lawyer to be able to get this lease and that's what's wrong with it.
>
> (Informant SHF14)

Farmers talked about coastal landowners feeling that shellfish farms were "ruining the beauty of the river", and they tended to blame amenity migrants for delaying leases. As one farmer put it, "the major problem is all of our shorefront properties are ending up in the hands of rich property owners from out-of-state. They don't need to earn an income. They don't have any intention of working. They just want to use this area to play and that is our big problem." To some extent we heard these sentiments across the state, but they were more prevalent in the Blue Hill Bay region.

Anecdotally, as part of our outreach and communication efforts for the larger SEANET research project, we held community meetings in both Damariscotta and Blue Hill Bay regions; in those meetings, we faced a more confrontational audience in Blue Hill Bay compared to Damariscotta region, reflecting the same sentiments heard in interviews and the lease hearing transcripts.

5 Discussion and conclusions

In the literature, social carrying capacity is often described as being a level or amount of activity from a specific sector that has undesirable social impacts or that society is willing to accept. This implies that there might be some specific number of farms or perhaps acreage, or even level of production, that represents the limits of aquaculture acceptable in a region due to social factors. Such prediction may be possible for the more biophysical aspect of carrying capacity with advanced modelling techniques. However, social carrying capacity is potentially a tricky subject to assess (e.g., McKindsey *et al.*, 2006) and we feel that it is difficult, if not impossible, to predict a definitive level or amount of activity that is acceptable. However, we can infer that a site is nearing its carrying capacity if high levels of conflict or low social acceptance are visible. We feel it is especially not practical to frame social carrying capacity in simple count terms, in areas where aquaculture is only just beginning. Our study suggests that clearly context matters and so we must not only be concerned with how much farming occurs in an area, but we should also consider what kinds of farms, their size, where they are located and who owns and run them. More useful and pragmatic is to try to understand the factors that influence a system's social carrying capacity. Nevertheless, settling on a way to empirically measure it, even through a proxy, is helpful because it makes analysis more concrete.

In this chapter, we rely on government documents to create an empirical measure of social carrying capacity by measuring conflicts. The value of this source is its extensive record and rich detail. Other coastal areas likely have similar sources. Legislative records and testimony are examples. This source had several limitations, too, discussed below, and so we felt it necessary to use additional sources to corroborate the hearing transcripts. Yet creating an empirical measure allowed us to focus our efforts and gave us a clear object of explanation. Certainly, the data set may miss key facts about aquaculture debates at a given site. DMR is bound by law to only disqualify leases based on certain criteria and this may shape public comments because those opposing leases know that only those criteria can derail a lease. A person may be worried about a lease's impact on property value, but there is no way this can stop a lease, so instead they protest environmental impact. This would seem to only have an impact on the types of concerns and not the intensity of concern. We can assume the level of conflict is accurate, as is the difference in social carrying capacity, but we are less sure whether the type of concerns we see are an accurate reflection of the NIMBY hearing speakers' real motivations. Uncertainty hearing speakers have no reason to lie (the opposite is true – they want their questions answered), so we can assume their statements reflect their actual motivations. Another issue is that some aquaculture farmers withdraw lease applications before public hearings because they realise there is intense opposition. Our data set misses cases like this. This is why it is important to supplement our lease-hearing analysis with other data sources, as we did with historical research and stakeholder interviews.

In this study gentrification affects social carrying capacity of aquaculture in some areas, but not others. Amenity migration has led to many conflicts in the Blue Hill Bay region, which reduces the social carrying capacity of aquaculture, but this is not the case for the Damariscotta area. Therefore, in restructuring coastal regions undergoing sustainability transitions, amenity migration need not conflict with marine resource development, but clearly relationship-building is key. Stakeholders need better information on what processes produce trust and mutual understanding and geographers are well-equipped to study and inform this. We suggest that where there are formal and informal opportunities for dialogue among stakeholder groups regarding aquaculture decisions, especially between riparian landowners and farmers, communities are afforded more opportunity to negotiate spatial conflicts, develop trust and mutual understanding and ultimately increase the social carrying capacity of aquaculture in a place. Although we recognise that public processes like mandatory lease hearings create delays and are burdensome to both farmers and government staff, they do offer opportunities for these groups to convene to share their values, interests and help to identify how to integrate these new aquaculture uses into the coastal system in a more socially sustainable way. In the absence of these meetings, the onus is placed on the farmers to reach out to the broader community if they wish to build trust and support.

We see public lease hearings as a bottom-up process more broadly applicable to coastal planning. The state's goal seems to be to find a way to make aquaculture fit amenity migrant priorities so that Maine has fair development that works for all community members. This benefits the industry too. It helps ensure that they do not drive off much-needed amenity migrant dollars. They know that they need to fit into amenity migrants' priorities because that helps to improve social acceptance. But the state could mandate how to do that much more explicitly. Instead they hold public hearings and then place site-specific conditions on leases. This helps individual farms fit better and arguably that helps the industry as a whole to do the same. The downside is that the overall process is slower, involves more work, etc., and sometimes it increases conflict. But if done right, as seems to happen in Damariscotta, our findings suggest that bottom-up processes can improve social carrying capacity.

While our research contributes to an understanding of why some amenity migrants and rural resource users find common ground, this remains an under-studied topic. Why do some people who oppose aquaculture or other marine resource uses later turn into supporters? What works for building trust and agreement and what does not? The literature focuses extensively on conflict between groups. Studies on co-operation between rural groups show that landowners getting involved in resource use can build relationships, at least when they feel that they are taking on a stewardship role (Abrams, 2013). Face-to-face contact and getting to know resource users are important factors in explaining co-operation as well (Sturtevant & Lange, 2003). Much more could be done on this.

Sustainable ecological aquaculture, viewed as a socio-technical transition, offers an opportunity to enhance coastal resilience in rural, natural resource-dependent regions like Maine. Our findings underscore the importance of paying attention to place and history when analysing such transitions. Conflicts and social acceptance of aquaculture are not homogenous across space, but are geographical processes (Hansen & Coenen, 2015). Aquaculture can diversify livelihoods for those living within coastal economies and can keep working waterfronts "alive", thereby giving some protection from displacement often arising from amenity migration. However, this resilience strategy will not work if common ground cannot be found among those who live and work in coastal areas. Thus, finding ways to not exceed and when possible perhaps even increase, the social carrying capacity for aquaculture in a given place is important for the sustainability and resilience of these coastal regions.

Acknowledgements

Funding for this research came from the National Science Foundation award #1355457 to Maine EPSCoR at the University of Maine. This research was supported also by the USDA National Institute of Food and Agriculture, #ME021509 through the Maine Agricultural & Forest Experiment Station and through the Maine Agricultural and Forest Experiment Station Publication #3581. We thank K. Beard-Tisdale and M. Kimble for assistance with cartography and C. Cleaver and M. Miller for interview data.

Notes

1 Gentrification is typically defined as an urban or rural location where higher-income persons have moved in and displaced lower-income persons. Researchers employ a wide variety of measures to assess its presence and extent. As a practical example, if wealthier persons move into an area in Maine and drive up home values, the local government must assesses taxes at less than 90% of their new value estimated by the Maine Revenue Service or state government may withhold funds for that area. Places raising taxes to comply with this law are widely referred to as "gentrifying" in Maine.

2 The Maine aquaculture data layer was compiled from the Maine Office of GIS by Kate Beard (UMaine). Cartography by Melissa Kimble (UMaine). Map design: T. Johnson.

Bibliography

Abrams, J. (2013). Amenity landownership, land use change and the re-creation of "working landscapes." *Society and Natural Resources, 26*(7), 845–859.

Bunting, S.W. (2013). *Principles of Sustainable Aquaculture: Promoting Social, Economic and Environmental Resilience.* Routledge.

Byron, C., Bengtson, D., Costa-Pierce, B., & Calanni, J. (2011). Integrating science into management: ecological carrying capacity of bivalve shellfish aquaculture. *Marine Policy, 35*(3), 363–370.

Colburn, L.L., & Jepson, M. (2012). Social indicators of gentrification pressure in fishing communities: a context for social impact assessment. *Coastal Management, 40*(3), 289–300.

Costa-Pierce, B.A. (2010). Sustainable ecological aquaculture systems: the need for a new social contract for aquaculture development. *Marine Technology Society Journal, 44*(3), 88–112.

D'Anna, L., & Murray, G. (2015). Perceptions of shellfish aquaculture in British Columbia and implications for well-being in marine social-ecological systems. *Ecology and Society, 20*(1), 57–67.

Donahue, M. (2014). Maine's working waterfront: preserving coastal access for the future of commercial fishing and other water-dependent businesses. *Ocean and Coastal Law Journal, 19*(2), 297–321.

Food and Agricultural Organization of the United Nations (FAO). (2016). *The State of World Fisheries and Aquaculture.* Rome: Food and Agricultural Organization of the United Nations.

Gabe, T., & McConnon, J. (2016). Economic Contribution of Maine's Aquaculture Industry [Report]. University of Maine, School of Economics.

Geels, F.W. (2002). Technological transitions as evolutionary reconfiguration processes: a multi-level perspective and a case-study. *Research Policy, 31*(8–9), 1257–1274.

Gibbs, M.T. (2004). Interactions between bivalve shellfish farms and fishery resources. *Aquaculture, 240*(1), 267–296.

Gibbs, M. (2009). Implementation barriers to establishing a sustainable aquaculture sector. *Marine Policy, 33*, 83–89.

Hale Group. (2016). *Maine Farmed Shellfish Market Analysis.* Gulf of Maine Research Institute.

Hanes, S. (2018). Aquaculture and the Post-Productive Transition on the Maine Coast. *Geographical Review, 108*(2), 185–202. doi: 10.1111/gere.12247

Hansen, T., & Coenen, L. (2015). The geography of sustainability transitions: review, synthesis and reflections on an emergent research field. *Environmental Innovation and Societal Transitions, 17*, 92–109.

Henry, A.M., & Johnson, T.R. (2015) Understanding social resilience in the Maine lobster industry. *Marine and Coastal Fisheries, 7*(1), 33–43.

Johnson, T.R., & McCay, B.J. (2012). Trading expertise: the rise and demise of an industry/government committee on survey trawl design. *Maritime Studies, 11*(1), 14.

Maine Department of Marine Resources (ME DMR). (2017). Maine commercial landings data. Boothbay, Maine.

Mazur, N.A., & Curtis, A.L. (2008). Understanding community perceptions of aquaculture: lessons from Australia. *Aquaculture International, 16*(6), 601–621.

McGinnis, M., & Ostrom, E. (2014). Social-ecological system framework: initial changes and continuing challenges. *Ecology and Society, 19*(2), 30.

McKindsey, C.W., Thetmeyer, H., Landry, T., & Silvert, W. (2006). Review of recent carrying capacity models for bivalve culture and recommendations for research and management. *Aquaculture, 261*(2), 451–462.

Mills, K.E., Pershing, A.J., Brown, C.J., Chen, Y., Chiang, F.S., Holland, D.S., Lehuta, S., Nye, J.A., Sun, J.C., Thomas, A.C., & Wahle, R.A. (2013). Fisheries management in a changing climate: lessons from the 2012 ocean heat wave in the Northwest Atlantic. *Oceanography, 26*(2), 191–195.

Murray, G., & L. D'Anna. (2015). Seeing shellfish from the seashore: the importance of values and place in perceptions of aquaculture and marine social-ecological system interactions. *Marine Policy, 62*, 125–133.

National Oceanic and Atmospheric Administration (NOAA). (2013). *National Coastal Population Report: Population Trends from 1970–2020*. Washington, DC: National Oceanic and Atmospheric Administration.

Pianka, K.E. (2016). Social and ecological factors affecting the adoption of aquaculture. Unpublished Masters thesis. Orono: University of Maine.

Rector, A. (2015). *Maine Population Outlook to 2030*. Augusta, Maine: Governor's Office of Policy and Management.

Reilly, C., & H. Renski. (2008). Place and prosperity: quality of place as an economic driver. *Maine Policy Review, 17*(1), 12–25.

Ridler, N., Wowchuk, M., Robinson, B., Barrington, K., Chopin, T., Robinson, S., Page, F., Reid, G., Szemerda, M., Sewuster, J., & Boyne-Travis, S. (2007). Integrated Multi- Trophic Aquaculture (IMTA): a potential strategic choice for farmers. *Aquaculture Economics & Management, 11*(1), 99–110.

Sherwood, R., & D. Megeean. (2004). Demography and Maine's destiny. In Barringer, R. (Ed.), *Changing Maine, 1960–2010*, 3–28. Portland: University of Southern Maine Press.

Sturtevant, V., & J. Lange. (2003). Getting from "Them" to "Us": The Applegate Partnership. In Kusel, J., Adler, E. (Eds.), *Forest Communities, Community Forests*, 117–134. New York: Rowman and Littlefield.

Suryanata, K., & Umemoto, K. (2005). Beyond environmental impact: articulating the "intangibles" in a resource conflict. *Geoforum, 36*, 750–760.

Thompson, C., Johnson, T., & Hanes, S. (2016). Vulnerability of fishing communities undergoing gentrification. *Journal of Rural Studies, 45*, 165–174.

Walters, B. (2007). Competing use of marine space in a modernizing fishery: salmon farming meets lobster fishing on the Bay of Fundy. T*he Canadian Geographer, 51*(2), 139–159.

Whitmarsh, D., & Palmieri, M.G. (2009). Social acceptability of marine aquaculture: the use of survey-based methods for eliciting public and stakeholder preferences. *Marine Policy, 33*(3), 452–457.

Wiber, M., Rudd, M., Pinkerton, E., Charles, A., & Bull, A. (2010). Coastal management challenges from a community perspective: the problem of 'stealth privatization' in a Canadian fishery. *Marine Policy, 34*, 598–605.

United States Census Bureau (USCB). (2016). American Community Survey: 2016 Maine Demographic and Housing Estimates. Available at https://factfinder.census.gov/faces/nav/jsf/pages/index.xhtml

Vail, D. (2004). Prospects for a rim population rebound. In Barringer, R. (Ed.), *Changing Maine, 1960–2010*, 429–449. Portland: University of Southern Maine Press.

11 Deltas in transition

Climate change, land use and migration in coastal Bangladesh

Boris Braun and Amelie Bernzen

1 Introduction

Countries with coastlines that are merely a few metres above sea level are highly concerned about the impacts that climate change and sea level rise are likely to have on their population, infrastructure and economy. The question of how the resilience and sustainability of affected populations can be strengthened and managed in times of rapidly changing coastal environments is of great importance. This is particularly true for countries of the Global South, whose highly vulnerable populations have very limited scope for local adaptation. Migration is an important strategy in this context for two main reasons. First, it can be understood as an efficient means of adaptation to changing environmental conditions. Second, it may result in a loss of human capital and therefore a reduced potential to build up local resilience. It also can cause considerable distress to the people, families and communities involved.

Small island states and large delta areas in particular have been in the spotlight in debates about the consequences of climate change, sea level rise and migration. More often than not, direct causal relationships are assumed between rising sea levels, catastrophic floods, tropical cyclones, coastal erosion and extensive migration flows. Thus, the migration of millions of "climate refugees" appears as an unavoidable consequence of climate change and the resulting environmental changes (e.g. Bierman & Boas, 2010; Myers, 2002). Apart from the fact that the term "refugee" is a legally defined category according to the UNHCR Convention and Protocol Relating to the Status of Refugees (so-called Geneva Convention) and must hence be used with deliberation, many reports in the mass media (and some scientific journals) are using morally charged terminology as well as simplifying causal assumptions about the relationship between environmental changes and migration.

At least four fundamental problems are related to these kinds of simplifications and argumentation patterns: first, the complexity of individual decision-making processes concerning the question of whether or not to migrate is not sufficiently considered; second, the ability of people to cope with changes is underestimated; third, permanent migration over long distances is overemphasised while temporary migration over short distances is often neglected; and fourth, migration often has a negative connotation – it is seen as an exception to the rule and as something that should be avoided if possible, rather than as a context-specific and flexible means of adaptation to changing circumstances.

The Ganges-Brahmaputra Delta is one of the largest deltas on earth, with a total area of 87,000 km^2. The nation state of Bangladesh, the territory of which makes up the majority of the delta, is a country which is in transition to a state where its natural resources are increasingly overexploited due to high pressure on land, widespread poverty, inadequate technologies and limited long-term sustainable planning. This pressure is going to increase due to demographic changes and competing interests by private and public entities in resources necessary to sustain rural livelihoods. Signs of climate change and related environmental hazards further exacerbate the existing challenges, most visible in the country's coastal regions. Migration is one way for individuals to respond and adapt to local environmental change and resource scarcities.

Bangladesh is also an often-cited example for the projected emergence of millions of climate refugees (e.g. Myers, 2002, p. 611). The mass media, and also non-governmental organisations, refer to so-called "expert opinions" warning of 20 million, 25 million or even more climate refugees in Bangladesh by 2050 (e.g. Heidegger, 2008, p. 6; Karim, 2008, p. 1). These forecasts, however, must be treated with caution as they are often based on simplified assumptions about the impact of climate change and even more so the dynamics of migration within Bangladesh.

Over the last few years, scholars have been analysing processes of climate-induced migration with much higher levels of differentiation and sophistication (Foresight, 2011). This chapter aims to give an overview of recent debates dealing with the interrelations of environmental changes – some of which are an outcome of economic transitions and land-use change in the area – and migration in coastal Bangladesh. We focus on migration decisions at the household level and their different drivers. Two important issues we shed light on are: a) the relative importance of environmental stressors to migration decisions vis-à-vis socio-economic and other factors; and b) the influence of household resources or assets on the ability to migrate. To this end, we discuss the major findings of recent research projects in this field, including the authors' own empirical results that are based on a large-scale household survey conducted in coastal districts in late 2014. This allows us to present a more realistic view on migration as a factor in sustainability transition strategies.

2 Coastal Bangladesh as a multi-risk environment

Bangladesh is one of the poorest and most densely populated countries on earth, with a total population of approximately 150 million in 2011, of which about 75% live in rural areas (BBS, 2016b, p. 60). Additionally, due to its distinct physiographical location, Bangladesh is highly exposed to climate-related extreme events. The mighty rivers Ganges, Brahmaputra and Meghna carry huge amounts of water from a vast catchment area. The catchment area of this river system is 12 times larger than the country of Bangladesh itself and includes the climate stations with the highest measured precipitation on earth (e.g. Cherrapunji in India's Meghalaya Hills). Moreover, almost all rainfall is recorded in the monsoon period between June and September. As a result, floods are very common in Bangladesh, covering up to more than half of the country's surface area (Braun & Shoeb, 2008). The "historic flood" of 1998 covered 68% of Bangladesh's total territory for more than two and a half months. Extreme weather events repeatedly result in dramatic changes in the population's environment in coastal Bangladesh. Regardless of the people's remarkable ability to cope with those catastrophic events, poverty, landlessness and lack of education tend to make households and individuals extremely vulnerable to environmental changes. Vulnerability will very likely increase with climate change, accelerated sea level rise and associated extreme events, including heavy and unseasonal rainfall, river and tidal floods, storms, heat waves, drought, etc.

Although Bangladesh has been experiencing a slightly lower frequency of devastating floods in more recent years, the overall frequency and intensity of flood events has clearly increased since the 1970s. In particular, the variance of the extent of floods from year to year has increased considerably (FFWC, BWDB, 2014, p. 7). This is critical for agriculture and the local population because both too much water but also too little water can cause enormous hardships. If flooding does not occur at all during the monsoon, problems with drought are to be expected during the following dry season, and a lack of fine-grained sediment deposits on the fields has a negative impact on soil fertility. However, when more than 20% of the total land area is flooded, living conditions for millions of people quickly deteriorate. Human lives are threatened, many families become homeless and water-borne diseases can easily spread. Moreover, crop failures, destroyed houses and villages, and infrastructure problems occur as a result of severe flooding.

A massive problem in many parts of Bangladesh is riverbank erosion. Every year this process destroys valuable agricultural land, crops and settlement areas (Mirza *et al.*, 2015). About 5% of the total area under cultivation is at risk of riverbank erosion, mainly along the Brahmaputra River and in the Meghna estuary (Figure 11.1). It is estimated that since 1973 about 1.6 million people have lost their homes due to riverbank erosion (Nishat & Mukherjee, 2013, p. 28).

The consequences of tropical cyclones hitting the Bangladesh coast are even more dramatic (Figure 11.1). Wind speeds of more than 200 kmph as well as tidal waves regularly cause massive destruction along the coast. However, the death toll of cyclones has significantly decreased over time, largely thanks to improved early warning systems, the erection of thousands

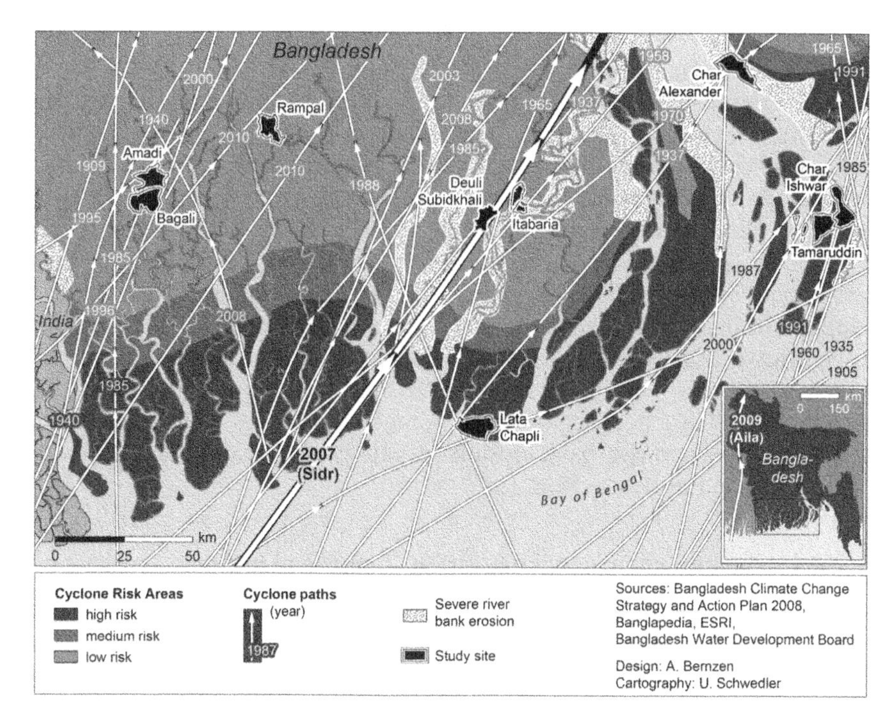

Figure 11.1 Major cyclones since 1900, cyclone and riverbank erosion risk areas, and the case study areas

of cyclone shelters and increasingly effective disaster management. In 1970, the Bhola cyclone was responsible for more than 500,000 deaths. Twenty years later, in 1991, another catastrophic cyclone cost 140,000 lives (Khalil, 1992, p. 15). Death tolls were significantly lower for the more recent cyclones Sidr (November 2009) and Aila (May 2009). Still, Sidr and Aila caused significant damage: the number of people heavily affected by these cyclones has been estimated at several hundred thousand (Shamsuddoha *et al.*, 2013, p. 9). Many of the affected cyclone victims remained homeless for an extended period of time because their homes as well as their agricultural lands were destroyed.

According to satellite radar altimeter measurements, absolute sea level rise along the Bangladeshi coast is estimated to amount to several millimetres per annum in the recent past. Long-term predictions suggest a sea level rise of 30 to 50 mm per annum in the Bay of Bengal by 2050 (Kusche *et al.*, 2016). In view of the fact that about 6,170 km^2 of settlement areas (including the approximately 140 polders) are less than 2 m above average sea level, these forecasts are alarming. In 2001, 38 million people were living in the 19 coastal districts of Bangladesh. This number increased to about 50 million people by

2011, and according to official projections the estimated population will reach about 61 million by 2050 (PDO-ICZMP, 2005). This population of millions will face major challenges in the decades to come.

Over centuries, the people living in the Ganges-Brahmaputra-Meghna Delta have learned to cope with permanent and dynamic changes in their immediate environment and recurring extreme natural events. They have historically developed various adaptation strategies such as forms of land use which are adaptable to regular floods (e.g. mixed forms of agricultural cultivation, seasonal brickfields, etc.) and elevated platforms for houses in rural settlements. Well-established social networks foster mutual assistance and support during and immediately after natural disasters (Braun & Aßheuer, 2011). However, if coping capacities are overstressed and the adaptability of people to changing conditions is no longer sufficient, out-migration seems to be inevitable. The extent to which environmental changes like sea level rise, floods, cyclones and riverbank erosion are causes of migration processes will be discussed in the next section.

3 Environmental change and migration – empirical insights

The academic and political debates about the interrelations of climate change, extreme natural events and migration in coastal Bangladesh have intensified in recent years (Call *et al.* 2017, p. 162). There are empirical studies on (potential) rural areas of out-migration as well as work on immigrant communities in the cities. Even if some of the empirical investigations are based on relatively small samples and studies with differing research designs are hard to compare, there is now more systematic knowledge about the motives of migrants than some years ago. There is an increasingly rich literature on environmental migration in Bangladesh and many other developing countries.

Some of the empirical studies focus specifically on the victims of extreme natural events. Mallick & Vogt (2012), for instance, conducted a survey of 280 households in Bangladesh's Satkhira district. They found that in the aftermath of cyclone Aila, average income was reduced by half, and almost 80% of the local workforce lost their jobs. Consequently, especially young and middle-aged men from poorer families moved to other regions and bigger cities in order to earn money to sustain their families' livelihoods.

Istahique & Ullah (2013) interviewed migrants who moved to the slums of Dhaka from surrounding rural areas. Only 7% of the 263 interviewees stated natural external events or disasters as major reasons for migration. Economic and family-related reasons played a much more decisive role in migration. Similarly, Gray & Mueller (2012), who conducted a comprehensive panel study, found that natural disasters may explain short-term migration, but have only little influence on long-term migration. For this study they used longitudinal data over a 15-year period from 1,680 households. Work by Chen *et al.* (2017), analysing official migration data of the Bangladesh Bureau of Statistics and remote sensing applications, basically confirms these findings.

Chen *et al.* estimated that about 5% of the households in the coastal region of Bangladesh had at least one family member who lived in a different place for at least six months a year. Based on multivariate statistical analysis, they showed that flooding events usually have a negative impact on out-migration rates. This finding appears to be paradoxical, but might be explained in two ways. On the one hand, extreme flooding might be an impediment to people who were planning to migrate (e.g. they might have lost their savings). On the other hand, (moderate) floods tend to improve soil fertility and increase agricultural yields. This leads to an increasing demand for labour and better income opportunities, largely counteracting rural out-migration. Of course, the latter line of argumentation does not hold for tropical cyclones and river-bank erosion.

Call *et al.* (2017) combined demographic surveillance data of 200,000 individuals over an 18-year period with biophysical data on riverine flooding, precipitation and temperature in a comprehensive study in the Matlab region in south-central Bangladesh. They were able to examine temporary migration at a very fine temporal scale. Their results clearly indicate that temporary migration quickly declines after major flood events. Environmental variability typically results in short-term disruption of long-term livelihood strategies (which may include temporary migration), rather than in a push to permanent mass out-migration in affected areas. Accordingly, they conclude "that climate change is much more likely to disrupt current livelihood-oriented migration flows than to directly induce mass displacement" (Call *et al.*, 2017, p. 164). Lu *et al.* (2016) argue along similar lines based on their study of call records from mobile phones before, during and after cyclone Mahasen in May 2013. Their analysis documented intensive short-term mobility in the hours before the storm, but only very little longer-term changes of migration patterns.

While the findings of Mallick & Vogt (2012) indicate that especially men from poorer families migrate after destructive cyclones, other studies suggest that the better-off and the better qualified are more likely to migrate (Call *et al.*, 2017). There are thus two opposing hypotheses which argue in different ways. The resource hypothesis assumes that successful migrants need financial, educational and social resources (e.g. savings, school education, social networks). Conversely, the poverty hypothesis argues that particularly poorer people who are highly vulnerable to natural or economic forces have to leave their home communities because migration is their last resort in times of crisis.

In order to better understand the motives and characteristics of migrants and non-migrants, we conducted a household survey in nine different rural regions in coastal Bangladesh (study regions shown in Figure 11.1). The survey took place in late 2014 in the context of an American-French-German-Bangladeshi co-operation project (BanD-AiD). The study areas cover three different ecological zones as well as regions with more and less dynamic land-use changes (mostly rice paddies to shrimp farms; Bernzen *et al.*, 2016; Braun *et al.*, 2016). In total, data of 1,188 households and 6,132 individuals were collected. About 6.3% of the adults (>=15 years), predominantly men,

Table 11.1 Time period and destination of migration and main reasons for migration

Time period and destination of migration			*Main reasons for migration*		
	Absolute number of migrants	*Share of all migrants (%)*		*Absolute number*	*%*
Temporary migration within Bangladesh	224	77.0	Better employment opportunities in the city	95	56.9
Temporary migration abroad	33	11.3	Better employment opportunities in agriculture	4	2.4
Permanent migration within Bangladesh	31	10.7	Education	28	16.8
Permanent migration abroad	3	1.0	Marriage or family reunification	34	20.3
Total	291	100	Political conflicts and riots	4	2.4
			Other reasons	2	1.2
			Total	167	100

Source: authors' own research

were migrants. Most of them had migrated temporarily, while only a few had settled permanently in another region or country (Table 11.1); 16.5% of all households had at least one family member with migration experience. This figure is considerably higher than the percentage mentioned by Chen *et al.* (2017). This difference, however, can be explained by the longer period of measurement in our study. In general, both studies show very similar percentages of migrants and non-migrants. Both indicate that temporary migration over mostly shorter distances is indeed a common phenomenon in Bangladesh's coastal districts. However, they do not indicate mass out-migration.

Where the cyclones Sidr (2007) and Aila (2009) caused massive damage, migration rates are higher than in other areas (between 8 to 12% compared to 5% on average, according to our study). However, when household representatives in our sample were asked for their major reasons to migrate, none of our respondents identified environmental factors, climate change or extreme natural events as the main cause. The majority of people migrated due to economic reasons, especially better opportunities to work and superior education facilities in larger towns and cities of the coastal region (such as Khulna or Barisal) or in the capital city Dhaka. Another important reason for migration was marriage. This especially applies to women who, according to Bangladeshi traditions, are expected to live in their husband's families after marriage.

Our data also revealed some interesting results with regard to the resource and poverty hypotheses mentioned above. Based on logistic regression models, we were able to identify possible factors influencing out-migration from rural communities in coastal Bangladesh. These drivers of migration include: education (e.g. secondary school or higher degree, English language proficiency), primary occupation in the off-farm sector, greater resources (e.g. measured by the access to land, living in permanent housing, etc.) as well as social capital, reflected through strong local support networks. These results support the arguments of the resource hypothesis. People leave their rural coastal communities if they have reason to expect better job opportunities or educational facilities in the chosen destination and are thus able to improve the income situation of their family. That said, our regression analysis also showed that permanently losing cultivable land also significantly increases the likelihood of migration. Both environmental and socio-economic causes contribute to the loss of land. While the majority of affected households cited two important environmental factors of erosion and cyclones, economically driven large-scale land-use changes dominate the landscape in the southwestern areas of Bangladesh. Here, this transformation has significantly reduced the necessary resources for sustainable and viable agricultural operations and has therefore led to increasing out-migration from these regions.

4 Shrimp production, land-use change and migration

Bangladesh's coastal belt is dominated by agriculture. About half of the total area in the 19 coastal districts is agricultural cropland with single or multiple annual harvests (BBS, 2016a, p. 262). This situation holds some risks and future challenges; the livelihoods of the predominantly rural population are based on natural resources such as water and land, and the latter is becoming increasingly scarce. According to estimations, in 2005 more than 80% of the people living in the coastal districts did not have access to agricultural land. Forecasts project the size of land available per capita will be reduced by more than half over the next few decades, from 0.056 ha to 0.025 ha (PDO-ICZMP, 2005 in Islam, 2006, p. 238).

Driven by an increasing demand for tropical shrimps in Europe, North America and Japan, large-scale conversion of rice paddies to shrimp farms has been observable in the southwestern districts of Bangladesh since the late 1970s (Falk, 2015). One of the reasons is the greater profitability of shrimp farming vis-à-vis rice production. Unlike the formerly practised traditional brackish-water shrimp farming, ponds for this type of production were much larger. The 139 polders, constructed since the 1950s and co-financed by the World Bank, were attractive locations for this undertaking, also due to increased waterlogging which had made rice farming less profitable in the first place (Bernier *et al.*, 2016). Consequently, the areas that had previously been protected from saltwater intrusion by embankments and dykes were flooded with saltwater to create large ponds for shrimp cultivation. Not only was existing farmland converted to shrimp farms, but new additional area was created by clearing mangrove forests, in the process destroying a natural protective barrier from cyclone and tsunami impacts. Today, many regions in the southwest, but especially the districts of Khulna, Bagerhat and Satkhira, are almost exclusively dedicated to shrimp farming, constituting up to 70% of the total area in some villages (Ministry of Land, 2011). The total area covered by shrimp farms is further increasing. Over the past decade, it rose from about 165,000 ha in 2005 to nearly 275,600 in 2015 (Azad *et al.*, 2009; BBS, 2016a, p. 465). Today, shrimps are in mass production and constitute the second most important export commodity after garments and textiles in terms of export value.

The integration of Bangladeshi shrimp farming into international value chains entails not only economic advantages for the national economy and the coastal region, but also some very serious social and ecological problems (Dietsche, 2009; Páez-Osuna, 2001; Paul & Vogl, 2011; Sohel & Ullah, 2012). Furthermore, the expansion of the sector has been largely unplanned, lacking systematic co-ordination, state control and regulatory oversight (Bernier *et al.*, 2006). In a number of cases, investors of large shrimp farms in Dhaka or overseas have extended their production areas by violent appropriation of smallholder properties (Islam, 2008; Kabir *et al.*, 2016). Over the past five to ten years, state involvement has increased. Efforts have been made by the government's National Land Zoning Project to define Land Use Zones, a planning instrument that is expected to support the government to regulate and implement ecologically and socio-economically sustainable land use (Ministry of Land, 2016).

Although there has been an increase of small and family-operated shrimp farms in recent years (Bernier *et al.*, 2016), the fundamental problem remains: shrimp farming is much less labour-intensive than the cultivation of rice or other traditional crops, resulting in reduced local employment opportunities. This is a significant problem for landless people like day labourers, harvesters and fishermen. In addition, shrimp farming-related ecological problems such as decreasing biodiversity, water pollution, shortage of drinking water and in particular soil salinisation are accumulating. The latter makes agricultural land use practically impossible, as crops cannot tolerate the salt levels which are often too high even for more salt-tolerant varieties. This applies to most types of grains, vegetables and fruits. Shrimp cultivation also affects plots of neighbouring lands as saline water seeps into the soil (Boem & Richter, 2010, p. 15). Consequently, these farmers are forced to change their farming to shrimp cultivation as well.

This situation will exacerbate the already existing challenge of improving food and nutrition security in coastal Bangladesh, as many rural households have insufficient means to sustain diverse diets due to low incomes or lack of land for self-subsistence farming. Another problem is that Bangladesh is already highly dependent on food imports, and food price fluctuations on global markets trickle down also to local markets. Price increases for staples and other food items are painfully felt by many households and represent one key factor contributing to the decision to migrate out of rural areas (Ahmed *et al.*, 2012, p. 25).

All things considered, economically driven land-use changes are one of the major reasons for rural out-migration in coastal Bangladesh. In fact, they seem to weigh more than environmental changes or the impacts of climate change and sea level rise, as evidence from empirical studies in communities and regions dominated by shrimp farming suggests (Amoako-Johnson *et al.*, 2016; Azam, 2011; Falk, 2015). Areas with a high share of shrimp production record a declining population, while many communities located closer to the coastline feature significant population increases due to both natural population growth and in-migration (Braun *et al.*, 2016).

Rural-urban migration has significantly increased in Bangladesh in the past three decades, but this also applies to migration between different rural communities. Over two-thirds of permanent migration flows are rural-to-urban (BBS, 2015; Braun *et al.*, 2016; Chen *et al.*, 2017). The rural-urban migration rate increased from 1.2 per thousand in 1984 to an impressive 22.5 per thousand by 2010 (BBS, 2015; Afsar, 2003, p. 1). This has created burgeoning urban slums, where 40 to 50% of all residents are rural immigrants (Ishtiaque & Ullah, 2013). Likewise, circular and more temporary migration of individual family members between rural and urban locations has been a common practice in many Bengali families for a long time. Distances covered by this latter type of migration vary greatly, however. Overall, a substantial impact of climate or environmental changes on migration can hardly be derived.

5 Conclusion

In conclusion, existing evidence from empirical studies on migration processes in coastal Bangladesh shows that economic reasons (e.g. pull factors like better job opportunities or educational facilities in the larger cities, or push factors like reduced demand for local agricultural labour due to the expansion of shrimp farming in rural areas) dominate migration decision-making in this area. In contrast, climate change, sea level rise and related extreme events have had only secondary effects, based on evidence collected to date. People tend to rather rationally "translate" changes affecting their immediate living environment into economic motives. These may or may not eventually result in a decision to migrate, more often in a circular fashion than permanently. Moreover, a large proportion of environmentally induced migration processes are only across short distances.

The hypothesis of (future) large-scale and long-distance migration flows attributed to climate change or sea level rise, often dramatically reported as oversimplified predictions of millions of "environmental refugees" in the media, cannot be supported by existing scientific findings. The implicit assumption of a causal relationship between changing climatic conditions and extensive flows of migrants across large distances cannot be justified based on what we know about migration processes and individual migration motives. On the one hand, from an empirical perspective, environmental and climate change effects can hardly be clearly distinguished from other migration motives. Simple, deterministic cause-and-effect approaches consider neither the multiplicity and complexity of migration drivers nor the enormous adaptability of people to changing circumstances. On the other hand, there is no clear link between geographical exposure to natural hazards, socio-economic vulnerability and migration. Most migrants are well-educated, have access to financial, social and physical resources and strive for better income opportunities, mainly in larger cities but also abroad. People from lower socio-economic backgrounds are considered to be particularly vulnerable, but are simultaneously restricted in their means to migrate as they lack the financial as well as social resources to support them. In consequence, this may mean that they miss out on an individual adaptation strategy which could potentially improve their livelihoods. We argue in line with Black *et al.* (2011) and Hartmann (2010) that debates should move away from treating migration as a threat or a problem. Rather, it should be seen as an efficient and reasonable way of dealing with changing environments for a society as a whole. Ultimately, future research should focus on the question of why people do *not* migrate (or cannot migrate), and on the development of so-called *trapped populations* that are stuck in their difficult circumstances without having the choice to migrate as an adaptation strategy.

Acknowledgements

The findings in this chapter are based on research that was conducted within the Belmont Forum BanD-AiD project "Collaborative Research – **Ban**gladesh **D**elta: **A**ssessment of the Causes of Sea-level Rise Hazards and **I**ntegrated **D**evelopment of Predictive Modeling Towards Mitigation and Adaptation". Research was funded by the German Research Foundation (DFG #BR1678/13-1). The authors wish to thank Craig Jenkins (State University of Ohio) for his valuable comments and their colleagues at Rajshahi University, in particular Raquib Ahmed and Abdullah Al-Maruf, for their priceless support during the fieldwork in coastal Bangladesh.

Bibliography

Afsar, R. (2003). Internal migration and the development nexus: The case of Bangladesh. Dhaka. Paper presented at the regional conference on migration, development and pro-poor policy choices in Asia. Available at: www.migrationdrc.org/publications/working_papers/WP-C2.pdf

Ahmed, A.U., Hassan, S.R., Etzold, B., & Neelorm, S. (2012). *"Where the Rain Falls"-Project. Case study: Bangladesh. Results from Kurigram District, Rangpur Division.* Report No. 2. Bonn: United Nations University Institute for Environmental and Human Security (UNU-EHS).

Amoako-Johnson, F.C.W., Hutton, D., Hornby, A., & Lazar, A.M. (2016). Is shrimp farming a successful adaptation to salinity intrusion? A geospatial associative analysis of poverty in the populous Ganges-Brahmaputra-Meghna Delta of Bangladesh. *Sustainability Science, 11*(3), 423–439.

Azad, A.K., Jensen, K.R., & Lin, C.K. (2009). Coastal aquaculture development in Bangladesh: unsustainable and sustainable experiences. *Environmental Management, 44*(4), 800–809.

Azam, M. (2011). *Factors Driving Environmentally Induced Migration in the Coastal Regions of Bangladesh: An Exploratory Study.* M.S. Thesis in Environmental Governance. Freiburg: Albert-Ludwigs-Universität.

BBS, Bangladesh Bureau of Statistics (2016a). Yearbook of Agricultural Statistics 2015. 27th Series. Dhaka.

BBS, Bangladesh Bureau of Statistics (2016b). Statistical Yearbook Bangladesh. 36th Series. Dhaka.

BBS, Bangladesh Bureau of Statistics (2015). *Population and Vulnerability: A Challenge for Sustainable Development of Bangladesh.* Population Monograph Vol. 7. Dhaka: BBS.

Bernier, Q., Sultana, P., Bell, A.R., & Ringler, C. (2016). Water management and livelihood choices in southwestern Bangladesh. *Journal of Rural Studies, 45,* 134–145.

Bernzen, A., Al-Maruf, A., Lin, A., & Ahmed, R. (2016). Landnutzungswandel im Küstenraum von Bangladesh: Management von Land auf Gemeinde- und Haushaltsebene. *Geographische Rundschau, 68*(7–8), 16–22.

Biermann, F., & Boas, I. (2010). Preparing for a warmer world: towards a global governance system to protect climate refugees. *Global Environmental Politics, 10*(1), 60–88.

Black, R., Adger, W.N., Arnell, N.W., Dercon, S, Geddes, A., & Thomas, D. (2011). The effect of environmental change on human migration. *Global Environmental Change, 21*(1), 3–11.

Boem, I., & Richter, N. (2010). Fader Beigeschmack. Die Folgen der industriellen Garnelenzucht in Bangladesch. *NETZ, 4,* 15–17.

Braun, B., & Aßheuer, T. (2011). Floods in megacity environments: vulnerability and coping strategies of slum dwellers in Dhaka/Bangladesh. *Natural Hazards, 58*(2), 771–787.

Braun, B., Chen, J.J., Dotzel, K.R., & Jenkins, J.C. (2016). Klimawandel und Migration im Küstenraum von Bangladesh. *Geographische Rundschau, 68*(7–8), 10–15.

Braun, B., & Shoeb, A.Z.M. (2008). Naturrisiken und Sozialkatastrophen in Bangladesch – Wirbelstürme und Überschwemmungen. In Felgentreff, C., & Glade, T. (Eds.), *Naturrisiken und Sozialkatastrophen* (381–393). Berlin, Heidelberg: Spektrum.

Call, M.A., Gray, C., Yunus, M., & Emch, M. (2017). Disruption, not displacement: environmental variability and temporary migration in Bangladesh. *Global Environmental Change, 46,* 157–165.

Chen, J.J., Mueller, V., Jia, Y., & Tseng, S. K.-H. (2017). Validating migration responses to flooding using satellite and vital registration data. *American Economic Review, 107*(5), 441–445.

Dietsche, C. (2009). Networking against stakeholder risks: a case study on SMEs in international shrimp trade, *Belgeo, 1*, 27–42.

Falk, G.C. (2015). Land use change in the coastal regions of Bangladesh: a critical discussion of the impact on delta-morphodynamics, ecology, and society. *ASIEN, 134*(1), 47–71.

FFWC, BWDB – Flood Forecasting and Warning Centre, Bangladesh Water Development Board (2014). *Annual Flood Report 2014*. Dhaka: FFWC/BWDB.

Foresight (2011). *Migration and Global Environmental Change. Final Project Report.* London: The Government Office for Science.

Gray, C.L., & Mueller, V. (2012). Natural disasters and population mobility in Bangladesh. *Proceedings of the National Academy of Sciences of the United States of America, 109*(16), 6000–6005.

Hartmann, B. (2010). Rethinking climate refugees and climate conflict: rhetoric, reality and the politics of policy discourse. *Journal of International Development, 22*, 233–246.

Heidegger, P. (2008). Die Welt ist aus dem Gleichgewicht. Der Klimawandel und seine Folgen. *NETZ, 2*, 6–9.

Ishtiaque, A., & Ullah, M.S. (2013). The influence of factors of migration on the migration status of rural-urban migrants in Dhaka, Bangladesh. *Human Geographies, 7*(2), 45–52.

Islam, M.R. (2006). Managing diverse land uses in coastal Bangladesh: institutional approaches. In Hoanh, C.T., Tuong, T.P., Growing J.W., & Hardy, B. (Eds.), *Environment and Livelihoods in Tropical Coastal Zones* (237–248). Wallingford: CAB International.

Islam, M.S. (2008). In search of "white gold": environmental and agrarian changes in rural Bangladesh. *Society & Natural Resources, 22*(1), 66–78.

Kabir, M.J., Cramb, R., Alauddin, M., & Roth, C. (2016). Farming adaptation to environmental change in coastal Bangladesh: shrimp culture versus crop diversification. *Environment, Development and Sustainability, 18*(4), 1195–1216.

Karim, M. (2008). Bangladesh faces climate change refugee nightmare. Retrieved 12 April 2017, from www.reuters.com/article/us-bangladesh-climate-islands-idUSDHA23447920080414 [

Khalil, G.M. (1992). Cyclones and storm surges in Bangladesh. Some mitigative measures. *Natural Hazards, 6*, 11–24.

Kusche, J., Uebbing, B., Rietbroek, R., Shum, C.K., & Khan, Z.H. (2016). Sea level budget in the Bay of Bengal (2002–2014) from GRACE and altimetry. *Journal of Geophysical Research-Oceans, 121*(2), 1194–1217.

Lu, X., Wrathall, D.J., Sundsøy, P.R., Nadiruzzaman, M., Wetter, E., Iqbal, A., & Bengtsson, L. (2016). Unveiling hidden migration and mobility patterns in climate stressed regions: a longitudinal study of six million anonymous mobile phone users in Bangladesh. *Global Environmental Change, 38*, 1–7.

Mallick, B., & Vogt, J. (2012). Cyclone, coastal society and migration: empirical evidence from Bangladesh. *International Development Planning Review, 34*, 217–240.

Ministry of Land (2011). Land zoning report of Rampal Upazila of Bagerhat District. Retrieved 29 February 2016, from www.landzoning.gov.bd/main1/report.php

Ministry of Land (2016). National Land Zoning Project, project brief. Retrieved 29 February 2016, from www.landzoning.gov.bd

Mirza, A.T.M, Rahman, T., Islam, S., & Rahman., S.H. (2015). Coping with flood and riverbank erosion caused by climate change using livelihood resources. *Climate & Development, 7*(2), 185–191.

Myers, N. (2002). Environmental refugees. A growing phenomenon of the 21st century. *Philosophical Transactions of the Royal Society of London B, 357*, 609–613.

Nishat A., & Mukherjee, N. (2013). Sea level rise and its impact in coastal areas of Bangladesh. In Shaw, R., Mallick, F., & Islam, A. (Eds.), *Climate change Adaptation Action in Bangladesh* (43–50). Tokyo: Springer.

Páez-Osuna, F. (2001). The environmental impact of shrimp aquaculture: causes, effects, and mitigating alternatives. *Environmental Management, 28*(1), 131–140.

Paul, B.G., & Vogl, C.R. (2011). Impacts of shrimp farming in Bangladesh: challenges and alternatives. *Ocean & Coastal Management, 54*(3), 201–211.

PDO-ICZMP, Program Development Office for Integrated Coastal Zone Management Plan (2005). *Living in the Coast: Urbanization.* Dhaka: PDO-ICZMP, Water Resources Planning Organisation.

Shamsuddoha, M., Islam, M., Haque, A.H., Rahman, M.F., Roberts, E., Hasemann, A., & Roddick, S. (2013). *Local Perspective on Loss and Damage in the Context of Extreme Events: Insights From Cyclone-Affected Communities in Coastal Bangladesh*. Dhaka: CPRD.

Sohel, M.S.I., & Ullah, M.H. (2012). Ecohydrology: a framework for overcoming the environmental impacts of shrimp aquaculture on the coastal zone of Bangladesh. *Ocean & Coastal management, 63*, 67–78.

Part III

Applied management

12 Coastal sediment management as a response to intensifying storms and sea level rise

A case study

James Tait, Ryan Orlowski, Jessica Brewer and Matthew D. Miller

Dedication: to our muses Irene and Sandy

1 Beaches and storm wave damage

In a period of rising sea levels and intensifying storms, enhancing coastal resilience and finding sustainable ways of living with the coast have become necessary concerns. Investigations of storm wave damage to coastal structures on the Connecticut shoreline during Hurricane Irene and Superstorm Sandy revealed that the width and height of frontal beaches were the most common denominator in mitigating such damages (Tait & Akpinar Ferrand, 2014). Three towns along the Connecticut coast were subject to particularly severe wave damage: Fairfield, Milford and East Haven. Along the western half of Cosey Beach Avenue in the town of East Haven, for example, the majority of homes with direct waterfront exposure were either severely damaged or were condemned and demolished. These homes were relatively new, were typically robust in their construction, often had small seawalls, but all shared the circumstance of having little or no beach at high tide. During Irene and Sandy, waves at high tide broke directly on to these properties. In comparison, houses along the adjacent West Silver Sands Beach typically experienced only minor wave damage despite being generally older and less robustly built than the houses on western Cosey Beach Avenue. The most common reason for this is their general position farther landward on the beach profile. There is a group of beach condominiums that escaped wave damage altogether despite also lacking a beach at high tide. The reasons for this are that they were built on an artificially elevated ridge and that they have a substantial seawall in front of the complex. It is interesting to note that Sandy, which was the larger storm, had its peak surge arrive just after low tide for a total local storm tide (height of the astronomical tide plus the height of the storm surge) elevation of 2.74 m (9 ft). If the peak surge had arrived at high tide, the storm tide elevation would have been 3.66 m (12 ft). In this case, the storm waves would have topped the seawall at the condominiums and wave damage would have been uniformly severe due to their proximity to the water.

2 How beaches buffer wave energy

Each individual wave can be regarded as a "packet" of wave energy. The energy per unit crest length of each wave is given by the equation

$$E = 1/8\, \rho g H^2$$

Where ρ is the density of the water, g is the acceleration of gravity and H is the height of the wave (Komar, 1998). Each packet of energy is capable of doing work in the sense of physics. To the extent that work is done by the wave, its total energy is diminished and, because the wave energy is proportional to the square of the wave height, the height of the wave is reduced. The work that each wave can do at the shoreline could be driving turbulence, moving sand or moving houses. At the western end of Cosey Beach Avenue, much of the wave energy was used to do the latter while on West Silver Sands Beach, most of the wave energy was used to do the first two things.

3 Connecticut's systematically eroding beaches

Typical open-ocean beaches exhibit a seasonal dynamic in which they are eroded during stormy periods, the sand being stored in offshore bars, and are rebuilt during fair-weather periods by moderately high, long period waves transporting the sand shoreward. This behaviour has been documented extensively (e.g., Shepard, 1950; Inman & Frautchy, 1966; Sonu & van Bleek, 1971; Griggs & Tait, 1988; Dean, 1991; Dalrymple, 1992; Griggs, Tait & Corona, 1994; Larson, Capobianco & Hansen, 2000). The fair-weather waves that restore the beach are known as *swell*. Swell waves typically emanate from distant storms and, during travel from their point of origin, they sort themselves into coherent groups via a process known as velocity dispersion. For most of Connecticut's beaches, the moderately high swell waves that are capable of transporting eroded sand back to the beach are filtered out by the presence of Long Island to the south. Long Island shelters the entire Connecticut coast. The fair-weather waves that do impinge upon the Connecticut shoreline are locally generated within Long Island Sound. They are typically on the order of 0.25 m high and lack both the threshold and cumulative power to transport sand lost to local storms shoreward again. This is because Long Island Sound is a fetch-limited body of water and the height of waves is a function, in part, of the fetch or the distance available for the waves to travel as they gain energy.

As a result of this seasonal energy imbalance, most of Connecticut's beaches are systematically erosive, losing sand offshore during storms but not recovering that sand during periods of fair-weather waves. Due to this physical dynamic, and as a result of extensive shoreline development, the Connecticut coast is highly vulnerable to the impacts of large storms such as Irene and Sandy.

4 Beach nourishment and beach replenishment as solutions

It is probably not a surprise that Connecticut has an extensive history of beach nourishment (import of sand to replace sand lost to erosion). Haddad & Pilkey (1998), in their publication on the New England beach nourishment experience, identified 44 beach replenishment episodes in Connecticut alone between 1938 and 1996. It is useful, at this point, to make a distinction between beach nourishment and beach replenishment. *Replenishment* is considered a single sand replacement *episode*, whereas *nourishment* applies to a series of *replenishment episodes* at a particular location. A beach that is being nourished is referred to as a *project.* Although some of these replenishments were part of a project, these 44 episodes amount to one replenishment for every 4.5 km of Connecticut's 154 km coast. The number of replenishment episodes is undoubtedly higher today.

In general, Connecticut's beaches tend to be relatively short and segmented by rocky headlands. Many of them are privately held. As a result, many of them come up short in cost/benefit analyses (performed by the US Army Corps of Engineers) that precedes authorisation of a beach nourishment (or replenishment) project. This aggravates the vulnerability of the Connecticut coast to systematic beach erosion and consequent storm wave damage. Although damage to coastal structures was intensive in shoreline towns such as Fairfield, Milford and East Haven, many of the developed stretches of shoreline remain exposed to future storm wave damage with little to no beach at high tide. This is due mainly to the lack of federal funds to implement the necessary projects. It may also be due to the perception that storm impacts and beach erosion are perceived to be random events rather than a systematic process on the Connecticut coast.

One of the major problems, particularly for a state like Connecticut that has a highly developed coastline with systematically eroding beaches, is that the cost of beach nourishment projects is not only high but is increasing rapidly. The cost of beach nourishment at Prospect Beach in the town of West Haven is instructive (Table 12.1).

Table 12.1 Major beach replenishment episodes at Prospect Beach, West Haven, Connecticut, and statistics for replenishment dimensions and costs

Replenishment episode	*1957*	*1973*	*1987*	*1994*	*2014*
Volume (m^3)	338,697	19,114	***	99,392	69,345
Length (m)	1,972	***	***	1,311	1,372
Cost ($)	358,507	166,000	2,268,000	1,700,000	3,790,000
Cost per m^3 ($)	1.06	8.68	***	17.10	54.65
Cost per m ($)	181.80	***	***	1296.72	2762.39

Source: Haddad & Pilkey (1998) and Town of West Haven Department of Public Works

The cost of beach replenishment at Prospect Beach went from \$1.06 per m^3 in 1957 to \$54.65 per m^3 in 2014. This is a 5,056% increase. Similarly, the cost per linear metre of beach (which is a less reliable metric) went from \$181.80 in 1957 to \$2,762.39 in 2014. This is a 1,419% increase. Meanwhile, the value of the US dollar since 1957 has increased by just 757%. The rate increase in the cost per cubic metre has been exponential.

In the half century between 1935 and 1996, \$48.2 million (Trembanis & Pilkey, 1998) was spent nourishing Connecticut beaches (expressed in 1996 dollars). As of 1996, the authors estimate that \$3.388 billion have been spent on beach nourishment in the nation as a whole. This would amount to \$75.8 million and \$5.33 billion respectively in 2017 dollars. The two major conclusions are that the cost of beach nourishment is increasing and so is the need for beach nourishment in a time of intensifying storms and sea level rise. This is particularly true for the state of Connecticut's beaches and its problem with systematic erosion and extensive coastal development. In addition to the cost of replenishment, there is another related variable: the availability of suitable sand. Aggregate (sand and gravel) is the most widely consumed solid natural resource in the world (UNEP, 2014). The rate of extraction far exceeds the rate of renewal. Between 47 and 59 billion metric tons are extracted every year. The world's use of aggregates for concrete is estimated at \$25.9–\$29.6 billion for the year 2012 alone. This represents enough concrete to build a wall that is 27 metres wide and 27 metres high around the world at the equator. As land-based sources of sand are depleted, the acquisition of sand and gravel for construction and other purposes has shifted to the marine environment, placing enhanced stress on offshore supplies. Therefore, beach nourishment projects are competing with other uses for the supply of sand, and that supply is dwindling.

In particular, sand sources that have the right characteristics, e.g., mean grain size, to be stable on a particular beach are becoming harder to locate (Daniel, 2001). It seems reasonable that beach sediments, and sand supplies in general, be managed. This emphasis on sediment management is reflected in the Bureau of Ocean Energy Management's outer continental shelf marine minerals (sand) studies (Bureau of Ocean Energy, 2017). For local communities with beach sand supply problems, a potential management alternative might be *sand reclamation* as opposed to replenishment in which sand with suitable characteristics is imported from a proximal or distant source. Sediment reclamation entails studies of sediment dispersal and deposition patterns so that sand that is eroded during storms can be reclaimed, i.e., the same sediment is recycled rather than replaced by new supplies of sand.

5 The town of West Haven: a case study

The town of West Haven, Connecticut (Figure 12.1a), has the longest stretch of public beaches in the state (5.3 km). These beaches provide recreational opportunities, promote economic activity and protect coastal structures and

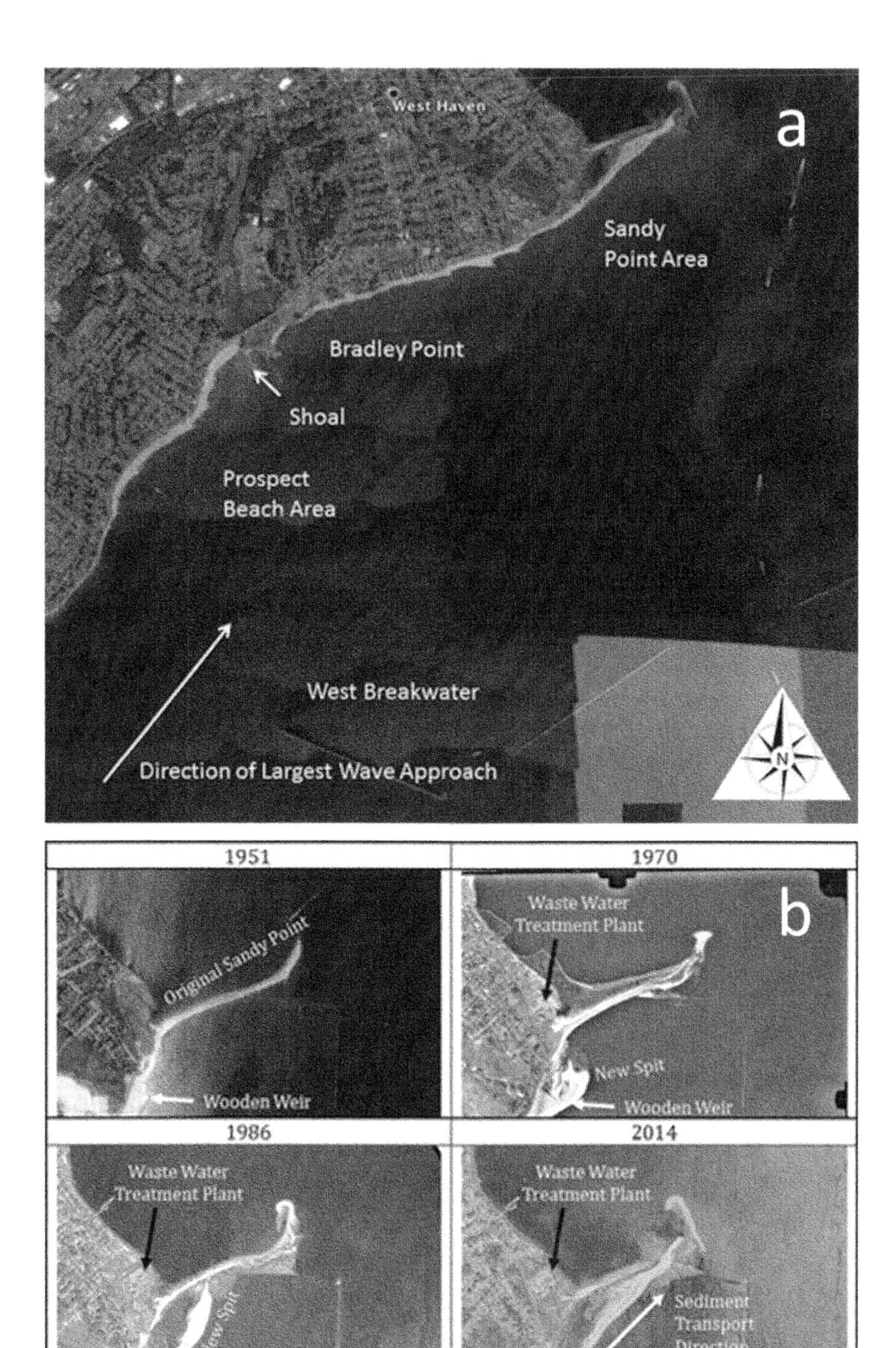

Figure 12.1 a) The town of West Haven has over 5 km of public beaches that have to be replenished on a regular basis. In many areas, beaches protect structures and infrastructure against storm wave damage; b) Sand transported to the northeast from several beach replenishment episodes was diverted seaward by a wooden "weir", prompting the growth of a second spit (also known as Morse Point), which threatens at this time to become a hazard to navigation and to interfere with a waste-water treatment plant

infrastructure, such as roads, sewer lines and utilities, from damage during storms. Like many other Connecticut beaches, the West Haven beaches have had to be nourished. The series of beaches known in US Army Corps of Engineers records as Prospect Beach comprise the southwestern portion of West Haven's beaches (Figure 12.1a). These beaches have undergone at least four replenishment episodes since 1957. The last episode was in the autumn of 2014 and involved the placement of 69,345 m^3 of sand imported from a glacial quarry in the nearby town of Wallingford at a total replenishment cost of $3.79 million.

Given that this investment would likely be lost to erosion (Tait & Akpinar Ferrand, 2014), scientists at Southern Connecticut State University's Werth Center for Coastal and Marine Studies measured bi-monthly beach profiles in order to document spatial and temporal sediment erosion, dispersal and depositional patterns. These profiles were subsequently assembled into sequential 3D images by J. Brewer at the State University of New York College of Environmental Science and Forestry. In a separate but related study, an aerial photography analysis was conducted on the northeastern, or Sandy Point, beaches (Figure 12.1a).

The results showed erosion along portions of the Prospect Beach beaches both in terms of beach width, measured at mean higher high water, and beach volume, measured above mean lower low water. In particular, erosion occurred in areas where the angle of the shoreline was at a maximum with respect to the incoming wave field. The direction of approach for the largest waves was controlled by the placement of breakwaters across the entrance to New Haven harbour (Figure 12.1a). In their present configuration, there is a gap between the western breakwater and the shoreline that provides waves coming from the southwest a fetch of 47 km. Storm waves entering New Haven harbour from this direction transported sand in the alongshore direction to the Bradley Point area. Bradley Point acts as a barrier to alongshore transport causing a shoal to form (Figure 12.1a). This shoal grew by approximately 40,000 m^3 over the study period from May 2015 to March 2016.

Aerial photography analysis of the Sandy Point beaches showed the same northeasterly transport of beach sediment driven by waves from the southwest. Much of this sediment appeared also to have originated as beach replenishment material. Beach nourishment in this area occurred via a number of replenishment episodes that were not well documented, according to the West Haven Department of Public Works. In a sort of perfect storm of unintended engineering consequences, nourishment of these beaches, plus the construction of a wooden "weir" intended to prevent a tidal channel from being closed by littoral drift, resulted in the creation of a sand spit that extends out into New Haven harbor, creating a hazard to navigation and threatening a waste-water treatment plant outfall pipe (Figure 12.1b). This artificially supplied spit overwhelmed a natural sand spit in the process. A shallow sandy embayment was converted into a saltmarsh with presumed ecosystem alteration.

The town of West Haven, over the years, has undertaken repeated beach replenishment episodes in order to maintain its beaches. Most of the sediment emplaced has eroded and been transported elsewhere. In the case of Prospect Beach, much of that sand has ended up in a large and still growing sand shoal. In the case of the Sandy Point beaches, the sediment placed during replenishment has ended up contributing to a new spit. Current Department of Energy and Environmental Protection policies discourage mining sand from local nearshore areas because of concern for nearshore ecosystems. However, considering the importance of beaches and the costs of replenishment, rethinking these policies in favour of instituting regional and local sediment management practices may be in order.

Using the data provided by the authors of this paper, the Town of West Haven has developed a Harbor Management Plan, a Coastal Resilience Plan and a Conservation and Development Plan that all incorporate the concept of *sand reclamation*. In this concept, the fate of sand used in replenishment projects is studied in terms of its spatial dispersal pattern. This sand can then be reclaimed and returned to the beaches from which it was eroded. This can be done at a much lower cost than that of beach replenishment in which sand is purchased and imported. Moreover, the timing of reclamation, the volumes reclaimed and other aspects of reclamation can be modulated in order to minimise impacts on the local ecosystem.

6 Conclusion

The need to sustain the presence of wide beaches as buffers against storm activity is clear. Damage to coastal structures and infrastructure during Irene and Sandy were the most severe in the towns with the most eroded beaches. Moving from a strategy of replacing lost sand to one of managing sand within the local system would appear to be a transition toward coastal sustainability that makes sense, at least in the time-frame that sea level rise allows. The possibility of doing so in consultation with marine ecologists and biologists to minimise harm to benthic ecosystems is even more enticing.

Acknowledgements

Resources for this research were provided by the Werth Center for Coastal and Marine Studies, Southern Connecticut State University.

Bibliography

Bureau of Ocean Energy Management Marine Minerals Program (n.d.). Retrieved 5 March 2017 from www.boem.gov/Marine-Minerals-Program/

Dalryple, R. (1992). Prediction of storm/normal beach profiles. *Journal of Waterway, Port, Coastal and Ocean Engineering, 118* (2).

Daniel, H. (2001). Replenishment versus retreat: the cost of maintaining Delaware's beaches. *Ocean and Coastal Management, 44* (1–2), 87–104.

Dean, R. (1991). Equilibrium beach profiles: characteristics and applications. *Journal of Coastal Research, 7*(1), 53–84.

Griggs, G., Tait, J., & Corona, W. (1994). the interaction of seawalls and beaches: seven years of monitoring, Monterey Bay, California. *Shore and Beach, 62*(3), 21–28.

Griggs, G., & Tait, J. (1988). The effects of coastal protection structures on beaches along northern Monterey Bay, California. *Journal of Coastal Research*, Special Issue 4, The Effects of Seawalls on the Beach, 93–111.

Haddad, T., & Pilkey, O. (1998). Summary of the New England beach nourishment experience (1935–1996). *Journal of Coastal Research, 14*(4), 1395–1404.

Inman, D., & Frautschy, J. (1966). Littoral processes and the development of shorelines. *Proc. Coast. Eng. Conf., Am. Soc. Civ. Eng.*, Santa Barbara, California, 511–536.

Komar, P. (1998). *Beach Processes and Sedimentation.* Upper Saddle River, New Jersey: Prentiss-Hall, Inc.

Larson, M., Capobianco, M., & Hanson, H. (2000). Relationship between beach profiles and waves at Duck, North Carolina, determined by canonical correlation analysis. *Marine Geology, 163*(1–4), 275–288.

Shepard, F. (1950). *Beach Cycles in Southern California. Corps of Engineers, Beach Erosion Board.* Washington DC, Technical Memo.

Sonu, C., & van Bleek, J. (1971). Systematic beach changes on the Outer Banks, North Carolina. *Journal of Geology, 79*(4), 416–425.

Tait, J.F., & Akpinar Ferrand, E. (2014). Observations of the influence of regional beach dynamics on the impacts of storm waves on the Connecticut Coast during Hurricanes Irene and Sandy. In Bennington, J.B., & Farmer, E.C. (Eds.) *Learning from the Impacts of Superstorm Sandy* (pp. 69–88). Cambridge, MA: Academic Press.

Trembanis, A.C., & Pilkey, O.H. (1999). Comparison of beach nourishment along the U.S. Atlantic, Great Lakes, Gulf of Mexico, and New England Shorelines. *Coastal Management, 27*(4), 329–40.

UNEP Global Environmental Alert Service. (2014). Sand, rarer than one thinks. Retrieved 5 March 2017, from the United Nations Environment Programme website: http://na.unep.net/geas/archive/pdfs/GEAS_Mar2014_Sand_Mining.pdf

13 Data and policy scale mismatch in coastal systems

The potential of μUAS as new tools for monitoring coastal resilience

Scott M. Graves

1 Introduction

An important challenge in monitoring, planning and evaluating coastal saltmarsh resources in the face of predicted sea level rise (*SLR*) using numerical modelling is that in some cases, perhaps many, we are at the nexus of potential "scalar mismatch" scenarios in the application of available data (elevation data, tidal range and descriptions of marsh species zonation – i.e., landcover). Johnson *et al.* (2012) write of "Social-Ecological Scale Mismatches" in their work on better understanding social, biophysical conditions at multiple scales when designing programs and projects that inform institutions on resource sustainability. Crowder *et al.* (2006) states that spatial/scale mismatches for resource planning can result from either jurisdictional boundaries being either too large or too small, or where administrative boundaries traverse natural ecological boundaries. Where these scalar mismatches are incongruent with the scale of ecological systems under investigation, either incomplete or ambiguous results/feedback make it difficult to learn from and/or adapt to expected ecological or social changes imposed, thus hindering efforts to effectively manage and sustain the resource in question (Berkes & Folke, 1998; Wilson, 2006). The particular scale mismatch scenarios highlighted here are in reference to a widely employed numerical modelling and visualisation system known as Sea Level Affecting Marshes Model (*SLAMM*). In *SLAMM* I identify scale mismatch as a serious challenge in interpreting model results due to imprecise and/or inappropriate scales/resolutions in the remotely sensed data used as model input for resolving the Cove River (West Haven, CT) saltmarsh responses to *SLR*. The data in question are Digital Surface Models (*DSM*) or elevation models derived from *LiDAR* (Light Detection And Ranging, a form of radar elevation mapping), and tidal inundations projected from tide gauges that may be miles from the wetland of interest, as well as "assumed marsh landcover classifications" across a wetland complex based solely on predicted tidal inundation and a generalised knowledge of marsh species' preference for living/thriving at different marsh elevations. Along with better *DSM* data and tidal inundation information – the initial conditions set in such modelling algorithms – the identified marsh species that are present and indicative of tidal range and inundation levels, need to be directly observed, rather than "assumed" by the elevation and tidal data alone. In this case study of the 20-hectare Cove River coastal saltmarsh/wetland complex in West Haven, Connecticut (see Figure 13.1), there appears just such a scalar (resolution) mismatch scenario in the *SLAMM* modelling products that have been produced and made available to coastal planners. It is the intent of this article to both provoke and challenge the coastal planning, engineering and management communities, as well as fellow academicians and researchers, to think more deeply about modelling programs such as *SLAMM* and to strongly suggest that any decision-making process includes continued on-ground observation and data gathering in addition to consideration of new and emerging technologies such as micro Unmanned Aerial Systems (*μUAS*/drones) and Structure from Motion (*SfM*) mapping and 3D modelling as appropriate alternative model inputs. It is my argument that by spending significant time in the field with new and traditional tools, and engaging multiple stakeholders in field work, we can see a more democratised process of coastal resilience and sustainability planning.

With this in mind, I believe that low-altitude, high-resolution *μUAS*/drone mapping can augment or even replace more traditional *LiDAR*-based modelling work by introducing a more precise scale/resolution-matching set of information to consider as model inputs. Thus, saltmarsh resilience, sustainability and upland migration potential in the face of future sea level rise can be more precisely and accurately determined.

2 Study site: Cove River Wetland complex, West Haven, Connecticut

The Cove River watershed (West Haven, CT) encompasses ~14 km^2, extending ~8 km inland, and drains through suburban, residential and commercial properties. The Cove River estuarine complex is a 20-hectare microtidal (~2 m tide range) coastal wetland extending 1.6 km north from the river mouth where waters exchange with Long Island Sound, and is bounded on all sides by residential and commercial development (again, see Figure 13.1). The Cove River wetland complex is broken into two distinct saltmarsh regions by a bridge, with its attendant channel constriction beneath. The lower saltmarsh covers ~9.5 hectares while the upper saltmarsh covers another ~9.5 hectares. The upper portion of the wetland complex includes a small freshwater marsh (~1 hectare) bound on the west by development and on the east by the Cove River Historical Site (*CRHS*), a designated Open Space which formerly was a working farm with orchards and pastures (early 1900s) and is now a completely reforested area. This freshwater wetland drains directly into the main saltmarsh complex at the nose of the *CRHS* forest. At that point, there is a half-metre drop from the fresh to the saltmarsh surface, though in recent years this distinct step-down area has been reduced to a low sloping surface, indicating some measure of marsh surface deflation or erosion.

The Cove River Historical Site has been monitored by the author with undergraduate and graduate students for a decade. Over the years, the author and his students have shared water quality data collected and high-resolution *μUAS* image/map mosaics with the City Public Works officials to augment their own field data. For many years, the trek to the water sampling locations was a frustrating scramble through thick stands of tall invasive *Phragmites australis* vegetation which began taking over the marsh system in the late 1970s. By the 1990s the entirety of the Cove River wetland complex was dominated, and exclusively so, by the invasive *Phragmites*. In the early 2010s, the entire wetland was mowed and herbicides applied to remove the invasive *Phragmites*. Since then the native marsh species have been slow to recolonise and there remains a high likelihood of the wetland not surviving future sea level rise. This small wetland complex has been modelled using the *SLAMM*; however, the results are questionable at the scale/resolution of input data used. It is my firm belief that as a case study of potential scale mismatch, this site represents an important example of how new technologies – *μUAS* and *SfM* – as well as diligent on-ground observations can significantly improve *SLAMM* modelling and any interpretations thereof.

3 SLAMM (Sea Level Rise Affecting Marshes Model) modelling

SLAMM (Sea Level Affecting Marshes Model) is a numerical modelling program whose output is rendered in a geographic information system (*GIS*) based visualisation tool for coastal planners and engineers, as well as public stakeholders and citizens. The purpose of *SLAMM* is to provide realistic projections of the anticipated effects of sea level rise on marsh complexes while highlighting potential upland/inland migration pathways. The *SLAMM* modelling algorithms were first developed with funding from the US Environmental Protection Agency in the 1980s (Park *et al.*, 1986). Warren Pinnacle Consulting (2016) describe the genesis and iterative upgrades of the *SLAMM* modelling process and its use as a tool to predict wetland response to long-term sea level rise. The model simulates the effect of long-term sea level rise (*SLR*) on the dominant processes that affect shoreline modifications and it has been applied in every coastal US state including Connecticut (Craft *et al.*, 2009; Galbraith *et al.*, 2002; Glick *et al.*, 2007, 2011; National Wildlife Federation, 2010; Park *et al.*, 1993; Titus *et al.*, 1991). *SLAMM* predicts where existing marshes are likely to be vulnerable to *SLR* as well as possible/probable upland migration pathways as *SLR* increases. *SLAMM* predictions focus on scenarios wherein irregularly flooded mid- to high-marsh habitats will, with *SLR*, become regularly flooded and transition to low-marsh habitat, as constant inundation pushes out the less-salt and inundation-intolerant species. And yet, as with many modelling processes, *SLAMM* has its limitations. In some cases, an overly generalised *LiDAR* footprint (1 m^2 to 3 m^2 resolution), and variable vertical resolution (centimetre to decimetre or more) lead to less-than-optimal validity in the modelled results. Vertical errors in *LiDAR* result when the *LiDAR* signal penetration to the ground surface is impeded. Often multiple *LiDAR* signal returns from marsh grass canopy, mid-level signal scattering/returns from the stalks, and the signals returning from the actual ground surface combine. Overall, the errors in determining the marsh ground surface can vary on the order of 10 cm to 40 cm+ when the return signal is a mash-up of these different targets (Schmid *et al.*, 2011).

SLAMM model output is provided in an ESRI web-map product online. Users can manipulate parameters in the calculated *SLR* scenarios based on IPCC (2007) projections for the years 2055, 2085 and 2100, with *SLR* forcings at Generalized Current Rates, as well as Minimum and Maximum Projected Rapid Ice Melt scenarios. To be fair, the *SLAMM* model documentation does include a caveat in stating that model results alone may not provide the best or most complete descriptions and thus should be considered a "starting point" in designing further studies; especially if and when marsh surface collapse may be occurring and where the result is that irregularly flooded marsh is converted to regularly flooded, and on toward tidal mudflat (Warren Pinnacle Inc., 2016). While reruns of the *SLAMM* modelling are being undertaken in New Hampshire and Massachusetts to accommodate just such updates on landcover vegetation transitions, this has not been done yet in Connecticut. If and when this is done, along with inclusion of better marsh elevation data (with micro Unmanned Aerial Systems/drones – *μUAS*- instead of or along with *LiDAR*), and where direct field observations of species distributions (landcover), salinity and inundation depths are included as valid model input, then the *SLAMM* modelling of Cove River and other similar fragmented and tidally impeded coastal wetlands can be better evaluated and understood. The following discussion covers the Cove River saltmarsh complex and the scale/resolution mismatch scenario that is apparent in the current representations in the available online *SLAMM* visualisation. Further, new tools (niche innovations) including *μUAS* and *SfM* as well as descriptions of necessary on-ground field observations including direct measures of tidal inundation and marsh vegetation mapping will be outlined.

In the current *SLAMM* model runs for the Cove River wetland complex there are a number of complicating factors that were not taken into account when the model run was initially set up. Most importantly, the "Initial conditions" set for the marsh were incorrect. The most recent run of the model assumes that the dominant vegetation over the entire marsh is associated with "Infrequent flooding". The species that are typical of this supra-tidal zone include *Spartina patens*, *Juncus gerardi* and other species that can only tolerate brief periods of brackish water inundation at shallow depths. In truth (from direct field observations), the actual vegetation present, and dominant across the entire wetland complex is *Spartina alterniflora*, a low-marsh species that does well in higher salinity environments and throughout nearly daily tidal inundations to depths of 10 cm or more. The left-hand side of Figure 13.1 shows a screen from the online *SLAMM* visualisation tool, centred on the Cove River Wetland complex, along with some notes on potential scale

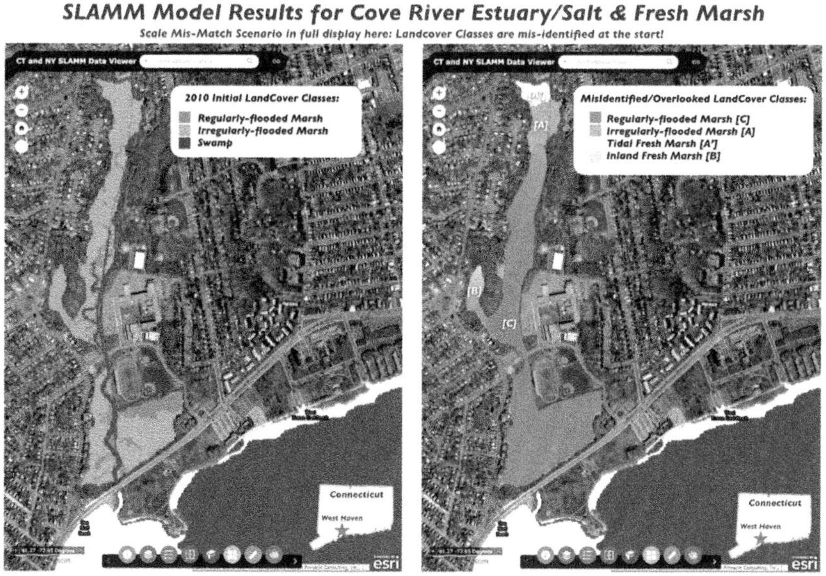

Figure 13.1 Left: *SLAMM* web interface showing Cove River wetland complex and spuriously identified landcover classes (marsh vegetation zones); right: author's overlays more precisely and correctly identify actual marsh vegetation zones: (A, A', B & C)

Source: http://ctdeep.maps.arcgis.com/apps/webappviewer/index.html?id=205df00b30de40c7b84626e4a77fb914

(resolution) mismatch scenarios and why coastal planners, engineers and other stakeholders must consider *SLAMM* as a starting-point, not the final analysis of *SLR* effects on marsh migration potential. Further, the current *SLAMM* model results do not sufficiently differentiate saltmarsh from tidal fresh or upland fresh marsh landcover classes at the Cove River site as shown in the right-hand side of Figure 13.1.

My critique of the Cove River *SLAMM* model results can be summarised as follows: what the current *SLAMM* modelling misses are some important nuances of the wetland complex whose fate might be better predicted with more accurate *DSM* (from *µUAS*) and through on-ground observations (more accurate landcover/marsh species classification) along with actual marsh inundation observations. To address this last issue, the author has developed a Marsh Inundation Camera System – *MICS*, comprising a set of simple GoPro Hero cameras with extended battery capacity and set to time-lapse with a view of the marsh surface and a meter staff anchored into the marsh surface in the field of view. On-ground and regular observations of the entire marsh complex by the author yield very different interpretations and landcover class (marsh species) assignments than those indicated as "Initial conditions" in the current *SLAMM* model results (personal observations, 2006–2017). The currently modelled landcover/marsh species class identified as irregularly flooded marsh over most of the wetland complex, is actually regularly flooded marsh. In Figure 13.1 I identify the correct/actual landcover classes as added overlays: the northern reach [A & A'] is flooded at every spring high tide yet functions as a tidal fresh marsh whose flow is stalled as tidal pulses approach but its salinity rarely exceeds 5–8 ppt and it is currently still dominated with the salt-intolerant common reed *Phragmites australis* (a deleterious invasive species); the western limb [B] stands ~40–50 cm above the rest of the saltmarsh complex [C] and is actually a ~1 hectare upland fresh marsh only partially flooded at its southernmost extent during proxigean spring tides and/or by hurricane surge. These two sub-components of the overall wetland complex are "Transitional tidal fresh" and "Inland fresh marsh", respectively, and should not be put in the class of "Irregularly-flooded" saltmarsh as indicated in the initial *SLAMM* results. It appears that the Digital Surface Model (*DSM*) used, derived from high-altitude *LiDAR*, and/or the "assumptions" on tidal inundation derived from tide gauges miles away, are of insufficient resolution (vertically) to identify the subtle, but important, elevation changes within the marsh complex, and the tidal inundation differences between the northern reach and western limb wetland areas. Thus, with potential errors in, or spurious interpretations of, the *LiDAR DSM*, along with an initial mis-identification of landcover classes, the details of the *SLAMM* results are misleading. Looking at parcel scale (few hectares) marsh systems, the limitations imposed by *LiDAR* accuracy and tide levels measured from afar, as well as assumptions rather than actual observations of marsh vegetation, result in a mismatch of the model input parameters for the given location.

In subsequent *SLAMM* model scenarios shown in the online visualisation tool (found at *http://ctdeep.maps.arcgis.com/apps/webappviewer/index.html?id=205df00b30de40c7b84626e4a77fb914*), the sequence of *SLR* scenarios presented retain these significant problems of inaccurately identifying landcover and thus marsh zone classifications and possible migration pathways… at least in the near term and under moderate *SLR* predictions. More accurate *DSM* and tidal inundation data – from *μUAS* and direct measures of inundation by time-lapse cameras (*MICS*) – would help sort out the nuances here and provide coastal planners with a much better starting-point in their deliberations on what to do about conserving and preserving marsh migration pathways. There does appear to be a couple of very limited potential migration pathways identified in the northern reach [A and A'] and western limb [B] areas in particular, which currently are very infrequently, if at all, inundated by normal tides. Any other potential marsh migration is limited to possible encroachment into the Cove River Historical Site's forest complex, located between the western limb fresh marsh and the Cove River saltmarsh to the east. Virtually all other migration pathways are currently blocked by hardened and elevated surfaces and/or development.

It must be said, however, that for this entire wetland complex, including the northern and western limb area (which stands some 10–50 cm above the majority of the lower saltmarsh), it is reasonable to conclude that the entire marsh/wetland complex, including its fringe tidal fresh and upland fresh marshes may turn to regularly and/or deeply flooded low marsh and tidal mudflat if the *SLR* predictions of 0.43 m to 1.7 m are realised by 2055, 2085 and 2100. In this case, the current *SLAMM* visualisation at maximum *SLR* at year 2100 gives a reasonable prediction, regardless of potential scale/resolution mismatch of input data for lower *SLR* scenarios.

4 New tools for coastal resilience and sustainability monitoring

The coastal transitions literature describes a number of novel and emerging innovations to enhance coastal management planning (Malone *et al.*, 2010). Multi-level perspectives (*MLP*) including "niche-innovations" afford space for new ideas, techniques and technologies to be developed and deployed. While *MLP* has been operationally defined as pertaining to socio-philosophical approaches, for my purposes, I adapt/adopt the term Multiple Spatial/Resolution Perspectives (*MS/RP*) to refer to work in introducing new tools (*μUAS* and other on-ground observations) to help eliminate the potential for scalar/resolution mismatches that may result from over-reliance on *LiDAR* and distant Tide Gauge data, and the other assumptions that are typically made with respect to marsh vegetation zonation. Further, I believe that low-altitude, high-resolution *μUAS* (micro Unmanned Aerial Systems) mapping can significantly augment or even replace more traditional *LiDAR*-based digital surface modelling by introducing a more precise scale and resolution-matching set of information to assist in determining saltmarsh elevations, and thus help better assess coastal wetland resilience, sustainability and upland migration potential in the face of future sea level rise.

Previous work by this author proved the viability and utility of engaging *μUAS* for coastal wetland monitoring at Cove River (Graves, 2017). This was, and is, a clear example of a "niche-innovation" (Heidkamp & Morrisey, 2017); an application of emerging technology that will certainly become more and more indispensable in future small scale (parcel and/or <100 hectare) coastal resilience planning. *μUAS* aerial mapping and Structure from Motion *(SfM)* processing represent real "niche-innovations" with the potential to "disrupt" (in a positive way) or enhance/compliment traditional ground-based surveying methods (Total Station, Precision *GPS* or *LiDAR*). *μUAS* aerial mapping and *SfM* has proven to be as, or more accurate, than *LiDAR* and is far less expensive and faster in gathering land surface elevations and nadir view images than either traditional ground-based surveying and/or traditional aerial photography (Leberl *et al.*, 2010; Fritz *et al.*, 2013; Dandois & Ellis, 2010; Fonstad *et al.*, 2013; Wallace *et al.*, 2016). With just a few select surveyed ground control points (*GCP*) collected using traditional surveying methods, entire hectares to km^2 scale image/map scenes with attendant digital surface models can be produced. Structure from Motion software allows for the creation of broad area coverage from a series of aerial images, leading to the creation of entirely new image/maps and digital surface modelling products. *μUAS*/drone mapping in conjunction with *SfM* processing will lead to new applications for ecological monitoring through the production of inexpensive, yet highly detailed map/mosaics and 3D landscape models. These can be used to supplement and/or "ground-truth" larger scale studies of landscapes, and as in the case explored here, visualisations of numerical modelling for investigations of coastal wetlands.

4.1 Micro Unmanned Aerial Systems (μUAS/drones)

UAS (Unmanned Aerial Systems) or drones come in a variety of sizes and configurations. Fixed wing *UAS* are often used when there is a need to map large ~100s of hectares to many km^2 regions of interest. Fixed wing *UAS* are rarely appropriate for locations in Connecticut due to the high density of development and the lack of free airspace for fixed wing flight manoeuvres. *UAS* also come in multi-rotor versions, from QuadCopters to OctoCopters capable of carrying up to 10 pounds of sensing payloads. For our purposes, off the shelf <5lb multi-rotor platforms such as those available from DJI and 3D Robotics, flying camera payloads capable of capturing 4K images and video, are most appropriate and rather low cost ($1,000 to $2,000 complete). These are the *μUAS* of (micro Unmanned Aerial Systems) of interest.

It is a fair certainty that in the near future, more instruments and sensors will be sufficiently miniaturised to allow far greater *μUAS* imaging and sensing capability, including multi- and hyperspectral imaging, RTK GPS *μUAS* positioning in flight, and perhaps even onboard *LiDAR*. As it stands, low-altitude flights conducted with *μUAS* platforms yield ground resolutions of millimetres to centimetres – depending on camera settings and flight altitude. In any case, a much higher resolution in the images gathered is achieved as compared to other higher flying aerial systems. An added benefit above traditional *LiDAR* surveys is that *μUAS* can be deployed as needed for on-demand and frequent mapping missions. Without this new low-altitude aerial mapping and the much-enhanced resolution provided, there is a strong potential for scale (resolution) mismatch and spurious interpretation of *SLAMM* model results due to the resolution limitations and inherent error involved in *LiDAR* data as initial inputs. As mentioned earlier, these errors can be as great as 4–6 cm even in perfect conditions, and upwards of 40 cm when the surface is covered with saltmarsh grasses – depending on species and height of grass canopy, *LiDAR* returns off the canopy, stalks and base, in addition to the actual marsh surface result in significant combined error (Schmid *et al.*, 2011).

4.2 Structure from Motion (SfM)

The roots of *SfM* can be traced back to two key fields, photogrammetry and computer vision. Those fields and the emergence of *SfM* algorithms and techniques are well described in the literature (Küng *et al.*, 2011; Lowe, 2004; Stretcha *et al.*, 2008; Agarwal *et al.*, 2010; Furukawa & Ponce, 2010; Snavely, 2008; Snavely *et al.*, 2006, 2007; McGlone & Lee, 2013; Sharp, 1951) and so will not be reiterated in detail here. In brief, *SfM* is a relatively new technique for processing large numbers of individual aerial images to generate seamless image/map mosaics of very high resolution – *SfM* is an example of a "Big Data" challenge (point clouds of billions of pixels) solved by the ever-increasing power of readily available computer platforms. The emerging industry-standard *SfM* software is Pix4D which costs ~$2,000 for an educational licence and $8,000 for a commercial licence. The Pix4D interface is rather intuitive and each run of the algorithms generates a Quality Report detailing all aspects of the processing run. This is a key innovation in mapping and monitoring landscapes from parcel-scale (hectares or smaller) up to km^2. It is important to understand that this process is *not* analogous to simply "stitching" photos together to create a mosaic. Rather, it is a completely new approach and it is very powerful. *μUAS* imagery/data processed through *SfM* are interrogated in each and every image frame taken, camera positions in space are calculated, individual pixel information is extracted where objects on the landscape are seen in consecutive frames, and further iterative processes of all information in a "cloud" of up to many billions of pixels, results in the generation a new seamless orthomosaic image and digital elevation models. *SfM* processing demands relatively high computational power – fast and multi-GB RAM and graphics processors. Until recently, *SfM* was impossible to run on anything but very large workstations or even super computers. Today, however, the processing power of even a modestly configured laptop or desktop computer is sufficient for *SfM*. There are a variety of commercial *SfM* packages available, some open source, but the emerging industry standard is from Pix4D, as mentioned, which has aligned with ESRI and AutoCad for geo-referencing orthomosaics and *DSM* output for use in a wide variety of commercial contexts. *SfM* and *μUAS* imagery are now becoming the tool of choice in many industries. *SfM* processing has been used to render scenes of many types, from archaeological sites, to eroded terrains, to mining operations, to forest and vineyard canopy studies, and submarine surfaces (Green *et al.*, 2014; Westoby *et al.*, 2012; Fitz *et al.*, 2013; Mathews & Jensen, 2013), and finally to wetland studies like that described here.

For many surveying projects *μUAS*/drones together with new Structure from Motion (*SfM*) software processing can significantly augment or even replace traditional ground-based (Total Station and GPS) or *LiDAR* survey techniques. *SfM* can be applied to resolve structure for almost any "scape" at any scale; e.g., subaerial landscapes, architectural scapes and structures, submarine seascapes, even microscopic and nano-scopic scapes. So *SfM* or the ability to determine structure from differing perspectives or camera angles is essentially a scale-independent analysis and rendering tool allowing researchers to document and model places and 3D structures at various scales in a multitude of environments.

5 Cove River Wetland revisited with new tools (*μUAS* and *SfM*)

In the autumn of 2012 the entire Cove River marsh system was treated with herbicide and mulched with an amphibious low-ground pressure mower to eradicate invasive *Phragmites australis*. I hypothesised that native marsh grasses (*Spartina sp.* and others) would recolonise the marsh-top and channel margins. Unfortunately that has not happened to the extent or rate anticipated. In addition, it appears that since the removal of the invasive marsh grasses, the marsh-top has begun deflating and the channel margins have started to collapse. Figure 13.2 shows the initial *SLAMM* model results (note the low resolution of the calculated landcover classes) compared to a high-resolution image/map mosaic produced from *μUAS* flight imagery and *SfM* processing (Figure 13.2b) for a detail study area of the Cove River marsh. In Figure 13.2a the *SLAMM* visualisation identifies regularly flooded marsh only along the channel margins and mis-identifies the rest of the marsh-top as irregularly flooded. In fact, the entire marsh surface in this area, and extending for significant distances both up and downstream, is regularly flooded. The current state of recovery of the marsh-top, since the eradication of the invasive *Phragmites a.*, is extremely problematic. The newly recolonising native marsh species (*Spartina a.*) have only taken hold along the channel margins with a few scattered clumps occurring across an otherwise barren marshtop mudflat. It remains to be seen whether further recolonisation of native salt-tolerant marsh species will ever result in a completely covered marsh-top. So, at present, the existing marsh surface is dominated by regularly inundated (and rather deeply so ~25 cm+) mudflat that may well remain or become the dominant landcover class. In essence, the well-intentioned initiative to remove the invasive grasses may have set the stage for an unrecoverable wetland complex. Only time will tell.

Deployment of *μUAS* can overcome frustrations of ground-level views of the wetland/marsh complex which can be misleading when it comes to attempts to determine vegetation coverage. At Cove River, the newly regrowing marsh grasses have colonised just the margins of the marsh channels with most all of the flat marsh-top unvegetated. This is difficult to see at ground level because the channel margin revegetated grasses block views of the marsh-top behind. With just a few tens of metres of altitude, the true story emerges. Multiple *μUAS* mapping missions conducted over the study area in the autumn of 2015 and spring and summer 2016 and 2017, and image processing with *SfM* produced parcel-scale (2–3 hectare) mosaics/maps and 3D landscape models that allowed us to investigate the recovery of the native saltmarsh vegetation. Figure 13.2b shows one of our *SfM* processed *μUAS* orthomosaics of our

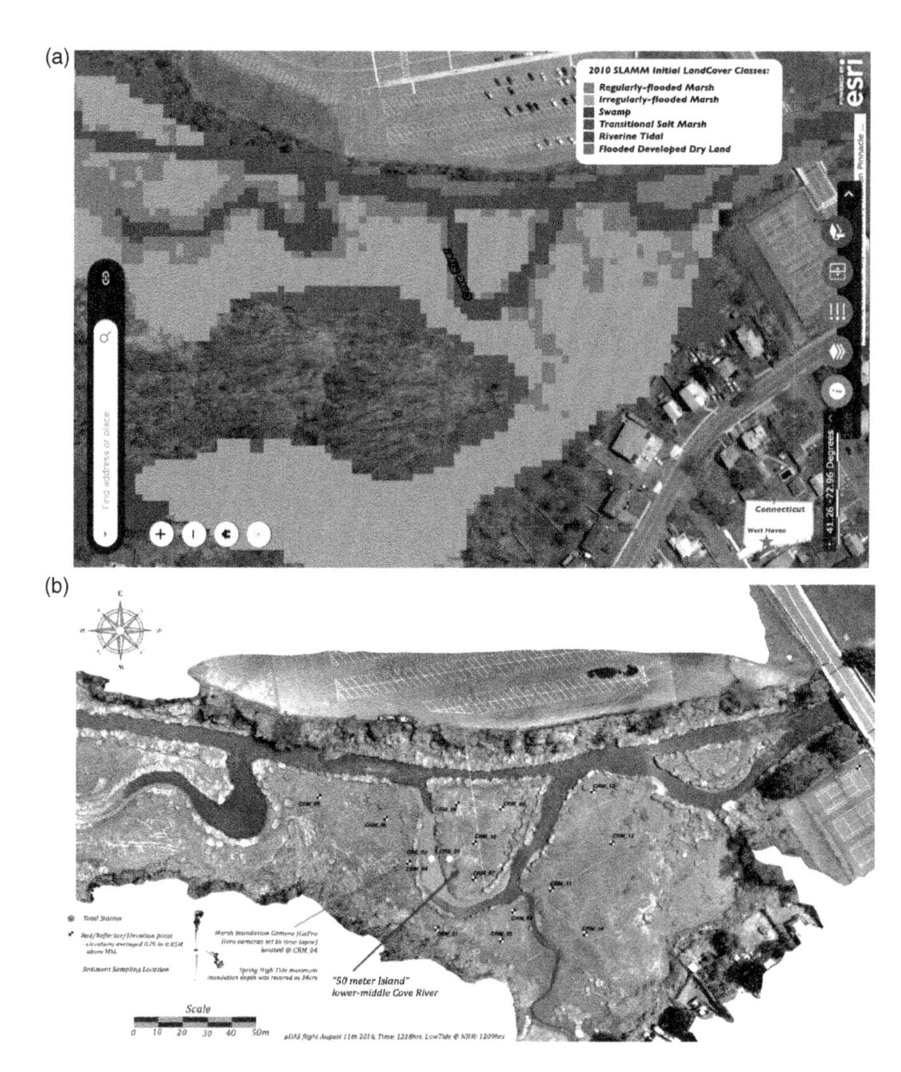

Figure 13.2 a) Initial *SLAMM* model results of Middle Cove River (detail study site) showing low resolution/scale of calculated landcover/marsh species identification and *DSM*; b) *SfM* rendered scene of the same Middle Cove River study site, with GCP, sediment sampling, and inundation camera locations (*μUAS* flight 08/11/2016)

parcel-scale detail study site in the upper half of the Cove River complex. Measuring the aerial extent of marsh vegetation and mudflats, yields some interesting numbers.

The visually approximated ratio (Google Earth tracing) of UnVegetated vs. Vegetated Marsh (*UVVR* for the area measured) is ~79%. These numbers are especially troubling, given the findings of Ganju *et al.* (2017) where they found such *UVVR* ratios correlated and scaled with net sediment (accretion) budget deficits and threats to projected marsh complex lifespans in eight Atlantic and Pacific coast saltmarshes. In the case of the Cove River, the entire marsh complex was intentionally converted to open water through mowing and poisoning of the invasive *Phragmites a.* So, sometime shortly after the downed *Phragmites* stalks completely decomposed, the *UVVR* was virtually 100%. After five years the recovering native grasses now occupy just 21% of the entire marsh complex area, indicating the future of the marsh may be seriously threatened unless significant marsh grass colonisation occurs with attendant suspended sediment trapping allowing marsh-top accretion. Again, if future sea level rise rates are realised (an additional 0.4–1.7 m by 2100), and *UVVR* correlated lifespan relationships prove accurate, the entire Cove River marsh system may not exist as a viable wetland complex and may not provide valuable ecosystem services beyond the year 2100, if not much sooner.

6 Cost/benefit analysis approach:

A brief cost/benefit analysis is worth considering for updating *SLAMM* model inputs and rerunning *SLAMM* to produce a more accurate prediction of the effects of *SLR* on the Cove River wetland complex. Costs are considered for new *LiDAR* vs. *μUAS/SfM* digital elevation models, as well as updated tidal inundation measures and redefined landcover classification for "Initial conditions".

LiDAR service providers, Maser Consulting (www.maserconsulting.com) and Flight Evolved (http://flight-evolved.com), quote traditional manned airborne (helicopter) *LiDAR* flight service costs as ranging from $20,000 to well over $50,000. This appears as a fair price range considering the logistics required. Conventional aircraft require pilots, fuel, airports, fuel trucks, hangar fees and a multitude of other things before they can even take off. Only recently, new *UAS*-based *LiDAR* service consultants have emerged, and they state that they can provide considerable cost savings purely on the basis of logistics. All these *LiDAR* services use industry-grade high-resolution *LiDAR* sensors that can cost more than $250,000 each, with the larger *UAS* platforms costing $10,000 to $80,000. Typical contracted small *LiDAR* mapping jobs can cost upward of $5,000 to $10,000 where *LiDAR* is deployed on a larger *UAS*, with processing costs running to another $500–$1,000. However the *LiDAR* data is acquired, a cost of ~$5,000 to $50,000 or more might be expected; and that is just for a one-time *LiDAR* data acquisition. Further processing requires an additional consultation fee or investment in software. Deploying smaller *μUAS* and *SfM* to derive new *DSM* also has a cost. A typical *μUAS* costs ~$2000. *SfM* software runs at ~$2,000 for an education licence and $8,000+ for a commercial licence. Again, as with *LiDAR* data, there is a time-cost requirement in formatting and processing the *UAS/SfM DSM* data. In the case of contracting with a *LiDAR* service provider, any follow-up *LiDAR* flights would incur another $10,000 to $50,000 cost. With *μUAS* and *SfM*-derived elevation models, the cost of the aerial platform is a one-time expenditure. This allows for multiple and on-demand flight operations, so surveys could be done at any time there is fair weather and light winds. One important aspect of the use of *μUAS* and the other low-cost alternatives mentioned here, is that considerably more field time is needed; but this is important time spent at the wetland site and it affords opportunities for multiple stakeholders and interested parties to collaborate in gaining a better understanding of the wetland, actual landcover classes (flora and fauna present), and hydraulic regime, as well as its ecosystems services. Gathering new and more precise tide inundation data is another important aspect of remodelling. Traditionally, the tide information is simply extracted from the average of data for a particular nearby tide station, and the cost is nil (except time spent on the task). The tide data acquired this way may not be accurate enough for precise modelling, and so more direct measures can be pursued. Portable tide and flow meter systems can cost anywhere from $1,000 to $5,000 each; cheaper if simply rented. As commonly practised, multiple gauges would be required in order to get data from across the wetland complex. An alternative (cheap) method of directly measuring tidal inundation on the marsh-top (rather than only channel flow) is to deploy Marsh Inundation Camera Systems, like that described elsewhere in this narrative. The cost is rather small: $300-$500 for GoPro or other time-lapse camera, plus an extended battery – another $50–$100; and the system can be deployed wherever desired. All this newly acquired data can be swapped in for traditional data input sources and the *SLAMM* algorithms rerun to generate new and more locally appropriate, scale-correct, and more believable *SLAMM* output. Finally, the cost of *SLAMM* remodelling with new input data can be $5,000 to $10,000 where farmed out to a consulting service, or it can be done in-house at little monetary cost, if sufficient expertise exists. The *SLAMM* platform itself is open source and thus free, but the expertise needed to redevelop the new input strategies and run the actual numerical simulation may be above the level of casual users.

The overall cost/benefit analysis here offers insight into what it would take to rerun *SLAMM* for the Cove River example to generate a better and scale-matching result that would be far more believable that the original results shared online at present. The least expensive, but most time-consuming (which can be a good thing as it gets more stakeholders involved), is to deploy *μUAS* with RTK GPS positioned Ground Control Points (*GPC*) on the marsh-top and in view of the drone, collect the imagery and process in *SfM*. Also, deploying Marsh Inundation Camera Systems can yield much better direct measures of tidal inundation across the marsh. Finally, while in the field, identifying the actual marsh vegetation present across the marsh complex will help in establishing the true "Initial conditions" (landform classification). *SLAMM* can then be rerun with new *GPC*-constrained *SfM* elevation models, along with new tide inundation data and more accurate/actual vegetation classifications and the output re-entered into a GIS visualisation. Likely the overall cost would be ~$5,000 at the least expensive configuration, and far in excess of $20,000 as the least cost to outsource the *LiDAR* survey, etc.

7 Conclusions and recommendations

New spatial perspectives (multiple spatial or multi-scalar perspectives) are required for full application of socio-technical understanding and in addressing many emerging challenges in the coastal zone (Morrisey & Heidkamp, 2017), especially when considering the fragmented nature of coastal wetland systems on an urban coastline; e.g., West Haven, CT has only ~2% of its coastal zone designated as open space wetlands, where municipalities work with non-profits to manage the ecosystems. (Ryan & Welchel, 2015). In particular, for the coastline of Connecticut, "parcel-level" monitoring, on-ground as well as low-altitude aerial mapping is required to best understand the habitat-level scale of ecosystem challenges. Our studies at the Cove River Historical Site in West Haven, CT are just such an example of multi-scalar, multiple spatial perspectives (*MSP*), parcel-level monitoring.

The critique offered here of the *SLAMM* modelling results is primarily one where I identify a definite scale mismatch in the modelling data input relative to the scale of the wetland in question. It is clear that the *LiDAR*-derived *DSM* along with the tidal inundation data used in the *SLAMM* modelling, and as importantly, the mis-characterisation of the actual landcover classification (wetland vegetation species and their zonation across the saltmarsh), are valid objections. Any interpretation of *SLAMM* modelling and the attendant visualisations provided online must thus be tempered with an understanding of the limitation of numerical modelling in general, and the need for more accurate and precise scale-matching input. With newly developed expertise in *μUAS* flight planning and execution, together with *SfM* image/map mosaic and 3D model production, I intend on routinely monitoring the Cove River marsh in future seasons. I also plan on increasing on-ground monitoring of channel and over marsh-top flow rates, along with salinity and tidal inundation depths (Marsh Inundation Camera System). With new and more precise scale-matching data acquired through *μUAS* deployment, I hope to initiate a rerun of *SLAMM* for the Cove River coastal wetland complex and produce more reasonable and believable output.

Morrisey & Heidkamp (2017) state that "Effective functional local alliances between disparate stakeholders may well represent the most important resource in determining local and regional transitions potential.". I believe my work on the Cove River estuary represents just such an important alliance between university research, community citizen science and delivery of resulting image maps/mosaics and field data to the local municipality (West Haven Public Works Department). Our research has shown clear benefits to our academic goals (new knowledge and skills developed by both faculty and students) while delivering much-needed and valuable information to our primary stakeholder/partner such that they can make data-driven decisions for future coastal resilience planning.

Acknowledgements

The author sincerely thanks the following persons for their invaluable help and assistance in conducting field surveys and in the μUAS aerial-mapping endeavours: μUAS field team – graduate students Peter Broadbridge, Scott Thibault, Darryl Nicholson; marsh-top surveying team – undergraduates Shannon Bronson, Matthew Connors, and Dr. James Tait – all from Southern Connecticut State University, Department of the Environment, Geography and Marine Sciences.

Bibliography

Agarwal, S., Furukawa, Y., Snavely, N., Curless, B., Seitz, S.M., & Szeliski R. (2010). Reconstructing Rome. *Computer, 43,* 40–47. doi: 10.1109/MC.2010.175

Berkes, F., & Folke, C. (1998). Linking social and ecological systems for resilience and sustainability. In Berkes, F., & Folke, C. (Eds.), *Linking Social and Ecological Systems: Management Practices and Social Mechanisms For Building Resilience.* Cambridge University Press, New York.

Bertness, M.D. (1992). The ecology of a New England salt marsh. *American Scientist, 80,* 268.

Craft, C., Clough, J., Ehman, J., Joye, S., Park, R., Pennings, S., Guo, H., & Machmuller, M. (2009). Forecasting the effects of accelerated sea-level rise on tidal marsh ecosystem services. *Frontiers in Ecology and the Environment,* 7, 73–78.

Crowder, L.B., Osherenko, G., Young, O.R., Airame, S., Norse, E.A., Baron, N., Day, J.C., Douvere, F., Ehler, C.N., Halpern, B.S., Langdon, S.J., McLoed, K.L., Ogden, J.C., Peach, R.E., Rosenberg, A.A., & Wilson, J.A. (2006). Resolving mismatches in US ocean governance. *Science, 313*(5787), 617–618. Available at http://dx.doi.org/10.1126/science.1129706

Dandois, J.P., & Ellis, E.C. (2010). Remote sensing of vegetation structure using computer vision. *Remote Sens, 2,* 1157–1176.

Donnelly, J.P., & Bertness, M.D. (2001). Rapid shoreward encroachment of salt marsh cordgrass in response to accelerated sea-level rise. *PNAS, 98,* 14218–14223.

Fritz, A., Katterborn, T., & Koch, B. (2013). UAV-based photogrammetric point clouds – tree stem mapping in open stands in comparison to terrestrial laser scanner point clouds, international archives of the photogrammetry. *International Archives of the Photogrammetry, Remote Sensing and Spatial Information Sciences, XL-1/W2.* Rostock, Germany.

Fonstad, M.A., Dietrich, J.T., Courville, B.C., Jensen, J.L., & Carbonneau, P.E. (2013). Topographic structure from motion: a new development in photogrammetric measurement. *Earth Surf. Proc. Landf.* doi: 10.1002/esp.3366

Fritz, A., Katterborn, T., & Koch, B. (2013). UAV-based photogrammetric point clouds – tree stem mapping in open stands in comparison to terrestrial laser scanner point clouds. *International Archives of the Photogrammetry, Remote Sensing and Spatial Information Sciences*, XL-1/W2. Rostock, Germany.

Furukawa, Y. & Ponce, J. (2010). Accurate, dense, and robust multiview stereopsis. *IEEE Transactions on Pattern Analysis and Machine Intelligence, 32,* 1362–1376. doi: 10.1109/TPAMI.2009.161

Galbraith, H., Jones, R, Park, R.A, Clough, J.S., Herrod-Julius S., Harrington, B., & Page, G. (2002). Global climate change and sea level rise: potential losses of intertidal habitat for shorebirds. *Waterbirds*, *25*, 173–183.

Ganju, N., Define, Z., Kirwan, M., Fagherazzi, S., D'Alpaos, A., & Carniello, L. (2017). Spatially integrative metrics reveal hidden vulnerability of microtodal salt marshes. *Nature Communications, 8*, article 14156, Available at: http://www.nature.com/articles/ncomms14156

Glick, P., Clough, J., & Nunley, B. (2007). Sea-level rise and coastal habitats in the Pacific Northwest: an analysis for Puget Sound, Southwestern Washington, and Northwestern Oregon. National Wildlife Federation, Seattle, Washington, pp. 94.

Graves, S.M. (2017). Scale mismatch in marsh migration modeling: opportunities afforded by μUAS/Drone Mapping. *Regions*, 307(3), 17–19. Regional Studies Association.

Green, S., Bevan, A., & Shapland, M. (2014). A comparative assessment of structure from motion methods for archaeological research. *Journal of Archaeological Science*, 46, 173–181.

Heidkamp, C.P., & Morrisey J. (2017). Coastal Sustainability I: Challenges Methods and Opportunities. *Regions, 307*(1), 8–9.

IPCC. (2007). Summary for Policymakers. In Solomon, S., Qin, D., Manning, Z., Chen, M., Marquis, K.B., Avery, M., Tignor, M., & Miller, H.L (Eds.), *Climate Change 2007: The Physical Science Basis. Contribution of Working Group 1 to the Fourth Assessment Report of the Intergovernmental Panel on Climate Change*, Cambridge University Press, Cambridge.

Johnson, T.R, Wilson, J.A., Cleaver, C., & Vadas, R.L. (2012). Social-ecological scale mismatches and the collapse of the Sea Urchin Fishery in Maine, USA, *Ecology and Society*, *17*(2), 15. Available at: http://dx.doi.org/10.5751/ES-04767-170215/

Küng, O., Strecha, C., Beyeler, A., Zufferey, J-C., Floreano D., Fua P., & Gervaix F. (2011). The accuracy of automatic photogrammetric techniques on ultra-light UAV imagery. Conference on Unmanned Aerial Vehicles in Geomatics, Zurich.

Leberl, F., Irschara, A., Pock, T., Meixner, P., Gruber, M., Scholz, S., & Weichert, A. (2010). Point clouds: Lidar versus 3D vision. *Photogramm. Eng. Remote Sensing*, 76, 1123–1134.

Lowe, D.G. (2004). Distinctive image features from scale-invariant keypoints. *International Journal of Computer Vision*, *20*(2). Springer.

Malone, T., Davidson, M., DiGiacomo, P., Goncalves, E., Knap, T., Muelbert J., Parslow, J., Sweijd, N., Yanagai, T., & Yap, H. (2010). Climate change, sustainable development and coastal ocean information needs. *Procedia Environmental Sciences*, *1*, 324–341, World Conference, Elsevier. Available at: www.sciencedirect.com/science/article/pii/S1878029610000228

Mathews, A.J., & Jensen, J.L.R. (2013). Visualizing and quantifying vineyard canopy LAI using unmanned aerial vehicle (UAV) collected high density structure from motion point cloud. *RemoteSensing, 5*, 2164–2183.

McGlone, J.C., & Lee, G.Y. (Eds). (2013). *Manual of Photogrammetry*. 6th edition. American Society for Photogrammetry and Remote Sensing: Bethesda, Md.

National Wildlife Federation. (2010). Assessing the vulnerability of Alaska's coastal habitats to accelerating sea-level rise using the SLAMM model: a case study for Cook Inlet. Available at: www.nwf.org/~/media/PDFs/Regional/Alaska/Alaska%20SLAMM%20Summary%20Report.pdf

Nixon, S.W. (1982). The ecology of New England high salt marshes: a community profile. *Biological Report 81*, 55. Washington DC: Fish and Wildlife Service.

Park, R.A., Lee, J.K., & Canning, D. (1993). Potential effects of sea level rise on Puget Sound Wetlands. *Geocarto International,* 8(4), 99–100.

Park, R.A., Armentano, T.V., & Cloonan, C.L. (1986). Predicting the impact of sea level rise on coastal systems. In *Supplementary Proceedings for the 1986 Eastern Simulation Conference*, Norfolk, Virginia, pp. 149–153

Redfield, A.C. (1972). Development of a New England salt marsh. *Ecological Monographs*, 42, 201–237.

Ryan, A., & Whelchel, A.W. (2015). The Salt Marsh Advancement Zone Assessment of Connecticut. The Nature Conservancy, Coastal Resilience Program, Publication Series #1:A-W Final, New Haven, Connecticut. Available at: www.conservationgateway.org/ConservationPractices/Marine/crr/library/Documents/Connecticut%20Salt%20Marsh%20Advancement%20Zone%20Assessment%20Final%20Report_small.pdf

Schmid, K.A., Hadley, B.C., & Wijekoon, N. (2011). Vertical accuracy and use of topographic LiDAR data in coastal marshes. *Journal of Coastal Research*, 27(6A), 116–132. Available at: www.jcronline.org/doi/abs/10.2112/JCOASTRES-D-10-00188.1?code=cerf-site

Sharp, H.O. (1951). *Practical Photogrammetry*. Macmillan Company: New York.

Snavely, K.N. (2008). Scene reconstruction and visualization from internet photo collections, Ph.D., University of Washington.

Snavely, N., Seitz, S.M., & Szeliski R. (2006). Photo tourism, presented at the ACM SIGGRAPH 2006 Papers. Boston, Massachusetts, pp. 835. [online] Available at: http://portal.acm.org/citation.cfm?doid=1179352.1141964 (Accessed 14 December 2010)

Snavely, N., Seitz, S.M., & Szeliski, R. (2007). Modeling the world from internet photo collections. *International Journal of Computer Vision,* 80, 189–210. doi: 10.1007/s11263-007-0107-3

Strecha, C., Gool, L.V., & Fua, P. (2008). A generative model for true orthorectification. *Proceedings of the International Society for Photogrammetry and Remote Sensing (ISPRS).*

Titus, J.G., Park, R.A., Leatherman, S.P., Weggel, J.R., Greene, M.S., Mausel, P.W., Trehan, M.S., Brown, S., Grant, C., & Yoje, G.W. (1991). Greenhouse effect and sea level rise: loss of land and the cost of holding back the sea. *Coastal Management, 19*, 171–204.

Wallace, L., Lucieer, A., Malenovsky, Z., Turner, D., & Vopenka, P. (2016). Assessment of forest structure using two UAV techniques: a comparison of Airborne Laser Scanning and Structure from Motion (SfM) Point Clouds, 2016. *Forests*, 7(62). Available at: www.mdpi.com/1999-4907/7/3/62

Warren, R.S. (1995). Evolution and development of tidal marshes. In Dreyer, G.D., & Niering, W.A. (Eds.), *Connecticut College Arboretum Bulletin,* 34. The Connecticut Arboretum, New London, CT, pp. 72.

Warren Pinnacle Consulting Inc. (2016). Advancing existing assessment of Connecticut Marshes' Response to SLR, Final Report prepared for Northeast Regional Oceans Council.

Westoby, M.J, Brasington, J., Glasser, N.F., Hambrey M.J., & Reynolds, J.M. (2012). 'Structure-from-Motion' photogrammetry: a low-cost, effective tool for geoscience applications. *Geomorphology*, *179*, 300–314.

Wilson, J.A. (2006). Matching social and ecological systems in complex ocean fisheries. *Ecology and Society*, *11*(1), 9. Available at: www.ecologyandsociety.org/vol11/iss1/art9/

14 Scale mismatches

Old friends and new seascapes in a planning regime

Karen Alexander and Marcello Graziano

Dedication: this chapter is dedicated to the memory of the late Professor Laurence Mee. Without him, three years of great collaboration and friendship between the authors would have not been possible.

1 Introduction: marine spatial planning and scale mismatches

Increasing demand for marine resources, because of the finite nature of ecological capital (e.g., food and energy), has led to a growing number of resource extraction activities in our oceans and coastal areas. We only need to highlight the rise of economic strategies such as 'Blue Growth' and 'Blue Economy' to see a clear trend towards increased exploitation of these areas. These strategies, and the policies originating from them, are leading to increasing pressures on marine and coastal ecosystems as well as competition and conflict amongst the users of these environments. These are major challenges for governance, and new management strategies tools are required to address these threats. One such tool, marine spatial planning (MSP), is quickly becoming the dominant approach to address these issues of increased pressure and conflict, with a number of policy initiatives calling for the development of marine spatial plans (Jay *et al.*, 2012; Qui & Jones, 2013) in addition to an increasing academic literature on the topic (Flannery *et al.*, 2016).

Marine spatial planning has been defined as "a process that can influence where and when human activities occur in marine spaces" (Douvere, 2008, p. 765). MSP is a process that brings together multiple users of the ocean – industry, government, conservationists and recreational users – to make decisions about how to make best use of marine resources (including space), and is often conflated with ocean zoning. Furthermore, a key underpinning philosophy of MSP is ecosystem-based management, the aim of which is to support healthy and productive ecosystems (Borja *et al.*, 2010; Dominguez-Tejo *et al.*, 2016). In its broadest sense, MSP is about allocating parts of three-dimensional marine space to particular human activities in order to achieve ecological, economic and social objectives (Douvere, 2008).

It has been suggested that MSP may have economic benefits such as conflict resolution and greater certainty for long-term investment; that it has ecological benefits such as the allocation of space for conservation and the opportunity for biodiversity commitments to be at the heart of management; and that it benefits administration by improving data collection, decision-making and regulation (Ehler, 2008). However, it has also been noted that to date there has been little critical investigation of MSP and a call has been made for the articulation of a 'radical MSP' (Flannery *et al.*, 2016).

One particular aspect of MSP which has received little attention is the concept of 'scale mismatch'. Scale may be understood in different ways within the different sciences. For example, scale may mean the nature of social structures that govern the spatial and temporal extent of resource access rights and management responsibilities or it may refer to the spatial and temporal dimensions of an environmental pattern or process. Often, environmental and social scales are not in alignment and this leads to a mismatch (Cumming *et al.*, 2006). In this chapter we consider some of these scale mismatches as they relate to MSP.

2 From DPSIR to DPSWR and why scale matters

In this chapter, we utilise a causal framework to identify, highlight and explore the scale mismatches that plague the non-critical conceptualisation of MSP as a tool. This tool is the Driver State Impact Response Framework (DPSIR), in the declination conceptualised by Cooper (2013). DPSIR as a tool has been used to identify, analyse and address complex environmental problems within the context of sustainable development, offering a transdisciplinary framework (Gari *et al.*, 2015). In its original formulation, DPSIR recognises a chain of causal links starting with Driving forces (needs) leading to Pressures (because of human activity) and in turn to altered States (of the environment) and therefore Impacts on ecosystems and societies which eventually lead to political Responses (Table 14.1).

In its development of the framework, Cooper (2013) improved the conceptual formulation of DPSIR, making the framework easier to be used comparatively, and redefining each of the five categories to better match human activities and their consequences on human well-being. In Cooper's formulation, the 'Impact' category is replaced by 'Welfare', which formalises the relationship between the benefits accruing from Drivers and the environmental costs of Welfare changes caused by changes in environmental State, bringing the framework its acronym, DPSWR (Cooper, 2013). The specific information categories we utilise in this chapter are:

- Driver: an activity or process intended to enhance human welfare;
- Pressure: a means by which at least one Driver causes or contributes to a change in State;

- State (change): an attribute or set of attributes of the natural environment that reflect its integrity relative to a specified issue (or change therein);
- Welfare: a change in human welfare attributable to a change in State;
- Response: an initiative intended to reduce at least one Impact (State or Welfare change).

In applying DPSWR, we additionally incorporate a modified version of the classification of scale mismatches introduced by O'Higgins *et al.* (2014). In its original design, this classification identified two types of mismatch based on their scale relative to the spatially fixed scale of legislative Response. 'Extent' mismatches (EMs) were classified as those where aspects of the ecological system lay outside the jurisdiction of a Response, for example, where part of a fishery lies outside the Exclusive Economic Zone of a nation (Figure 14.1). 'Grain' mismatches (GMs) were those where ecological aspects occurred within the jurisdiction of the Response but at scales smaller than those where enforcement of legislation could be implemented, for example, where breaches in regulation are widespread within a national jurisdiction but there is insufficient resource to enforce the regulations on an individual basis. In our modification, we extend the scope of the determinants of 'scale mismatches' to include the *locus* of power relations and economic processes affecting each category. For example, in the case of large-scale offshore wind farms whose development occurs well within a single jurisdiction (e.g., within one county in the US system), major influence is exercised by state or national energy plans, which often operate on a larger scale than those considered by marine spatial planners.

In applying a scale-based tool (DPSWR) in combination with power relations, we recognise the existence of the concept of 'scale' in environmental

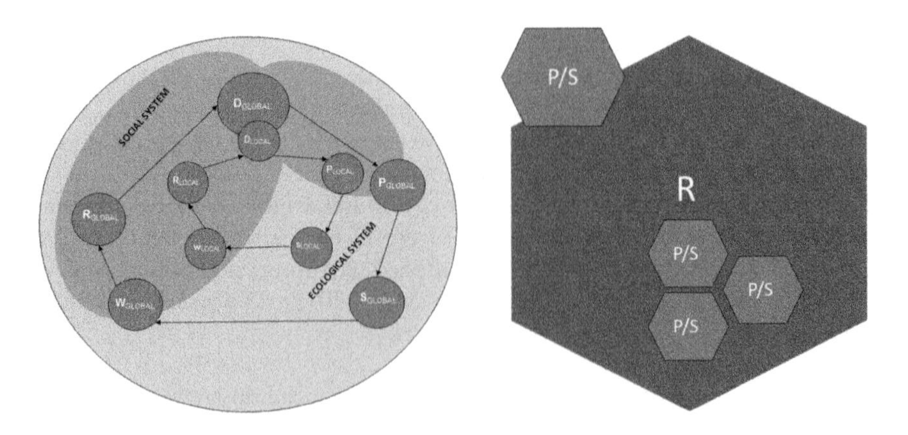

Figure 14.1 Where part of a fishery lies outside the Exclusive Economic Zone of a nation

management and human geography in line with that of Cumming *et al.*, (2006) and of Jonas (2006). We also recognise Geels' response in relation to levels and hierarchies within multi-level sustainability transitions, which does not necessarily reject 'flatness', but reformulates the role of levels as degrees of stability within the socio-technical landscape (Geels, 2011). Conceptually, 'scale' pervades MSP in two ways. First, MSP operates within man-made boundaries, which are arranged hierarchically in vertical structures defining the area where MSP operates (e.g., town-county-state-federal in the US system). In this sense, 'scale' operates conceptually closer to Neil Smith's version of it (Smith, 2008), where ecological, socio-political and economic processes are constructed throughout vertical levels. This first pervasive character of scale is, perhaps, the one more open to the critique raised in the mid-2000s by Marston *et al.* (2005; and Jones III *et al.*, 2016). This particular critique notes the confusion between scale as a matter of 'size', and scale as a matter of 'level', or the issues associated with binary distinction 'local-global' (Marston *et al.*, 2005).

In addition to the scales defining the area where MSP is implemented, other scales affect the agents of MSP, defining their ability to influence and be influenced by the decision-making process (e.g., trans-national corporations and local businesses). The second way in which scale operates is within the ecological realm(s). In this sense, ecological processes transcend human boundaries and interact with them, often forcing a rethinking of the areal scope of MSP itself (Jay *et al.*, 2016), and the scale process itself has a more spatial meaning (e.g., changes in fish stock patterns, Jensen *et al.*, 2015).

Within the context of MSP, scale has been accepted as one of the elements characterising the object of planning (i.e., the marine ecological processes to be managed) (Jay *et al.*, 2016), and the industries regulated or affected by MSP (Nursey-Bray, 2016). As for power relations, the recent literature emerging around critical MSP has recognised the limitations associated with MSP as a neutral 'forum' where stakeholders deliberate a-critically (Tafon, 2017), or without exogenous influences from their state outside of the forum itself (e.g., Flannery *et al.*, 2016). The combination of 'scale' and 'power' allows us to identify several shortcomings of MSP. It shows how transdisciplinarity, meant as a co-participatory (academic + public) circular process as defined by Mauser *et al.* (2013), can be used to address these shortcomings. It does this by recognising MSP as a tool and implementing it within a broader framework of existing practices in social sciences to achieve just and sustainable development in coastal areas. The 'transdisciplinary' character of these approaches is not new to academics and policy-makers, as it has become more and more integrated into another recent and broad framework for sustainable transitions: the water-energy-food nexus (Kurian, 2017).

In this chapter, we utilise DPSWR as a framework to identify scale mismatches in three MSP case studies: the Great Barrier Reef in Australia, the Massachusetts Ocean Plan in the United States of America, and Scotland's Marine Plan in the United Kingdom. Each of these MSPs has been conceived because one or more Drivers generated Pressures upon the social and ecological states of the regions. Furthermore, they have been studied both retroactively and during their development. After briefly presenting these cases, and identifying the mismatches within the DPSWR framework, we discuss the commonalities among these cases, and conclude by making several recommendations to policy-makers and researchers in the broader field of ecosystem-based management.

3 Scale mismatches and MSP

3.1 Australia: Great Barrier Reef

The Great Barrier Reef is the largest coral reef system in the world, located in the Coral Sea off the coast of Queensland, Australia. The reef stretches 2,300 km, extends over 14 degrees of latitude from shallow estuarine areas to deep oceanic waters, and is comprised of some 3,000 individual reefs. Within the reef are a unique range of ecological communities, habitats and species, including 600 types of soft and hard corals, more than 100 species of jellyfish, 3,000 varieties of molluscs, 500 species of worms, 1,625 types of fish, 133 varieties of sharks and rays, and more than 30 species of whales and dolphins (GBRMPA, 2017).

Historically, the Great Barrier Reef came to the attention of the Australian public in the 1960s due to limestone mining and oil drilling on the reef leading to two Royal Commissions, after which the Great Barrier Reef Marine Park (GBRMP) was established (Craik, 1992). Today, the key Drivers of change continue in the same vein, first and foremost economic and population growth. The GBR outlook report of 2014 notes that Queensland's economy is based principally on mining, construction, tourism and agriculture and the state's economy has had an average annual growth rate of 4.2% over the last decade. Growth in the resource sector has led to altered land use in the catchment area, unprecedented demands for water, power and new infrastructure including roads, railways and large-scale ports. Indeed, more than 80% of the Great Barrier Reef catchment supports grazing, cropping, dairy and horticulture (GBRMPA, 2014). In addition, the catchment is rich in mineral resources and has long supported significant mining activity with the area used for mining activities in the catchment almost doubling to 125,579 hectares between 1999 and 2009 (GBRMPA, 2014). Furthermore, tourism has become an increasingly prominent use of the GBRMP with 2.4 million visitors to the park in 2016 (GBRMPA, 2017). In addition, annual population growth is forecast at 1.6% or higher for much of the catchment, with the fastest growth in coastal regional centres (GBRMPA, 2014). This increasing population means an increase in coastal development and use of the marine environment.

These Drivers place Pressures on the Great Barrier Reef ecosystem in a number of ways. Increasing terrestrial run-off and pollution (sediments and associated nutrients, pesticides, heavy metals) caused by land-based activities has had clear detrimental impacts on freshwater aquatic systems and has put near-shore reefs and seagrass beds at risk (Productivity Commission, 2003). Marine debris pollution linked to marine park visitors has been found in the GBRMP with concerns particularly for seabirds and turtle species, which are prone to plastic ingestion (Wilson & Verlis, 2017). Diving tourism also affects coral reefs directly, particularly those reefs on which dive training is undertaken. Last but not least, expanding urbanisation, in coastal areas of the GBRMP in addition to port expansions, threaten coasts and estuaries which support the life histories of marine and freshwater species; species which often rely directly on connectivity between the reef and freshwater wetland features which are often lost to development (Waltham & Sheaves, 2015).

In combination with a changing climate which has seen 'marine heatwaves' (Hobday *et al.*, 2016), these Drivers and their subsequent Pressures have led to a deterioration in the state of the GBRMP. This is particularly evidenced in the 2016 and 2017 coral bleaching events which saw two-thirds of the world's largest reef system impacted. Such state changes are likely to have an impact on human welfare. It has been suggested that these bleaching events may lead to a reduction of up to 80% of visitors in the event of a decrease in coral cover or fish biodiversity (Kragt *et al.*, 2009) which will likely have knock-on economic impacts on the local and national economies. It is also likely that such state changes will have impacts on commercial fishing in the vicinity. The oceans are also now facing a serious threat from pollution, particularly plastic, which has environmental, economic, aesthetic and human health impacts.

As a response to several of these Pressures and the consequent changes to the coastal and marine environment, the Great Barrier Reef Marine Park (GBRMP) was established in 1975 as a multiple-use marine park. The Australian nation identified the need to address growing threats to a nationally and globally iconic environment from activities such as oil exploration and production, limestone mining, unsustainable tourism, unsustainable fishing and coastal development (Kenchington & Day, 2011). The aim of GBRMP was to provide long-term protection, ecologically sustainable use, understanding and enjoyment of the Great Barrier Reef Park (Day, 2008). Zones have been created for the purposes of general use, habitat protection, conservation, scientific research, preservation, marine national park, estuarine conservation and buffer zones. Different activities are then allocated to specific zones, with no uses whatsoever allowed in preservation zones. Spatial planning is considered the cornerstone of management of the GBRMP. This has included a comprehensive multiple-use zoning system which provides a high level of protection for some areas and allows a variety of uses in others, and which has evolved and changed considerably over time (Day, 2008).

In this case example, for this particular response mechanism, EMs are clear. The effects of point and diffuse source pollution on reef biodiversity, for example, cannot be mitigated through the establishment of habitat protection, conservation or preservation zones, due to the fluid and dynamic nature of the environment. This particular Pressure lies mostly outside the spatial domain of the Response, but similarly contamination from, for example, an oil spill or marine litter in a zone which can be used for commercial purposes, cannot be constrained to that zone. Climate change is another driver which creates an EM.

It should be noted, however, that marine spatial planning is only one of many tools that are used in management of the GBRMP with other non-spatial tools including community engagement, public education, industry partnerships, method or skill-based licensing, and economic instruments such as the Environmental Management Charge – a charge associated with most commercial activities (Kenchington & Day, 2011).

3.2 USA: Massachusetts OMP

Situated on the East Coast of the United States (US), the Commonwealth of Massachusetts has the fourteenth longest coastline of any US state (NOAA, 2017a), which hosts about 75% of the state's population (NOAA, 2013). The state's marine region is important for two main reasons: 84% of all economic activities are located in coastal counties (NOAA, 2016; BEA, 2017), and marine activities, as defined by NOAA, contribute to 2% of the state's Gross Domestic Product (ENOW, 2014; BEA, 2017). In addition to its ecological and economic capital, Massachusetts' coast also possesses a vast cultural capital, partly preserved within the Cape Cod National Seashore, and by two Marine Sanctuaries, a Management Area and several Restricted Areas (NOAA, 2017b).

Massachusetts initiated plans to manage human activities off its coast in 2008, following the promulgation of the Oceans Act, which "[...] defined the Commonwealth's goals, siting priorities, and standards for ensuring effective stewardship of ocean waters and resources [...] for activities, uses, and facilities allowed by the Ocean Sanctuaries Act [...]" (EEA, 2015, p. 1–1). The initial act was followed by the Ocean Management Plan (OMP) in 2009, which provided several regulatory directives, including a focus on offshore renewable energy technologies (ORETs). The plan identified three types of areas: prohibited (13%), renewable (2%), and multi-use (85%) (EEA, 2009). However, the revised version of the plan published in 2015 reduced the areas previously open for offshore wind sites (EEA, 2015; Bates, 2017). The development of a MSP in Massachusetts primarily attempted to regulate and incorporate the rising pressures from ORETs, specifically of offshore wind farms (Bates, 2017).

The promulgation of the state's Renewable Portfolio Standard (RPS) in 1997, established a target of 15% of electricity generated by Class I renewables by 2020 (of which 400 MW from solar, and 2,000MW from wind power), and a subsequent annual increase of 1% per year (NCSL, 2017; EEA, 2015). Within the Driver of the RPS, which is defined at state level, ORETs represent the most viable way to satisfy the state's mandate (Schwartz *et al.*, 2010; Beiter *et al.*, 2017). These technologies directly interact with Massachusetts' OMP, as they compete with other sea uses for space. Overall, the OMP has sped up the application process for sub-sea cables, and partly facilitated the interaction between fishers and wind-energy developers (Blau & Green, 2015). Nevertheless, what should have been the first utility-scale, offshore wind farm, Cape Wind, has faced considerable opposition, even after the OMP was put in to place. Since its initial inception and approval, the project has faced continuous delays and opposition (see e.g., Firestone & Kempton, 2007; Pasqualetti, 2011; Fischlein *et al.*, 2014), mainly concerned with visual impacts and, more recently, with fisheries access (Fischlein *et al.*, 2014). The pressure generated by the public has created the first major change in the updated version of the OMP, released in 2015. The pressure generated by the public has created the first major change in the updated version of the OMP, released in 2015, citing both new ecological data, as well as stakeholders' concerns to justify the reduction in the areas assigned for ORETs (EEA, 2015). The EEA cited the fact that no party presented bids for the areas temporarily identified as viable for offshore wind development, although it is difficult to imagine investors willing to commit resources in a framework of unsecured property rights and clear opposition, as faced by the Cape Wind project. In 2011, Massachusetts requested changes to be made to the federal Request for Interest for developing offshore wind farms in federal waters, to protect areas critical to commercial fisheries, marine fauna and navigation, thus implicitly advancing a series of priorities relative to ORETs.

This case identifies two EM scale mismatches. First, the overall pursuit of the RPS by the state generates a series of Pressures on the marine zones under the OMP that originate within the purview of the state, but are regulated by and are of interest to the state as a whole. As an example, the state has passed an Act to Promote Energy Diversity (H.4568) requiring 1,600 MW of offshore wind to be installed by 2027. Whether the EEA objectives are followed (2,000 MW by 2020) or the H.4568 (1,600 MW by 2027), current leased federal (not state) areas come short of these values (1,503MW projected if all leases are allocated, see Musial *et al.*, 2013). Assuming that the Drivers related to ecological protection will remain unchanged; will the current balance of uses remain intact? Furthermore, can public concerns change the content of the EEA, potentially infusing a sense of uncertainty for developers previously observed in other regions (Flannery *et al.*, 2016)? The Drivers and Pressures generating within the state, but outside the purview of the EMP, can have two types of State changes on the local marine ecosystem. Assuming no changes will affect sites excluded because of ecological concerns, in case of an expansion of the wind sites in future, EEA will signal a shift in the state's priorities based on the balance of powers and interests between coastal and non-coastal counties. This is particularly true as shown by the most recent version of the OMP, which resulted in a local balance of powers between ORETs developers and fishers (Blau & Green, 2015) whose roots are not necessarily based uniquely within the coastal region. Conversely, if the EEA remains fixed, the state will likely have to seek to meet its objectives through an expansion of renewable electricity supply, whether from the western portion of the state or through out-of-state suppliers, for example, the highly contested Northern Pass. This second option will effectively outsource any ecological and social impact from the marine space on to the land space, in a sense solving any MSP-related issue by simply moving it to a different planning realm, while implicitly recognising the structure of objectives (and power) in place within the region.

The second EM generate stems from the policy extent of the current regulatory framework. Essentially, community-level projects are the purview of regional agencies, whereas the state oversees the broader objectives for RETs. Above all, the federal government has a longer-term strategy to develop offshore wind power (US DoE and US DoI, 2016), one that is currently pursuing the development of new projects, which will occur in areas adjacent to the state's waters (Musial *et al.*, 2016). Questions remain, however, on how the expansion of new lease sites on both sides of Cape Cod could potentially impose new restrictions (Pressures) on the EEA, leading to the incorporation of new changes. Now, the fact that the EEA can change can seem, at first, as the ability of the MSP to absorb changes. However, this ability also shows the limits of any planning tool generated in a multi-scale context, where Drivers and Pressures often are borne away from the marine realm.

3.3 United Kingdom: the Scottish National Marine Plan

Scotland, including its numerous islands, has some 10,250 miles (16,500 km) of coastline. The seas around Scotland are almost six times the area of its land mass; 20% of Scotland's population live within 1 km of the coast and there are 130 inhabited islands. What's more, a quarter of Scottish businesses, accounting for 10% of Scottish turnover and 20% employment, are within 1 km of the coast (Scottish Government, 2005).

There are a range of anthropogenic activities occurring in Scottish waters. Historically, these have included exploitation of mineral and hydrocarbon resources, marine transportation and shipping, and commercial fisheries. In the financial year 2016–17, oil and gas production in Scotland is estimated to have been 74.7 million tonnes of oil equivalent, an increase of 2.9% on the previous year, valued at £17.5 billion (Scottish Government, 2017a). Ports and shipping provide for the transport of freight and passengers, including ferry services to the Scottish islands. The 11 major ports regularly handle over 1 million tonnes of freight per year. In 2015, 9.5 million passengers travelled by ferry and three million cars were transported on ferry routes (Transport Scotland, 2016). In 2016, Scottish-based vessels landed a total of 453,000 tonnes of sea fish and shellfish, with a value of £557 million, landed by 2,033 active Scottish-based vessels employing a total of 4,823 fishers (Scottish Government, 2017b). More recently, we have seen the development of new activities in the marine environment such as aquaculture and offshore renewable energy extraction. Production of the key aquaculture species, Atlantic salmon, during 2016 was 162,817 tonnes, undertaken by 15 businesses farming 253 active sites (Scottish Government, 2017c). For offshore renewable energy, a number of offshore windfarms are already in operation, mostly on the east coast. Although marine renewable energy requires further development prior to commercial operation, in August 2017, the MeyGen project, operating three tidal turbines in the Pentland Firth, generated 700 megawatt hours (MWh) over the course of the month (Coates, 2017). This increasing use of marine resources by different users of this environment has been driven largely by such needs as food and energy security, as well as economic development, and has the potential for substantial user conflict and significant impacts on Scotland's ocean and coastal areas.

These Drivers of change in the Scottish marine environment have led to, and may continue to lead to, a number of Pressures. These include: loss of biodiversity, degradation/destruction of habitats, introduction of harmful non-native species, unsustainable exploitation of fish stocks and potentially eutrophication through discharges of nutrients from aquaculture. Almost all (97%) of Scotland's coastal waters currently (2017) have a good or high status as assessed under the European Union's Water Framework Directive, but there are local impacts from commercial fishing, aquaculture and diffuse pollution. One example of this is changes to fragile and highly biodiverse habitats such as maerl beds or deep-water coral reefs, caused by mobile fishing gears (Hall-Spencer & Moore, 2000; Roberts, 2002), which may lead to changes in food-web structure and thus change the State of the whole ecosystem. Given the large reliance of Scotland's economy on marine resource use, changes in ecosystem state have large impacts on human Welfare, particularly seafood production due to changing species availability or water quality.

To manage the increasing, and often conflicting, demands on Scottish seas, the Marine (Scotland) Act 2010 brought into force a new statutory marine planning system, which is in accordance with the European Union's Maritime Spatial Planning Directive (2014/89/EU). A National Marine Plan (Scottish Government, 2015) has been created to provide a comprehensive overarching framework for all activity in Scotland's inshore and offshore waters (out to 200 nautical miles) in order to protect and enhance the marine environment at the same time as promoting existing and emerging industries. Additional Regional Marine Plans will be implemented at a local level within 11 Scottish Marine Regions, extending out to 12 nautical miles, and will be developed by Marine Planning Partnerships. Although a National Marine Plan would tend to focus more broadly than just on the geographical aspects, in the Scottish case there has been much focus on Scotland's Marine Atlas (Scottish Government, 2011) and National Marine

Plan interactive, both of which very much focus upon spatial components of the marine environment.

In this case, we can see some clear scale mismatches of both types. In the case of fisheries, we can see a GM where the Response to the problem may be too large to deal with the specifics. Although the banning of mobile gear fisheries in certain habitats such as maerl or deep-water corals, for example, will definitely contribute to the conservation of these species, it may also lead to displacement thus putting more pressure on other areas of the ecosystem (also within the marine spatial plan), as well as increased competition between fishers, thus leading to more conflict. In the case of the spread of invasive species, or of impacted water quality, and subsequent state changes to the marine environment, we can see a clear EM. For example, banning shipping in certain locations of an MSP may help to prevent the spread of marine invasives, while man-made infrastructure in other locations may also facilitate spread, by acting as stepping stones (Adams *et al.*, 2014). Also, as in the case example of the GBR, pollution will not remain within one area. Both Pressures may lie outside the spatial extent of the Response.

4 Discussion and conclusions

The scale mismatches identified in the case studies included here are mostly EMs (Table 14.1).

This is unsurprising: because of its planning nature, MSP is bounded by jurisdictional and domain extents. The former tends to highlight the inability of MSP to overcome the disconnect between coastal and marine environments; the most prominent example of this being the jurisdictional break-points within federal systems (i.e., local-state-federal). The latter highlights methodological limitations of MSP in dealing with broader, overarching policy drivers, such as decisions made within, for example, the domains of energy planning or regional economic development, where the marine environment is but one place where transitions will occur. In addition, and possibly because of these limitations, the EMs identified in this work arise from two different spatial domains. The first domain is ecological, and concerns primarily Drivers and States related to habitat and cumulative impacts (e.g., in Scotland and in Massachusetts). The second domain is policy-related, and links directly to Drivers, Pressures and even Responses arising from domains not directly (or solely) within the purview of MSP. An example of this would be aquaculture as a driver for an export-orientated, economic development policy for Scotland (Alexander *et al.*, 2014). Finally, EMs are further intensified by the inability of the MSPs to incorporate both power relations (Flannery *et al.*, 2018) and to move beyond a tokenistic selection of stakeholders (Flannery *et al.*, 2016), often rooted in the pre-existing policy and economic landscape of the coastal zone, which inevitably fails to incorporate representatives of the Drivers, Pressures and Responses non-traditionally (i.e., at the time of the MSP) within the marine domain.

Table 14.1 Overview of DPSWR components of case studies and whether they feature extent mismatches (EM) or grain mismatches (GM)

Case study	*Drivers*	*Pressures*	*State changes*	*Welfare impact*	*Response*	*EM*	*GM*
Australia	Agriculture and urban growth	Terrestrial run-off and pollution; marine debris; habitat destruction	Deterioration of coral reefs	Seafood production; tourism	MSP	✓	
United States	Offshore wind energy	Recreational and displacement of seafood production; space competition; potential habitat destruction	Changes in food-web structure	Seafood production; tourism; shipping	MSP	✓	
United Kingdom	Aquaculture and offshore wind energy	Biodiversity loss; habitat destruction; exploitation of fish stocks non-native species; eutrophication	Changes to food-web structure	Seafood production	MSP	✓	✓

4.1 MSP and scale mismatches in the context of sustainable coastal transitions

Due to the existing – and unaddressed – mismatches, MSP plays a major role in the context of sustainability transitions in coastal zones (STC). As a tool, MSP has the potential to be an effective forum for operationalising STCs, providing a context within which current incumbent and emerging sectors can communicate at various scales. In this way, it may potentially overcome the issues emerging from niche-centred processes outlined by Hansen & Coenen (2015). Further, as a tool, MSP can potentially build upon local forms of knowledge and capital, while connecting the micro, meso and macro levels of STCs, something identified as lacking in the current research (Coenen *et al.*, 2012; Hansen & Coenen, 2015; Murphy, 2015). This potential is currently undermined, however, by the existing scale mismatches, which, interestingly, generate similar issues as those found in STCs when it comes to operationalise transition processes, which take place across several levels, and beyond the original niche-to-regime initially conceptualised by Geels (2004, 2011).

4.2 Moving forward: people and the human scale as elements of a transitional MSP

As part of the broader transition experienced by coastal and marine zones, MSP can play an important role as one of several tools within the realm of ecosystem-based management (EBM). In this chapter, we have identified some of the tool's limitations, using DPSWR as a framework to identify scale mismatches within three case studies. A final, general element of mismatch also arises from our analysis: a temporal scale mismatch. Those implementing MSP must recognise that it cannot sanitise existing power relations, and, therefore, cannot prevent current priorities to be Drivers-Pressures-Responses of the future. In addition, and because of continuous changes in the Drivers-Pressures on the marine environment, MSP has possibly to recognise its limitations as a tool (rather than a 'discipline' or a 'framework') in providing a risk-reducing environment for stakeholders. This is particularly evident in the example of climate change which increasingly acts as a global-local Driver-Pressure (e.g., changing fish stock dynamics). As these changes modify the stakeholders' landscape at local level, so do the power relations and priorities (not to mention resource endowments) for the region.

We hope to offer a path forward for researchers and practitioners of EBM. It is evident from these case studies and their generalisation, that the human component (*lato sensu*) is a contributing limitation of MSP. It appears that jurisdictions, human welfare, power structures, human activities, etc., make it difficult for regional MSPs to avoid EMs. To overcome this, we suggest as a first step to more fully include humans in the MSP discourse, or, to paraphrase the geographer Torsten Hägerstrand, MSP practitioners and scholars should ask themselves 'What about people in MSP?' (Hägerstrand, 1989).

This question is non-trivial, in so far as it marks the importance to look at the human domains that determine the Drivers-Pressures-Responses beyond the marine region, and yet which influence it. Currently, MSP tends to approach and package objectives from a conservation and/or use-based perspective, often implicitly rejecting other factors, or reducing their importance in, for example, the choice of stakeholders and their relative powers during the planning process. To address this, a true nexus of disciplines is pivotal for incorporating existing knowledge and research from various fields such as political economy, regional studies/sciences, sociology, etc. Furthermore, within the operationalisation process of MSP, tokenistic bottom-down approaches, often limited in their socio-economic extents and driven by non-elected officials (planners), could be replaced with top-down-bottom-up approaches, where broader, inter-sectoral and society-wide objectives are determined at political level. This could reduce the risk of EMs or of continuous change in MSPs, or, worse, from MSP becoming a Driver (or roadblock) for emerging human activities. This is particularly important as Blue Growth or Blue Economy imperatives become more pressing globally, and national or supranational objectives put further Pressures on marine (and coastal) regions (Jones *et al.*, 2016).

To conclude, MSP is an important tool, one of many in the EBM toolkit. The positive news emerging from the recent literature on 'critical MSP' (see e.g., Jones *et al.*, 2016; Flannery *et al.*, 2016) and from our brief work is that many shortcomings of MSP have been studied and often corrected by other disciplines and across multiple human landscapes (marine or not). As such, transdisciplinarity, or a nexus, or any other research strategy capable of connecting practitioners and researchers in MSP with those studying the Driver-Pressures-Responses from other perspectives and using other tools (e.g. economic impact modelling), is imperative.

Acknowledgements

The authors would like to thank Dr. Tim O'Higgins for his insights into the DPSWR framework.

Bibliography

Adams, T.P, Miller, E.G., Aleynik, D., & Burrows, M.T. (2014). Offshore marine renewable energy devices as stepping stones across biogeographical boundaries. *Journal of Applied Ecology*, *51*(2), 330–338.

Alexander, K.A., Gatward, I., Parker, A., Black, K., Boardman, A., Potts, T., & Thomson, E. (2014). An Assessment of the Benefits to Scotland of Aquaculture. The Scottish Government. Retrieved from www.gov.scot/Topics/marine/Publications/publicationslatest/farmedfish/AqBenefits

Bates, A.W. (2017). Revisiting approaches to marine spatial planning: perspectives on an implication for the United States. *Agricultural and Resource Economics Review*, *46*(2), 206–223.

Beiter, P., Musial, W., Kilcher, L., Maness, M., & Smith, A. (2017). *An Assessment of the Economic Potential of Offshore Wind in the United States from 2015 to 2030*. NREL/TP-6A20-67675. Retrieved from www.nrel.gov/docs/fy10osti/45889.pdf

Blau, J., & Green, L. (2015). Assessing the impact of a new approach to ocean management: evidence to date from five ocean plans. *Marine Policy*, *56*, 1–8.

Borja, A., Elliott, M., Carstensen, J., Heiskanen, A-S., & van de Bund, W. (2010). Marine management – towards an integrated implementation of the European Marine Strategy Framework and the Water Framework Directives. *Marine Pollution Bulletin*, *60*(12), 2175–2186.

Bureau of Economic Analysis (BEA) (2017). Regional data – gross domestic product (GDP) by state. Retrieved from www.bea.gov/iTable/index_regional.cfm.

Coates, A. (2017). Blue Energy: the marine renewables sector starts to show promise. *The Independent*. 11 October. Retrieved from www.independent.co.uk/news/long_reads/blue-energy-marine-renewables-secto-show-promise-a7990726.html

Coenen, L., Benneworth, P., & Truffer, B. (2012). Toward a spatial perspective on sustainability transitions. *Research Policy*, *41*, 968–979.

Cooper, P. (2013). Socio-ecological accounting: DPSWR, a modified DPSIR framework, and its application to marine ecosystems. *Ecological Economics*, *94*, 106–115.

Craik, W. (1992). The Great Barrier Reef Marine Park: its establishment, development and current status. *Marine Pollution Bulletin, 25*(5–8), 122–133.

Cumming, G.S., Cumming, D.H.M., & Redman, C.L. (2006). Scale mismatches in social-ecological systems: causes, consequences, and solutions. *Ecology and Society, 11*(1), 14–34.

Day, J. (2008). The need and practice of monitoring, evaluating and adapting marine planning and management – lessons from the Great Barrier Reef. *Marine Policy*, *32*(5), 823–831.

Dominguez-Tejo, E., Metternich, G., Johnston, E., & Hedge, L. (2016). Marine Spatial Planning advancing the Ecosystem-Based Approach to coastal zone management: a review. *Marine Policy*, *72*, 115–130.

Douvere, F. (2008). The importance of marine spatial planning in advancing ecosystem-based sea use management. *Marine Policy*, *32*, 762–771.

Executive Office of Energy and Environmental Affairs (EEA) (2009). *Massachusetts Ocean Management Plan, Volume 1: Management and Administration*. Boston (MA), USA: Massachusetts Executive Office of Energy and Environmental Affairs.

Executive Office of Energy and Environmental Affairs (EEA) (2015). *Massachusetts Ocean Management Plan, Volume 1: Management and Administration*. Boston (MA), USA: Massachusetts Executive Office of Energy and Environmental Affairs.

Economics: National Ocean Watch (EOW). Dataset 2003–2014. Retrieved from https://coast.noaa.gov/digitalcoast/data/

Ehler, C. (2008). Conclusions: benefits, lessons learned, and future challenges of marine spatial planning. *Marine Policy*, *32*(5), 840–843.

Firestone, J., & Kmepton, W. (2007). Public opinion about large offshore wind power: underlying factors. *Energy Policy*, *35*(3), 1584–1598.

Fischlein, M, Feldpausch-Parker, A.M., Peterson, T.R., & Stephens, J.C. (2014). Which way does the wind blow? Analysing the state context for renewable energy deployment in the United States. *Environmental Policy and Governance*, *24*, 169–187.

Flannery, W., Ellis, G., Nursey-Bray, M., van Tetenhoven, J.P.M., Kelly, C., Coffen-Smout, S., Fairgrieve, R., Knol, M., Jontoft, S., & O'Hagan, A.M. (2016).

Exploring the winners and losers of marine environmental governance. *Planning Theory & Practice*, *17*(1), 121–151.

Flannery, W., Haley, N., &, Luna, M. (2018). Exclusions and non-participation in Marine Spatial Planning. *Marine Policy*, *88*, 32–40.

Gari, S.R., Newton, A., & Icely, J.D. (2015). A review of the application and evolution of the DPSIR framework with an emphasis on coastal social-ecological systems. *Ocean & Coastal Management*, *103*, 63–77.

Geels, F.W. (2004). From sectoral systems of innovation to socio-technical systems: insights about dynamics and change from sociology and institutional theory. *Research Policy*, *33*, 897–920.

Geels, F.W. (2011). The multi-level perspective on sustainability transitions: Responses to seven criticisms. *Environmental Innovations and Societal Transitions*, *1*, 24–40.

Great Barrier Reef Marine Park Authority (GBRMPA) (2014). GBRMPA Outlook Report 2014. Retrieved from www.gbrmpa.gov.au/managing-the-reef/great-barrier-reef-outlook-report. Last accessed on 08/11/2017

Great Barrier Reef Marine Park Authority (GBRMPA) (2017). Facts about the Great Barrier Reef. Retrieved from http://www.gbrmpa.gov.au/about-the-reef/facts-about-the-great-barrier-reef.

Hägerstrand, T. (1989). Reflections on "What about people in Regional Sciences?". *Papers of the Regional Science Association*, *66*, 1–6.

Hansen, T., & Coenen, L. (2015). The geography of sustainability transitions: review, synthesis and reflections on an emergent research field. *Environmental Innovation and Societal Transitions*, *17*, 92–109.

Hall-Spencer, J. M., & Moore, P.G. (2000). Scallop dredging has profound, long-term impacts on maerl habitats. *ICES Journal of Marine Science*, *57*(5), 1407–1415.

Hobday, A.J., Alexander, L.V., Perkins, S.E., Smale, D.A., Straub, S.C., Oliver, E.C.J., Benthuysen, J.A., Burrows, B.T., Donat, M.G., Feng, M.G., Holbrook, N.J., Moore, P.J., Scannell, H.A., Gupta, A.S., & Wernberg, T. (2016). A hierarchical approach to defining marine heatwaves. *Progress in Oceanography*, *141*, 227–238.

Jay, S., Alves, F.L., O'Mahony, C., Gomze, M., Rooney, A., Almodovar, M., Gee, K., de Vivero, J.L.S., Gonçalves, J.M.S., Fernandes, M. d.L., Tello, O., Twomey, S., Prado, I., Fonseca, C., Bentes, L., Henriques, G., & Campos, A. (2016). Transboundary dimensions of marine spatial planning: fostering inter-jurisdictional relations and governance. *Marine Policy*, *65*, 85–96.

Jensen, F., Frost, H., Thøgersen, T., Andersen, P., & Andersen, J. L. (2015) Game theory and fish wars: the case of the Northeast Atlantic mackerel fishery. *Fisheries Research*, *172*, 7–16.

Jonas, A.E. (2006). Pro scale: further reflections on the 'scale debate' in human geography. *Transitions (Institute of British Geographers)*, *31*(3), 399–406.

Jones, P.J.S., Lieberknecht, L.M., & Qiu, W. (2016). Marine spatial planning in reality: introduction to case studies and discussion of findings. *Marine Policy*, *71*, 256–264.

Jones III, J.P., Leitner, H., Marston, S.A., & Shepperd, E. (2016). Neil Smith's Scale. *Antipode*, *49*(S1), 138–152.

Kragt, M.E., Roebeling, P.C, & Ruijs A. (2009). Effects of Great Barrier Reef degradation on recreational reef-trip demand: a contingent behaviour approach. *Australian Journal of Agricultural and Resource Economics*, *53*(2), 213–229.

Kenchington, R.A., & J.C. Day. (2011). Zoning, a fundamental cornerstone of effective Marine Spatial Planning: lessons learnt from the Great Barrier Reef, Australia. *Journal of Coastal Conservation*, *15*(2), 271–278.

Kurian, M. (2017). The water-energy-food nexus trade-offs, thresholds and transdisciplinary approaches to sustainable development. *Environmental Science & Policy*, *68*, 97–106.

Marston, S.A., Jones III, J.P., & Woodward K. (2005). Human geography without scale. *Transactions of the Institute of British Geographers*, *30*, 416–432.

Mauser, W., Klepper, G., Rice, M., Schmalzbauer, B.S., Hackmann, H., Leemans, R., & Moore, H. (2013). Transdisciplinary global change research: the co-creation of knowledge for sustainability. *Current Opinion in Environmental Sustainability*, *5*, 420–431.

Murphy, J.T. (2015). Human geography and socio-technical transition studies: promising intersections. *Environmental Innovation and Societal Transitions*, *17*, 73–91.

Musial, W., Parker, Z., Filed, J., Scott, G., Elliott, D., & Draxl, C. (2013). Assessment of Offshore Wind Energy Leasing Areas for the BOEM Massachusetts Wind Energy Area – NREL/TP-5000–60942. Retrieved from www.nrel.gov/docs/fy14osti/60942.pdf

Musial, W., Beiter, P., Schwabe, P., Tian, T., Stehly, T., & Spitsen, P. (2016). 2016 Offshore Wind Technologies Market Report. Retrieved from www.energy.gov/sites/prod/files/2017/08/f35/2016%20Offshore%20Wind%20Technologies%20Market%20Report.pdf

National Conference of State Legislatures (NCSL) (2017). State Renewable Portfolio Standards and Goals. Retrieved from www.ncsl.org/research/energy/renewable-portfolio-standards.aspx

National Oceanic and Atmospheric Administration (NOAA) (2013). National Coastal Population Report: Population Trends from 1970 to 2020. Retrieved from https://coast.noaa.gov/digitalcoast/training/population-report.html

National Oceanic and Atmospheric Administration (NOAA) (2016). STICS: Total Economy of Coastal Areas. Retrieved from https://coast.noaa.gov/dataregistry/search/collection/info/coastaleconomy

National Oceanic and Atmospheric Administration (NOAA) (2017a). Shoreline Mileage of the United States. Retrieved from https://coast.noaa.gov/data/docs/states/shorelines.pdf

National Oceanographic & Atmospheric Agency (NOAA). (2017b). MPA Inventory Data. Retrieved from https://marineprotectedareas.noaa.gov/dataanalysis/mpainventory/mpaviewer/

Nursey-Bray, M. (2016). More than fishy business: epistemology, integration and conflict in marine spatial planning. *Planning Theory and Practice*, *17*(1), 121–151.

O'Higgins, T., Farmer, A., Daskalov, G., Knudsen, S., & Mee, L. (2014). Achieving good environmental status in the Black Sea: scale mismatches in environmental management. *Ecology and Society, 19*(3), 54.

Pasqualetti, M.J. (2011). Opposing wind energy landscapes: a search for common cause. *Annals of the Association of American Geographers*, *101*(4), 907–917.

Productivity Commission (2003). Industries, land use and water quality in the Great Barrier Reef catchment. Paper No. 0305001. EconWPA, Canberra, Australia.

Qiu, W., & Jones, P.J. (2013). The emerging policy landscape for marine spatial planning in Europe. *Marine Policy*, *39*, 182–190.

Roberts, C.M. (2002). Deep impact: the rising toll of fishing in the deep sea. *Trends in Ecology & Evolution*, *17*(5), 242–245.

Scottish Government (2005). Seas the opportunity: a strategy for the long term sustainability of Scotland's coasts and seas. Retrieved from www.gov.scot/Publications/2005/08/26102543/25444

Scottish Government (2011). Scotland's Marine Atlas: Information for The National Marine Plan. Retrieved from www.gov.scot/Publications/2011/03/16182005/0

Scottish Government (2015). Scotland's National Marine Plan: A Single Framework for Managing Our Seas. Retrieved from www.gov.scot/Resource/0047/00475466.pdf

Scottish Government (2017a). Oil and Gas Production Statistics 2016/2017, Release Date: 13 September 2017 Retrieved from www.gov.scot/Resource/0052/00524651.pdf

Scottish Government (2017b). Scottish Sea Fisheries Statistics 2016. Retrieved from www.gov.scot/stats/bulletins/01291

Scottish Government (2017c). Scottish Fish Farm Production Survey 2016. Retrieved from www.gov.scot/Publications/2017/09/5208

Schwartz, M., Heimiller, D., Haymes, S., & Musial, W. (2010). Assessment of Offshore Wind Energy Resources for the United States. NREL/TP-500–45889. Retrieved from www.nrel.gov/docs/fy17osti/67675.pdf

Smith, N. (2008). *Uneven Development: Nature, Capital, and the Production of Space*. Athens (GA), USA: University of Georgia Press.

Tafon, R. (2017). Taking power to sea: towards a post-structuralist discourse theoretical critique of marine spatial planning. *Environmental and Planning C: Politics and Space*, In press. doi.org/10.1177/2399654417707527

Transport Scotland (2016). Scottish Transport Statistics No 35: 2016 Edition. Retrieved from www.transport.gov.scot/publication/scottish-transport-statistics-no-35-2016-edition/

US Department of Energy and US Department of the Interior (2016). National offshore wind strategy. Retrieved from www.boem.gov/National-Offshore-Wind-Strategy/

Waltham, N.J., & Sheaves, M. (2015). Expanding coastal urban and industrial seascape in the Great Barrier Reef World Heritage Area: critical need for coordinated planning and policy. *Marine Policy*, *57*, 78–84.

Wilson, S.P., & Verlis, K.M. (2017). The ugly face of tourism: marine debris pollution linked to visitation in the southern Great Barrier Reef, Australia. *Marine Pollution Bulletin*, *117*(1), 239–246.

15 The varying economic impacts of marine spatial planning across different geographical scales

A Q methodology study

Madeleine Gustavsson and Karyn Morrissey

1 Introduction

Home to 40% of the global population and believed to be the next economic frontier (K. Morrissey, 2017), the coast is of strategic importance for future sustainability trends (J.E. Morrissey & Heidkamp, 2017). Traditionally seen as a sector dominated by the seafood industry (K. Morrissey, Donoghue & Hynes, 2011), 'emerging', often high-tech, research-led sectors such as marine renewable energy, offshore aquaculture and biotechnology have entered the core of the marine economy (K. Morrissey & O'Donoghue, 2012; K. Morrissey, O'Donoghue & Farrell, 2013). Effective social control of technology has been a constant concern for industrial societies (Berkhout, Smith & Stirling, 2004). In order to manage both incumbent and new sectors in the ocean space, the last two decades have seen a rapid increase in interest and action at varying political levels to implement spatially explicit management of marine resources (Halpern *et al.*, 2012; McLeod & Leslie, 2009).

Emerging from Integrated Coastal Zone Management, marine spatial planning (MSP) is increasingly promoted by governments and international bodies as a means to reduce sectorial conflicts and maintain the good environmental status (GES) of the marine environment (European Commission, 2011). As a management tool, MSP aims to move away from a traditional, failing, sectorial focus on the management of marine space to a more holistic approach which understands the full use of marine space (Kidd, Plater & Frid, 2011; White, Halpern & Kappel, 2012). In its broadest sense, marine spatial planning can be defined as:

> Analyzing and allocating parts of three-dimensional marine spaces to specific uses or non-use, to achieve ecological, economic, and social objectives that are usually specified through a political process.
>
> (Ehler & Douvere, 2007, p. 13)

Complementing the current spatial focus occurring in the transitions literature (Coenen, Benneworth & Truffer, 2012), MSP brings a spatial dimension to the regulation of marine activities by identifying which areas of the ocean are appropriate for different uses or activities in order to reduce conflicts and achieve ecological, economic and social objectives (see Jay, Ellis & Kidd, 2012). Within the Blue Economy agenda, MSP is seen as a means of creating "an optimal investment climate for maritime sectors and give operators more certainty as to what opportunities for economic development are possible[1]". A report by the European Commission in 2011 (p. 7) found that if MSP is managed properly economic benefits would arise from: a) "enhanced coordination and simplified decision processes"; b) "enhanced legal certainty for all stakeholders in the maritime area"; c) "enhanced cross border cooperation"; and d) "enhanced coherence with other planning systems". Through these gains, the study estimated that the effects of MSP have potential to cumulate in a saving of between €400 million to €1.8 billion due to the reduction of transaction costs and from €155 million to €1.6 billion due to the acceleration of activities such as wind energy and aquaculture, by 2030.

In the UK, the non-governmental organisation RSPB (2004) highlighted the economic potential of MSP, in particular to three areas:

1) "Facilitating sectoral growth – the MSP can provide a framework that facilitates the sustainable development of different economic activities, therefore helping to enhance incomes and employment;
2) Optimising the use of the sea – MSP can help to ensure that maximum benefits are derived from the use of the sea by encouraging activities to take place where they bring most value and do not devalue other activities;
3) Reducing costs – MSP can reduce costs of information, regulation, planning and decision-making" (RSPB, 2004, p. 69).

Furthermore, Jay (2013) suggests that expansion of new marine sectors and economic activities through MSP can lead to increases of state revenues by means of licensing fees and taxes on potential developers. With its focus on GES and the integration and co-evolution of emerging sectors such as marine renewable energy and biotechnology with traditional marine sectors, MSP offers an important tool in transitioning towards a sustainable coastal zone. However, from a sectorial perspective, Jay (2013, p. 519) notes that: "Newcomers to the marine environment, such as the wind energy industry, appear to be benefitting well from the allocation of space, whilst more traditional users, such as the fishing industry, feel more constrained as a result."

Studying five ocean plans – two in North America; two in Europe; and one in Australia – Blau & Green (2015) found that economic benefits were not shared equally. In particular, they found that capital-intense projects, such as wind farms, have gained the largest benefits because of the increased certainty and enhanced speed of regulatory processes. The same authors found that commercial and recreational fishing, tourism and shipping (so called 'incumbent' industries) did not receive any substantial economic benefits. However, they argue "a case can be made that they could have lost greater economic value without the plans (e.g., if wind farms were sited in spawning areas or shipping lanes)" (Blau & Green, 2015, p. 6). Drawing on these findings, the authors suggest that ocean planning has the potential to produce net benefits at little cost, but that the "distribution of these benefits [...] depends on the context, politics and goals underlying the plan" (Blau & Green, 2015, p. 7). In contrast, White *et al.* (2012) found that optimal planning create in Massachusetts Bay, USA, led to over $10 billion from wind-energy development whilst not compromising the commercial fishing industry. Lester *et al.* (2013) found similar results in their study of wave energy in Oregon, USA.

Although researchers have recognised diversity in economic impacts across sectors, very little attention has been paid to how these link to the onshore communities involved in the marine economy (St. Martin & Hall-Arber, 2008). In reality, all policies have spatially differentiated outcomes. In order to formulate effective management policies, it is necessary not only to understand the nature and the operation of policies at the national level but also to evaluate the likely impact of policies on activity at the local level (K. Morrissey & O'Donoghue, 2012). With regard to MSP, Flannery & Ellis (2016, p. 121) note that:

> While MSP is quickly becoming the dominant marine management paradigm, there has been comparatively little assessment of the potential negative impacts and possible distributive impacts that may arise from its adoption.

The spatial implications of MSP are particularly important within the context of a sustainable coastal transitions literature as many coastal areas are undergoing rapid sociotechnical change (J.E. Morrissey & Heidkamp, 2017), of which the impact on the local community is unclear. Drawing on these insights and using Q methodology, this paper seeks to elicit the perception of the potential economic impact of MSP across different scales, including households, coastal, rural versus urban communities, regional and national level.

2 Methods

This UK-based study uses a Q methodology approach to understand sectorial perspectives on the potential impact of MSP across different scales, including households, coastal, rural versus urban communities, regional and national level. The Q methodology has been described as the 'science of subjectivity' in that it examines the subjectivities of individuals in a systematic way (McKeown & Thomas, 2014b). Q methodology differs from other data-rich empirical (quantitative) methods in that it does not seek to identify traits across a population, nor provide results that are generalisable. As such, one of the advantages is that Q methodology does not rely on large numbers of participants. The focus of Q methodology is on identifying shared ways of

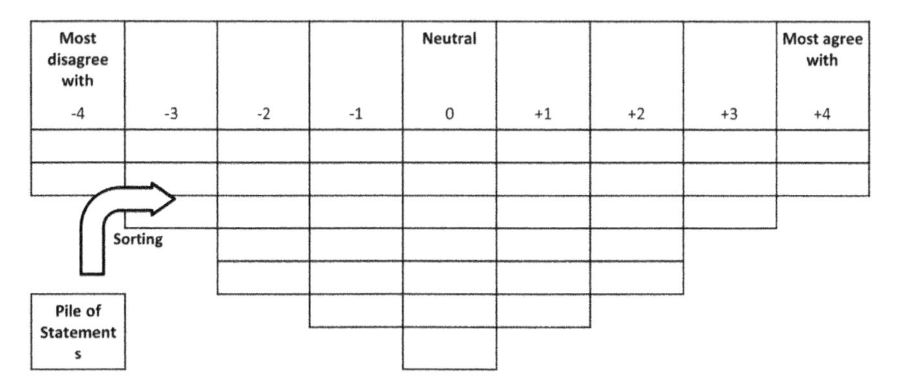

Figure 15.1 The Q-sort 'forced distribution' which is used in the current study; the total number of statements are 39, and one statement should be placed in each box

thinking about an issue through revealing a number of different discourses (Eden, Donaldson & Walker, 2005; Ellis, Barry & Robinson, 2007; McKeown & Thomas, 2014b).

In Q methodology, participants are asked to organise a pre-determined number of statements into a range of categories according to which they agree with the most, or the least. Only a fixed number of statements can be sorted in one particular category of agreement (see Figure 15.1). This means that participants need to choose carefully the number of statements they completely agree with (or completely disagree with, etc.) a process referred to as 'forced distribution' or "forced choice" (McKeown & Thomas, 2014a, p. 3). The sorted statements are the 'data' of Q methodology – also called 'Q-sort'.

To administer the survey, the online software 'Q sortware' (www.qsortware.net) was utilised and a total of ten participants were recruited for this study. The 39 statements presented to participants in this study (see Table 15.1) were derived from an extensive literature review (Eden *et al.*, 2005) on the economics of marine spatial planning, marine economic geography and the economics of specific marine sectors. The selected statements were chosen to represent a diverse set of positions on the economic impacts of MSP across our four stakeholder groups. To ensure the clarity and comprehensiveness of statements, the research used a piloting phase where three participants completed the online exercise (and measured the time needed for completion) and provided feedback and suggestions for improvements. As this study uses a fairly large number of statements, the sorting is preceded by an initial sorting where participants are asked to simply read the statements and store them into a small set of piles without restrictions (Agree, Neutral and Disagree) in order to become familiar with the content of each card before the final sorting. The final 'Q-sort' sets are analysed through factor analysis using the software PQMethod (Schmolck, 2014)

Table 15.1 Statements presented to participants. The Q-sort position for each statements in the respective factor (F1=Factor 1; F2=Factor 2; F3= Factor 3) is presented below (*dark grey indicates distinguishing statements with p-value <0.01 and light grey indicate p-values <0.05*). Consensus statements – that is, statements that are similar across factors – *are marked grey in the statement column*

	Statement	*F1*	*F2*	*F3*
1	MSP is an important process for equitably dividing space between different users	+3	+1	+1
2	MSP will reduce costs for development at sea	0	-3	-1
3	Allocation of space within MSP should be based on sound scientific principles and economic rationality seeking to maximise national economic revenues from the sea	0	+3	+4
4	Economic diversification can help traditional industries adapt to the negative economic impacts caused by MSP	+1	-2	+3
5	MSP will have positive economic effects as a result of better coherence between planning systems, such as between the sea and land planning systems	+4	0	+1
6	MSP should prioritise marine businesses and sectors who spend their money regionally	0	+1	-2
7	MSP is moving jobs from rural coastal communities to urban areas	-4	-1	0
8	MSP will have economic benefits as it will simplify decision-making	+3	0	-1
9	Stakeholder participation is crucial to reduce the negative economic impacts from MSP	+2	+4	+2
10	Small-scale businesses will benefit economically from MSP	+1	+1	0
11	I believe that maximising the national economic profits from the use of the sea will lead to economic benefits for my sector	0	0	+3
12	MSP will have positive economic impacts for my sector as it will enhance co-operation across regional and national borders	+2	-4	-3
13	Expansion of new marine industries will generate jobs in local coastal communities	+2	0	+2
14	Jobs will be created in cities and not in local coastal communities as a consequence of MSP	-4	-1	-2
15	Banks will grant loans much more easily because of MSP	-1	-1	-4
16	MSP will lead to economic growth of all marine-based sectors and will create jobs and income	-1	-1	-2
17	In cases of displacement of previously existing activities economic compensation should be paid	-1	+2	-4
18	MSP will have economic benefits at the regional level	+1	+1	-3
19	Low levels of negative economic impacts to already existing activities are acceptable to make space for new profitable activities	0	-4	+3
20	The negative economic impacts from MSP will be felt on the household and local level whilst benefits will be gained at the national level	-2	+1	0
21	MSP has economic benefits as it improves the investment climate by clarifying who has the right of use to areas at sea	+1	0	-1

Table 15.1 (Cont.)

	Statement	*F1*	*F2*	*F3*
22	The development of stationary objects at sea ruins the aesthetic value of the sea which will have negative impacts on the local economy	-3	-3	-1
23	MSP will reduce conflicts between users which will lead to economic benefits for all marine sectors	+2	-2	+1
24	Competition for areas at sea will be greatest in inshore areas as these are the most profitable areas	-1	+3	+4
25	Coastal communities and families need to economically benefit from new marine sectors, or such activities should not be allocated space at sea	0	0	+1
26	MSP is benefiting sectors with large-scale investments	-1	-1	0
27	MSP will speed up the process of investment in the marine sector	+1	-2	0
28	There will be no economic impacts (neither positive or negative) from MSP on any marine sector	-3	-3	-2
29	Rural coastal communities will benefit economically from MSP	+2	-1	0
30	Better legal certainty from MSP will provide economic benefits to my sector	0	-2	+2
31	It is important that the use of the sea contributes to sustaining vibrant coastal communities	+4	+4	+2
32	MSP should seek to plan for co-existence of activities as much as possible to maximise economic output from the sea	+3	+2	+2
33	MSP will benefit the region as a whole, but won't have any significant economic impacts on the local level	-2	0	+1
34	Development of new marine industries will lead to the displacement of jobs in other marine sectors which were there previously	-2	+2	0
35	Skilled labour for new marine sectors can be found in rural coastal areas	+1	+2	-1
36	The necessary economic burdens from MSP will be carried by all marine activities equally	-2	-2	-2
37	Those sectors which historically used the sea (previous to MSP) should be continuously allowed to do so	-1	+3	-1
38	New jobs in the marine economy have to be full-time jobs, not seasonal part-time jobs	-2	+1	+1
39	The biggest threat to the marine economy are marine conservation zones which is part of MSP	-3	+2	-3

to produce a number of 'ideal sort', or factors, which represent the different discourses identified.

Before engaging with the Q-sort, respondents were asked a number of background questions. These included the number of their employees, location of their business and importantly their 'position' on MSP and whether they had or were involved in the MSP process. This qualitative data was used to interpret and make sense of the results of the factor analysis of the Q-sorts.

2.1 Selection of participants

Q methodology relies on theoretical sampling rather than random sampling (Eden *et al.*, 2005; McKeown & Thomas, 2014b). Participants in this study were selected based on four sectors of interest: 1) recreational sea-angling fishing; 2) marine renewable energy; 3) aquaculture; and 4) commercial fisheries. The justification for choosing these specific sectors is as follows:

1) Pre-existing conflicts have been reported between recreational sea anglers and fishers. This conflict stems from the exclusion of sea anglers from fishing quotas which fishers deem unfair (Voyer, Barclay, McIlgorm & Mazur, 2017). Some studies have already touched on the relation between recreational fishing and MSP (Hooper, Hattam & Austen, 2017) and therefore we have some pre-existing knowledge on which to build our study.
2) Marine renewable energy (wave, tidal and offshore wind energy) is a recent addition to the marine economy (K. Morrissey & O'Donoghue, 2013) and as such poses particular challenges to already existing marine activities, as well as possibilities for growth of the marine sector.
3) Aquaculture currently comprises 50% of seafood production and it is one of the marine sectors that is expected to expand rapidly over the short to medium term. Licensing and planning for aquaculture sites is contentious across other marine sectors and among public stakeholders. Conflict between inshore fisheries and aquaculture is already evident, while the push to move aquaculture further offshore will mean that aquaculture will be competing for space with a wider range of marine sectors (Alexander, Wilding & Jacomina Heymans, 2013; Asche & Khatun, 2006).
4) Literature suggests that commercial fisheries can become displaced from areas used in the past as space is allocated to new marine sectors (Berkenhagen *et al.*, 2010).

Within each stakeholder group, the researchers sought to include diverse representatives from the small- and large-scale sectors, located in urban as well as rural areas, and operating on global, national, regional and local levels. Each of these sectors, for the reasons listed above, is likely to hold a specific perspective on the issues pertaining to MSP. The specific businesses and organisations were identified through online searches in databases held by national and regional associations representing these sectors.

2.2 Analysis

PQMethod (Schmolck, 2014) was used to analyse the Q-sorts. To begin with, principal component analysis (PCA) was used to calculate eigenvalues to identify the strength of each factor. Following Addams & Proops (2000), factors with eigenvalues greater than one were maintained. This final set of eigenvalue-selected factors was then rotated using a Varimax rotation. The built in add-on application PQROT was used for automatic 'flagging' (i.e., loading particular Q-sorts on to specific factors). The factor loadings represent the correlations between extracted factors and the variables (i.e., participants) (Farrell, Carr & Fahy, 2017). Following this, PQMethod was able to identify a number of 'ideal sorts' or 'factors', which represents the different discourses identified on the studied topic within the studied sample. These 'ideal sorts', similar to a Q-sort, organise statements on a scale from 'most agree with' (+4), to 'most disagree with' (-4) (see Figure 15.1 and Table 15.1). The PQMethod analysis also determines the strength of each factor within the final set, groups participants according to factor similarity of their Q-sorts, and calculates a z-score, which represents the relative rank-order of each statement, for each sorted Q statement for each factor. PQMethod also analyses which statements are distinguishing a factor (and produces a p-value for these), and/or, which statements are so-called 'consensus statements' – similar across all factors.

3 Results

All respondents loaded on a factor, which cumulatively explained 68% of the total variance within the data. The factor analysis revealed that there are three operating discourses (see Table 15.1). Factor Group 1 is composed of six individuals: all three representatives from the marine renewable energy (all limited companies less than ten years of age, located in urban areas or towns, with between four and 60 employees operating on a national to global level), two from aquaculture (one large-scale limited company of 25 years located in a town with ten employees operating on a national level; and one ten years old small-scale sole-trade registered company located in an 'isolated dwelling' operating on local scale) and one representative from a fisheries organisation (mainly representing the large-scale fisheries sector with regional to national scope). Factor Group 2 comprises two individuals, one from a charter boat for sea angling (one registered partnership, approximately 20 years of age, with two employees located in a village, operating on a local scale) and one from a fisheries producer organisation (limited company approximately 20 years old with only one employee which is operating on a regional scale). Factor Group 3 consists of two individuals, one from a sea-angling club (with 50 members, that has been around for over 60 years with local scope) and one from aquaculture (limited company with ten years of operation, located in an urban area with eight employees operating on a national level). As will be discussed, the discourse held by respondents loading on to Factor 1 will be referred to as 'place-makers', respondents loading on to Factor 2 will be referred to as 'place-holders' and respondents loading on to factor 3, 'place-less'. The eigenvalue of the place-maker discourse is 3.98; it is 2.1598 for the place-holder discourse and 1.15 for the place-less discourse. The composite reliability is 96%, 89% and 89%, respectively.

The questionnaire revealed that out of the ten participants in the study, seven were in favour of MSP and three were neutral, with no participant stating that they were 'against' MSP. Additionally, four companies had been involved in an MSP process. With regard to the percentage breakdown of the participants' sectors and their position on MSP:

- 100% of both renewable energy and recreational sea-angling business are in favour of MSP;
- 33% of aquaculture businesses and 50% of fisheries are in favour of MSP.

With regard to work location and the participants' positions on MSP,

- 66% of urban, town and village-based businesses are in favour of MSP;
- 100% of businesses located in isolated areas are in favour of MSP.

3.1 Factor 1 – the optimistic 'place-makers'

What is distinctive about this factor is that representatives are optimistic about MSP and its role in maintaining coastal communities in the future. Respondents loading on to Factor 1 assert that it is important for the marine economy to sustain coastal communities and that new marine economic activities will help to do so. Their optimism about using the sea to contribute to sustainable coastal communities, its capacity to equitably dividing space between activities and MSP capacity to harness positive economic effects as a result of better coherence between planning systems underpins a view of MSP as a 'place-maker'. Representatives loading on to Factor 1 do not agree that MSP will lead to negative consequences for households at the local level and they do not agree that jobs will be displaced from sectors historically present in local areas because of MSP. They disagree that MSP will result in significant negative economic impacts on the local level and that only seasonal (lower-quality jobs) will be created in local communities. Furthermore, they strongly disagree that jobs will be created in cities rather than rural areas. From this, it could be understood that representatives loading on to this factor are optimistic about MSP and the structural changes it can deliver, and therefore see MSP as an opportunity to 'make places', as a means of creating new opportunities in coastal communities.

3.2 Factor 2 – the sceptical 'place-holders'

Representatives loading on to Factor 2 are distinguished by their strong sense of maintaining the historic practices associated with the sea, believing that the historical use of the sea should be taken into account when planning and implementing MSP. Given their strong preference for maintaining historic practices and coastal communities, we refer to this group as 'place-holders'. This factor strongly agrees that stakeholder participation is crucial to reduce the negative economic impacts from MSP and that it is important that the use of the sea contributes to sustaining vibrant coastal communities and that those industries previously using the sea should be continuously allowed to use these areas. In line with this they strongly disagree that low levels of economic impacts are acceptable to make space for new profitable activities, and if businesses are displaced they should be compensated for any negative economic impacts. They do not agree that they will benefit from MSP and that most of the structural changes to marine governance that MSP brings about will benefit them economically. Also, they do not agree that MSP will reduce conflicts that will benefit them economically. Furthermore, they do not agree that economic diversification can soften the negative economic impacts of MSP. From this, it could be understood that representatives loading on to Factor 2 are less optimistic about MSP than Factor 1 representatives and their main concern is that MSP maintains or 'holds' current practices in the sea.

3.3 Factor 3 – the utilitarian 'place-less'

Representatives loading on to Factor 3 are distinguished by their strong sense of economic rationality and the need to maximise national economic revenues from the sea. Given their utilitarian, national level focus we refer to this factor as 'place-less'. Specifically, this factor strongly agrees that national economic gains will lead to benefits for their sector, and that at a sectorial level 'low levels of negative economic impacts to already existing activities are acceptable to make space for new profitable activities'. Representatives loading on to this factor strongly disagree that compensation should be paid to affected businesses, further demonstrating their utilitarian approach to MSP. With regard to location or 'place', Factor 3 representatives disagree that there should be some level of prioritisation given to businesses that benefit the local/regional economy. Interestingly, this factor does not tend to agree that the structural changes imposed by MSP will lead to specific economic impacts on their sectors. For instance, they do not agree that co-operation across regional/national borders will lead to economic benefits and do not agree that banks will more easily grant loans because of MSP. From this, it could be understood that representatives loading on to Factor 3 have a more utilitarian approach to MSP compared to Factor 1 or Factor 2 representatives and their main concern is that MSP be carried out in a scientific manner that focuses on national level rather than sub-national economic objectives for the marine resource.

3.4 Consensus statements

Though clear differences between the three groups can be seen, there are significant areas of consensus among the factors that can provide further insights on stakeholders' perception of the distributional impact of MSP across sectors and locations. First, all factors agree that stakeholder participation is crucial to reduce the negative economic impacts from MSP. They also agree it is important for MSP to seek the co-existence of activities to maximise economic output from the sea. They strongly disagree that there will be no economic impacts of MSP and that the economic burdens from MSP will be carried by all marine sectors equally. All factors disagreed that the development of stationary objects (such as offshore wind turbines) at sea ruins the aesthetic value of the sea, which has negative impacts on the local economy. All three factors also had some statements which they all felt neutral about. For instance, the factors do not highlight any conflicts between small-scale and large-scale businesses in terms of economic impacts of MSP. Also, the factors are neutral about the statement that coastal communities and families need to economically benefit from new marine sectors or that such activities should not be allocated space at sea.

4 Discussion

The blue economy agenda (European Commission, 2017; Koundouri & Giannouli, 2015) has highlighted the economic potential which the marine environment offers. Simultaneously, there have been concerns raised on the increased demand for – and potential conflicts over – the use of marine space and importantly the sustainability of marine resources in the face of these conflicts. Against the background of the sustainable transitions literature, this paper investigated the perceptions of stakeholders in the fishing, aquaculture, sea angling and marine renewable energy sector of a) the economic impact of MSP and b) the geographical scale of this impact, using Q-methodology. The factor analysis revealed that there are three operating discourses. Results emerging from this study indicate that all three factors agree that stakeholder participation is crucial to reduce the negative economic impacts from MSP; however, each of the three factors has a distinct sense of the distributional impact of MSP across sectors and places. Compared to Factor 3, representatives of Factor 1 and 2 are distinguished by their strong sense of location, both Factor (1, 2) representatives agree that MSP should ensure the local level to benefit from the marine economy. However, while Factor 1 focuses on the role of MSP in emerging marine activities and their benefits to coastal communities, Factor 2 is distinguished by a strong sense of maintaining the historic practices associated with the sea for the benefit of local, coastal communities. In contrast, representatives loading on to Factor 3 have a more utilitarian viewpoint and are distinguished by their beliefs that it is the overall economic benefit of MSP that is important and that low levels of negative economic impacts to already existing activities are acceptable to make space for new profitable activities.

Whilst previous studies have recognised how MSP will lead to growth of the blue economy, not many studies have sought to understand the commonly asked questions: growth for whom and of what? This study found that while each of the sectors represented in this study were either in favour or at least neutral on the implementation of MSP, only participants from the marine energy sector had the same perception of the economic impact of MSP at different scales. Representatives from fisheries, sea angling and aquaculture belong to a mix of the three identified discourses, the optimistic place-makers, the sceptical place-holders and the utilitarian place-less. Interestingly, business location was also not a strong indicator of the perception of the economic impact of MSP with participants from across the four identified locations, urban areas, towns, villages and isolated hamlets, each belonging to the three discourses. More participants are however needed to draw any further conclusions regarding this.

Through examining the perceptions held by stakeholders, this chapter has produced novel insights into the distributional effects of MSP across different geographical scales. Nevertheless, future studies need to engage more quantitative methods to measure the *de facto* economic impacts, rather than the perceptions held by stakeholders. Future research could also use the typology developed here to explore in more depth the underlying socio-cultural identities which underpin these positions. A limitation of this research is that the results only reflect businesses involved in the blue economy; policy-makers and other non-commercial entities were not surveyed. Regardless, the findings of this study have implications for sustainable coastal transition. Whilst participants of this study were not against MSP, they held varying positions in regards to the economic impacts of changed marine governance. Similar to previous research on sustainable transitions (Geels, 2011), differences between incumbent industries (fishing) and new industries (renewable energy) are identified.

Acknowledgements

The authors would like to thank the participants for generously volunteering their time. We are also thankful to the anonymous reviewer as well as the editors of this book for their constructive comments.

Note

1 Maria Damanaki (European Commissioner for Maritime Affairs and Fisheries).

Bibliography

Addams, H., & Proops, J. (2000). *Social Discourse and Environmental Policy: An Application of Q Methodology*. Cheltenham: Edward Elgar Publishing.

Alexander, K.A., Wilding, T.A., & Jacomina Heymans, J. (2013). Attitudes of Scottish fishers towards marine renewable energy. *Marine Policy*, *37*(0), 239–244. https://doi.org/http://dx.doi.org/10.1016/j.marpol.2012.05.005

Asche, F., & Khatun, F. (2006). Aquaculture: issues and opportunities for sustainable production and trade. *Natural Resources*, *5*, 1–52.

Berkenhagen, J., Döring, R., Fock, H.O., Kloppmann, M.H.F., Pedersen, S.A., & Schulze, T. (2010). Decision bias in marine spatial planning of offshore wind farms: problems of singular versus cumulative assessments of economic impacts on fisheries. *Marine Policy*, *34*(3), 733–736. https://doi.org/10.1016/j.marpol.2009.12.004

Berkhout, F., Smith, A., & Stirling, A. (2004). Socio-technological regimes and transition contexts. In *System Innovation and the Transition to Sustainability: Theory, Evidence and Policy* (pp. 48–75). Cheltenham: Edward Elgar.

Blau, J., & Green, L. (2015). Assessing the impact of a new approach to ocean management: evidence to date from five ocean plans. *Marine Policy*, *56*, 1–8. https://doi.org/10.1016/j.marpol.2015.02.004

Coenen, L., Benneworth, P., & Truffer, B. (2012). Toward a spatial perspective on sustainability transitions. *Research Policy*, *41*(6), 968–979. https://doi.org/10.1016/j.respol.2012.02.014

Eden, S., Donaldson, A., & Walker, G. (2005). Structuring subjectives? Using Q methodology in human geography. *Area*, *37*(4), 413–422.

Ehler, C.N., & Douvere, F. (2007). *Visions for a Sea Change. Report of the First International Workshop on Marine Spatial Planning*. Intergovernmental Oceanographic Commission and Man and the Biosphere Programme. Paris: UNESCO.

Ellis, G., Barry, J., & Robinson, C. (2007). Many ways to say ‘no’, different ways to say ‘yes’: applying Q-methodology to understand public acceptance of wind farm proposals. *Journal of Environmental Planning and Management*, *50*(4), 517–551. https://doi.org/10.1080/09640560701402075

European Commission. (2011). *Study on the economic effects of Maritime Spatial Planning. Framework*. Luxembourg: Publication Office of the European Union. https://doi.org/10.2771/85535

European Commission. (2017). Blue growth | Maritime Affairs. Retrieved 21 February 2017, from http://ec.europa.eu/maritimeaffairs/policy/blue_growth/

Farrell, D., Carr, L., & Fahy, F. (2017). On the subject of typology: how Irish coastal communities’ subjectivities reveal intrinsic values towards coastal environments. *Ocean and Coastal Management*, *146*, 135–143. https://doi.org/10.1016/j.ocecoaman.2017.06.017

Flannery, W., & Ellis, G. (2016). Exploring the winners and losers of marine environmental governance. *Planning Theory & Practice*, *17*(1), 121–151. https://doi.org/10.1080/14649357.2015.1131482

Geels, F.W. (2011). The multi-level perspective on sustainability transitions: responses to seven criticisms. *Environmental Innovation and Societal Transitions*, *1*(1), 24–40. https://doi.org/10.1016/J.EIST.2011.02.002

Halpern, B.S., Diamond, J., Gaines, S., Gelcich, S., Gleason, M., Jennings, S., … & Zivian, A. (2012). Near-term priorities for the science, policy and practice of Coastal and Marine Spatial Planning (CMSP). *Marine Policy*, *36*(1), 198–205. https://doi.org/10.1016/j.marpol.2011.05.004

Hooper, T., Hattam, C., & Austen, M. (2017). Recreational use of off shore wind farms: experiences and opinions of sea anglers in the UK. *Marine Policy*, *78*, 55–60. https://doi.org/10.1016/j.marpol.2017.01.013

Jay, S. (2013). From disunited sectors to disjointed segments? Questioning the functional zoning of the sea. *Planning Theory & Practice*, *14*(4), 509–525.

Jay, S., Ellis, G., & Kidd, S. (2012). Marine spatial planning: a new frontier? *Journal of Environmental Policy & Planning*, *14*(1), 1–5. https://doi.org/10.1080/1523908X.2012.664327

Kidd, S., Plater, A., & Frid, C. (Eds.). (2011). *The Ecosystem Approach to Marine Planning and Management*. London: Earthscan.

Koundouri, P., & Giannouli, A. (2015). Blue growth and economics. *Frontiers in Marine Science*, *2*, 94. https://doi.org/10.3389/fmars.2015.00094

Lester, S.E., Costello, C., Halpern, B.S., Gaines, S.D., White, C., & Barth, J.A. (2013). Evaluating tradeoffs among ecosystem services to inform marine spatial planning. *Marine Policy*, *38*, 80–89. https://doi.org/10.1016/j.marpol.2012.05.022

McKeown, B., & Thomas, D.B. (2014a). A concluding subjective-science postscript. In McKeown, B., & Thomas, D.B. (Eds.), *Q Methodology*. Thousand Oaks: SAGE Publications.

McKeown, B., & Thomas, D.B. (2014b). *Q Methodology*. Thousand Oaks: SAGE Publications.

McLeod, K., & Leslie, H. (2009). *Ecosystem-Based Management for the Oceans*. Washington DC: Island Press.

Morrissey, J.E., & Heidkamp, P. (2017). Coastal sustainability ii: frontiers for regional transition towards sustainability transitions in the coastal zone. *Regions Magazine*, *308*(4), 9–10.

Morrissey, K. (2017). *Economics of the Marine: Modelling Natural Resources*. London: Rowman and Littlefield International.

Morrissey, K., Donoghue, C., & Hynes, S. (2011). Quantifying the value of multi-sectoral marine commercial activity in Ireland. *Marine Policy*, *35*(5), 721–727. https://doi.org/10.1016/j.marpol.2011.02.013

Morrissey, K., & O'Donoghue, C. (2012). The Irish marine economy and regional development. *Marine Policy*, *36*(2), 358–364. https://doi.org/10.1016/j.marpol.2011.06.011

Morrissey, K., & O'Donoghue, C. (2013). The role of the marine sector in the Irish national economy: an input–output analysis. *Marine Policy*, *37*(0), 230–238. https://doi.org/http://dx.doi.org/10.1016/j.marpol.2012.05.004

Morrissey, K., O'Donoghue, C., & Farrell, N. (2013). The local impact of the marine sector in Ireland: a spatial microsimulation analysis. *Spatial Economic Analysis*, *9*(1), 31–50. https://doi.org/10.1080/17421772.2013.835439

RSPB. (2004). *Potential Benefits of Marine Spatial Planning to Economic Activity in the UK*. Plymouth: RSPB.

Schmolck, P. (2014). PQMethod v.2.35. Retrieved 1 March 2017, from http://schmolck.userweb.mwn.de/qmethod/

St. Martin, K., & Hall-Arber, M. (2008). The missing layer: geo-technologies, communities, and implications for marine spatial planning. *Marine Policy*, *32*(5), 779–786. https://doi.org/10.1016/j.marpol.2008.03.015

Voyer, M., Barclay, K., McIlgorm, A., & Mazur, N. (2017). Connections or conflict? A social and economic analysis of the interconnections between the professional fishing industry, recreational fishing and marine tourism in coastal communities in NSW, Australia. *Marine Policy*, *76*, 114–121. https://doi.org/10.1016/j.marpol.2016.11.029

White, C., Halpern, B.S., & Kappel, C.V. (2012). Ecosystem service tradeoff analysis reveals the value of marine spatial planning for multiple ocean uses. *Proceedings of the National Academy of Sciences*, *109*(12), 4696–4701. https://doi.org/10.1073/pnas.1114215109/-/DCSupplemental.www.pnas.org/cgi/doi/10.1073/pnas.1114215109

16 Steps towards the sustainable management of sediment in ports and harbours

Amélie Polrot, Jason R. Kirby, Jason W. Birkett, Ian Jenkinson and George P. Sharples

1 Introduction

Among the different coastal environments, ports and harbours hold an important place. Dominating the local economy, they permit trade, fishing and tourism or can shelter a naval base. A well-known challenge in world ports and harbours is to maintain their navigability. Indeed, sediments accumulate with time and activities, which creates significant problems for shipping as sedimentation decreases nautical depth. The most widespread method to tackle siltation problems is dredging, which consists of sediment excavation from the site, followed by its transport and disposal in a designated area. Dredging is a significant industry with a global turnover estimated as €5.02 billion (IADC, 2016). However, dredging is also associated with many detrimental issues including the fundamental non-sustainability of the practice, sediment loss from the coastal system (Manap & Voulvoulis, 2015) and the presence of contaminants in the targeted sediments (Birkett *et al.*, 2002). Indeed, polychlorinated biphenyls (PCBs), polycyclic aromatic hydrocarbons (PAHs), pesticides, heavy metals, dioxin or organotin compounds coming from industry, agriculture and urban run-off are commonly found in sediments in the world's ports and harbours (United States, 1999) in concentrations exceeding sediment quality standards (Al Sawai, 2015; Buruaem *et al.*, 2012; Nasr *et al.*, 2006; Sany *et al.*, 2013).

Currently, there are two options to deal with the contaminated sediment problem – leave the material *in situ* and close off the facility, or removal of sediment by dredging and costly waste disposal in a licenced Confined Disposal Facility (CDF). This usually involves relocating the mud to a terrestrial settling lagoon where it may be treated with active remediation. This procedure is often prohibitively expensive, and contributes to carbon emissions through release of greenhouse gases during removal and transportation.

Considering the issues linked to dredging, research has been carried out to find more sustainable and innovative ways to manage fine sediment problems in ports and harbours (Kirby, 2011). Different methods have been investigated, some of which aim to evacuate sediment through flushing techniques (e.g., harbour entrance current deflecting walls), others are designed to reduce the accumulation of sediment or manipulating the physical properties of the sediment to maintain sufficient nautical depth. The latter methods are called the Passive and the Active Nautical Depth (PND/AND) concepts. PND is basically a new way to define nautical depth, considering fluid mud as navigable above a critical threshold density value of 1,200 kg/m^3 (Welp & Tubman, 2017). By carefully monitoring mud rheology using density and shear strength measurements, the need for dredging can be greatly reduced using this concept, as better navigability in fluid mud is enabled. AND is a step further, where a navigable fluid mud is created *in situ* by mixing and aerating the sediment, thus increasing the nautical depth. This process of sediment conditioning manipulates not only the physical conditions but also the biological and chemical environment. After its application, AND has demonstrated promising results reducing the impact of organic micropollutants, including degradation of tributyltin (TBT) (Kirby, 2013), a highly toxic legacy contaminant found in ports from its application as an anti-fouling agent on ships. Preliminary studies demonstrate that this bioremediation activity is related to *in situ* micro-organisms that have been promoted by the aeration (Cruz *et al.*, 2015). AND may, therefore, provide the dual benefit of reducing dredging need as well as representing a novel bioremediation technique with the potential to resolve two significant problems facing the maritime industry.

In this chapter, the issues surrounding conventional dredging practice will be highlighted and the principle of AND outlined. The first results from the application of AND and its potential for the bioremediation of contaminated sediments in coastal ports and harbours will be presented. Thus, the potential to turn an intractable waste into a valuable ecosystem resource will be explored. More broadly, this paper explores the AND concept as a key socio-technical transformation towards the sustainable management of sediment in the maritime industry.

2 Dredging

Most ports and harbours in the world experience siltation problems that have hindered ship navigation since ancient times. In Ancient Egypt, workers used to manually drag mud until the method improved when the first dredging machine was developed in 1796 (Knight & Lacey, 1843). Dredging consists of the excavation of the sediment from the congested site, followed by its transport and disposal in a designated area, normally offshore. Both the excavation and the disposal are strictly regulated and subject to legislation aimed at minimising environmental impact, especially because of the potential presence of harmful chemical contaminants. In the United Kingdom, the Marine Management Organisation is the licensing authority for dredge disposal sites and operates under the OSPAR guidelines (OSPAR, 2004).

2.1 Environmental impacts of dredging

The negative impacts of dredging comprise effects related to the excavation method itself and also to the impact of contaminated sediment manipulation. These effects can be categorised into three types – physical, chemical and biological impacts – and are discussed below in relation to the dredging of non-contaminated and contaminated sediment.

When dredging non-contaminated sediment, different problems can be encountered. First, an increase in turbidity takes place at the excavation site and at the disposal site in the sea, which can affect photosynthetic activity. This can result in the widespread loss of seagrass vegetation (Erftemeijer & Lewis, 2006), a reduction in the production of phytoplankton and can also affect fish or membrane-feeding organisms through the clogging of gills and membranes (Balchand & Rasheed, 2000). A lethal effect on corals caused by turbidity and sedimentation at the disposal site has also been shown (Erftemeijer *et al.*, 2012). During the excavation, an abundance of nutrients is released in the water column. This causes a strong perturbation to the ecosystem, which can have an impact on the macrobenthic fauna by causing the population of native organisms to decrease in number (Ponti *et al.*, 2009). The habitat is also modified during the process, with a change of the seabed surface at the excavation site and a potential change in sediment properties at the disposal site. These changes can affect the ability of the benthic fauna to recover after the dredging perturbation (Cooper *et al.*, 2011). The overall consequence of these phenomena is a decrease in benthic faunal diversity after dredging operations (Barrio Froján *et al.*, 2011; Kenny & Rees, 1996). Another secondary impact of dredging is the emission of greenhouse gas that occurs mainly during the transportation phase (Choi *et al.*, 2016).

For dredging of contaminated sediment, the negative effects increase significantly (Manap & Voulvoulis, 2015). The re-suspension of sediment during excavation can result in the release of contaminants around the excavation site (Munawar *et al.*, 1989) and the excavation process may itself expose a new layer of potentially highly contaminated sediment. Some of these compounds, such as heavy metals, can even become more toxic after re-suspension through an oxidation process (Roberts, 2012). Organisms affected by contaminant exposure comprise three types: the organisms living in the sediment (benthic fauna), pelagic organisms (fish and plankton) and consumers (fish, birds, mammals and even humans) (Bridges *et al.*, 2010). Strong increases in the bioavailability (Eggleton & Thomas, 2004) and bioaccumulation of contaminants have been reported after dredging activities (Hedge *et al.*, 2009; Martins *et al.*, 2012; Winger *et al.*, 2000), which leads to the spreading of these toxic compounds through the entire food chain.

2.2 Regulatory framework

In recognition of the significant environmental impacts of dredging, a range of rules and regulations have been implemented at local, national and international level with the aim to control and reduce the negative effects of this process. Firstly, restrictions have been put in place by the London Convention (IMO, 1972) that "*prohibits the dumping of certain hazardous materials in the sea and requires a prior special permit for the dumping of a number of other identified materials and a prior general permit for other wastes or matters*". Several international convention agreements have followed (Abriak *et al.*, 2006) and, consequently, laws and directives have been created across the world with obligatory procedures in place before dredging is authorised. These include: evaluation of sediment contamination, framing of contaminated sediment disposal and remediation, justification of dredging methods used, agreement for the follow-up monitoring of the dredged site, etc.

In parallel to the implementation of restrictive laws, effort has been made to develop tools and methods of management to match the new regulations (Cooper, 2013). A wide range of concepts and decision-making frameworks have been proposed (Bates *et al.*, 2015; Manap & Voulvoulis, 2014; Palermo *et al.*, 2008) in an attempt to limit and reduce the environmental consequences of dredging. The complex legislation and the negative public perception of dredging make managing the process a challenge (Cutroneo *et al.*, 2014; Hamburger, 2002). Conflicts can appear between the different stakeholders and projects are consequently subjected to delays or cancellation.

A further significant issue with dredging is its high cost, comprising the cost for the operation and the cost for the disposal. The cost can vary depending on the technology and equipment used, as well as the volume of sediment targeted, the distance to the disposal site and the presence of contaminants. Moreover, since many ports and harbours are further developing, for example by adapting to enable the entry of larger vessels, the increased need for dredging comes with increases in consequence and with increased associated costs (Kirby, 2011; Manap & Voulvoulis, 2015).

Finally, even if modifications are proposed to limit its environmental impact, the principle of the method itself, removing the sediment from its initial environment, is fundamentally non-sustainable. The way in which we manage sediment in ports and harbours requires innovation and we urgently need new ways to deal with siltation and contamination that reduce costs and do not have an impact on marine ecosystems or contribute significantly to carbon emissions.

3 Port contamination: need for remediation

Port and harbour activities generate many types of pollution: sewage and wastewater, petroleum and its derivatives, greenhouse gas emission and release of compounds from antifouling paints. The multiple sources of contamination and the usual enclosed configuration of ports and harbours result in limited circulation leading to high levels of contaminants in sediments and subsequent negative impact to aquatic life. The main contaminants persisting in sediment are organotin compounds, heavy metals, PCBs and PAHs.

All of the methods designed for the remediation of contaminated sediment involve its dredging and placement *ex-situ* followed by a designated treatment.

Most of the available treatments are physical and chemical. Thermal treatment such as incineration, as an example of a physical treatment, is often used because of its efficiency but incineration consumes a lot of energy and has a high cost (Du *et al.*, 2014).

Efforts have been made to find more environmentally friendly and cost-effective ways for the remediation of dredged contaminated sediment and bioremediation is an encouraging process in this regard. Bioremediation consists of the degradation of a contaminant as a result of the activity of a living organism. It usually involves contaminant breakdown by micro-organisms (biodegradation) or by plants (phytoremediation). Bioremediation has been applied successfully as an *ex-situ* treatment for contaminated sediment (Novak & Trapp, 2005; Wu *et al.*, 2014). It represents a low-cost method with few environmental impacts. Used *ex-situ*, however, bioremediation is still associated with the negative effects of dredging described above (e.g., strong environmental impact, complex legislation, high cost) and remains unsustainable as the sediment is removed from its initial location. Consequently, developing *in-situ* solutions that do not require dredging for the remediation of contaminated sediment is most desirable.

4 Nautical depth: toward a sustainable way to manage sediment

4.1 Passive Nautical Depth

The issues with conventional dredging described above highlight the non-sustainability of currently applied methods and illustrate the need for developing new, more innovative ways to manage sediment in ports and harbours that generate less pollution, have a lower economic cost and provide an opportunity to reduce contamination and carbon emissions. The first step in reducing the need for dredging has been the adoption of a concept called Passive Nautical Depth (Kirby & Parker, 1974).

Passive Nautical Depth consists of changing the criteria defining the nautical bottom, defined as the level at which physical characteristics of the bottom can cause either damage or unacceptable effects on controllability and manoeuvrability by contact with a ship's keel (Kirby, 2011; McAnally *et al.*, 2016). Before the application of this concept, the depth was measured with a fathometer, which records the time for a sound pulse to be reflected from the channel bottom and back to the device. Depending on the rheological parameters (e.g., density, viscosity, etc.) of the sea bottom (especially in muddy bays and estuaries), the fathometer generates ghost echoes, that can either be associated with solid bed or with fluid mud that would be navigable. None of the instruments currently used is able to differentiate ghost echoes from real solid bed (McAnally *et al.*, 2007). By precaution, ghost echoes are always considered to be associated with solid bed, which leads to a potentially unnecessary dredging of the fluid mud, resulting in a waste of money and generation of additional pollution that could be avoided.

During Passive Nautical Depth, the depth should be defined by the parameters that permit discrimination between solid bed and fluid mud. The density criterion is generally used but density alone is not sufficient. Other parameters, such as shear stress, should be considered to establish whether the mud is fluid enough to be navigable (Wurpts, 2005). These parameters, however, are not easy to record routinely and different particle size arrangements (which are locally variable) also influence density, shear strength and therefore navigability. As a consequence, for each port the density at which the sediment is in a fluid mud state has to be determined. Most of the time the density threshold used is 1,200 kg/m^3 (Welp & Tubman, 2017). The concept of Passive Nautical Depth is now widely used in the world's ports and harbours and permits reduced application of dredging (McAnally *et al.*, 2016). Nevertheless, Passive Nautical Depth does not deal with the issue of chemical contaminants.

4.2 Active Nautical Depth

4.2.1 Principle

By derivation of the concept of Passive Nautical Depth, a new method to manage sediment in muddy ports and harbours has been developed, called Active Nautical Depth (Kirby, 2011; McAnally *et al.*, 2016). The principle (see Figure 16.1) is to manipulate the fluid mud cloud to perpetuate its navigability by mixing and aerating it. Aeration is a critical step that determines

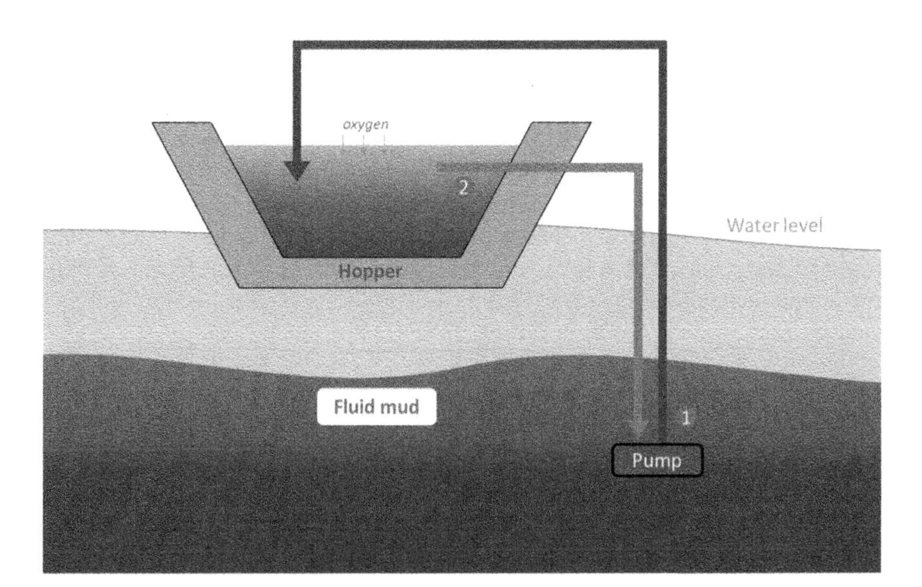

Figure 16.1 Active Nautical Depth principle

the sustainability of the method. Indeed, the new aerobic state of the mud promotes the growth of aerobic micro-organisms that start producing large amounts of extracellular polymeric substances (EPS). Without EPS production, the mud would rapidly go back to its initial non-navigable state but with EPS, the particles are kept in suspension longer (Pang Qi Xiu *et al.*, 2018) and the fluid remains navigable for weeks. The physical properties of EPS also permit the hulls of vessels to pass through with minimal friction, thus facilitating navigability through the fluid mud cloud. Moreover, this aerobic condition may allow another phenomenon to take place. Indeed, numerous biochemical processes occur aerobically, notably the biodegradation of chemical contaminants such as tributyltin (TBT). Consequently, as a beneficial side-effect, aeration of the mud may favour bioremediation of sediment pollutants while reducing the production of other pollutants such as methane, ammonia or hydrogen sulphide by anaerobic micro-organisms.

Figure 16.1 shows a process whereby the mud is pumped into a hopper dredger (2) where it is oxygenated, pumped back to the sea bottom (1) and remains in a fluid mud state for weeks.

4.2.2 Application

Emden port (Ems estuary, Germany) was the first to experiment with AND. The method has been successfully applied and is well described in the literature (Kirby, 2011; McAnally *et al.*, 2016; Wurpts, 2005). In this case, mixing is achieved by pumping the fluid mud with a low-power submerged dredge pump into a hopper dredger (see Figure 16.1). The pumping initially alters the physical conditions by breaking the inter-particle bonds and fluidising the mud. This mud goes in the hopper and is exposed to the atmosphere, thus rapidly becoming aerobic and ready to be returned to the sea-bottom. The fluid mud cloud remains in suspension during the three to four months before the mixing episode has to be repeated (Kirby *et al.*, 2008). In Emden's port configuration, the fluid mud cloud maintained by AND prevents exterior sediment entering into the basin, consequently reducing the need for dredging to zero where previously 4 million m^3 of sediment was dredged each year. Finally, as a result of the reduced need for maintenance dredging, the overall cost of sediment management has decreased from €12.5 million per year to €4 million per year (Kirby, 2013) since the initiation of AND in Emden port.

4.2.3 Worldwide applicability

Based on the successful results obtained following the implementation of AND in Emden port, an investigation of its potential to be upscaled and used in other ports and harbours worldwide has been performed (Wurpts, 2005). There are some critical conditions necessary for AND to be successful and these include grain size. A muddy substrate with low sand content is required in the targeted area. According to Wurpts (2005), AND should be easily applicable for a sand content of up to 10% with a grain size of between 60 and 200 μm. For sediment with a sand content exceeding 10%, however, the process can be refined. Indeed, the hopper dredger applied in Emden port has been designed in such a way that a sand extraction can be performed if needed.

These application conditions are technically viable for many ports in the world with muddy sediment problems, such as Liverpool, Bristol, Leer, *etc.* (Wurpts, 2005) and feasibility studies could be performed to evaluate the possibility of applying AND as a sustainable method for sediment management (to replace or reduce dredging).

4.2.4 Potential for bioremediation

TBT is one of the most toxic marine contaminants, which has been widely used in antifouling paints and has been globally banned since 2008 (Sonak *et al.*, 2009). Studies have shown, however, that TBT is highly persistent in anaerobic sediment with half-lives reaching decades (Langston *et al.*, 2015). Before AND, TBT remediation could only be achieved by treatment of dredged sediment through processes such as incineration or vaporisation (Du *et al.*, 2014), which are expensive and energy-consuming methods in addition to the negative impact of the dredging itself. TBT remediation *in situ* can be achieved with AND, which offers the potential for a new sustainable low-cost method for active bioremediation.

Many studies have focused on the isolation of TBT-resistant or TBT-degrading micro-organisms (Barug, 1981; Cruz *et al.*, 2007; Ebah *et al.*, 2016; Khanolkar *et al.*, 2015; Murthy *et al.*, 2007; Sakultantimetha *et al.*, 2009). Bacteria, as well as fungi, from a wide range of genera, are able to degrade TBT and have been successfully isolated and characterised. Some studies have also evaluated their potential for TBT degradation in pure culture media or microcosms, whether by adding a solution of TBT or by using contaminated sediment. Most of the isolated bacteria were able to grow at TBT concentrations as high as 3 mM, far exceeding the concentration found in contaminated areas. Only three studies, however, have estimated the potential for TBT bioremediation from an indigenous microbial community using contaminated sediment from a river (Sakultantimetha *et al.*, 2010; Suehiro *et al.*, 2006; Tessier *et al.*, 2007), and no large-scale field trials have been undertaken to assess the full potential of the technique in coastal or marine environments or to establish possible undesirable impacts.

4.2.5 Research perspectives

Although TBT biodegradation has been observed following sediment oxygenation (Kirby *et al.*, 2008; Prexl, 2002) and many TBT-degrading micro-organisms have been isolated, no field studies have been performed to evaluate *in-situ* biodegradation and the microbial communities involved in port and harbour environments. The only microcosm experiments reported have been conducted on freshwater sediment, whereas salinity is known to affect TBT adsorption and desorption in sediment (Fang *et al.*, 2017). Consequently the results of such experiments cannot be translated directly to the marine environment. The absence of microcosm and field-scale studies involving full native microbial communities thus constitute a gap in our knowledge and a priority for future research.

Furthermore, TBT is not the only contaminant found in ports and harbours, and studies could be applied to the potential aerobic bioremediation of other pollutants. One study has already highlighted the isolation of both methylmercury-degrading bacteria and TBT-degrading bacteria from the same seawater sample (Lee *et al.*, 2012), highlighting the fact that different communities of degraders cohabit in the same environments. Furthermore, another study has successfully investigated the remediation of heavy metals in the laboratory using a technique of re-suspension/aeration of contaminated freshwater sediment from a Quebec harbour (Pourabadehei & Mulligan, 2016). Further research on seawater sediments is needed especially since changing the availability of metals in oxygenated sediment could be an important issue because some metals can become more toxic after oxidation (Roberts, 2012).

5 Conclusion

Dredging is the conventional technique used for sediment management in ports and harbours but this is now recognised to have a highly negative impact on the environment in numerous ways. Alternative methods exist, however, for managing sediment more sustainably, and at lower cost. Such methods, including Active Nautical Depth, have so far only been applied at a few locations in Europe. Active Nautical Depth in particular shows promising results, representing a far more sustainable method for sediment management and the bioremediation of TBT contaminated sediment in ports and harbours. With AND the issues concerning the transport and disposal of sediment or the procurement of a dredging licence disappear, resulting in a smaller environmental impact, a facilitation of the procedures (e.g., no need for dredging licences) and a consequent decrease of the cost for sediment management. Significant carbon savings are also possible. Further research into the bioremediation potential of the AND method is necessary so that it can be optimised and adapted to other situations where it has the potential to revolutionise the ways in which we approach sediment management problems in ports and harbours worldwide. Once the technical feasibility of the concept has been fully scientifically trialled, this socio-technical approach to sustainably managing sediment in the marine and coastal zone would be globally applicable, especially if encouraged through policy shifts.

Abbreviations

PCBs Polychlorinated Biphenyls
PAHs Polycyclic Aromatic Hydrocarbons
CDF Confined Disposal Facility

PND Passive Nautical Depth
AND Active Nautical Depth
TBT Tributyltin
EPS Extracellular Polymeric Substance
Kg/m^3 Kilogram per cubic metre
m^3 Cubic metre
µm Micrometre
mM Millimolar

Bibliography

Abriak, N.E., Junqua, G., Dubois, V., Gregoire, P., Mac Farlane, F., & Damidot, D. (2006). Methodology of management of dredging operations I. Conceptual developments. *Environmental Technology*, *27*(4), 411–429. https://doi.org/10.1080/09593332708618653

Al Sawai, A. (2015). Determination of heavy metal contamination in the surface sediments at Sohar Industrial Port (SIP) and the nearby coastal regions (Thesis). Cardiff Metropolitan University. Retrieved from https://repository.cardiffmet.ac.uk/handle/10369/7527

Balchand, A.N., & Rasheed, K. (2000). Assessment of short term environmental impacts on dredging in a tropical estuary. *Terra et Aqua*, 79.

Barrio Froján, C.R.S., Cooper, K.M., Bremner, J., Defew, E.C., Wan Hussin, W.M.R., & Paterson, D.M. (2011). Assessing the recovery of functional diversity after sustained sediment screening at an aggregate dredging site in the North Sea. *Estuarine, Coastal and Shelf Science*, *92*(3), 358–366. https://doi.org/10.1016/j.ecss.2011.01.006

Barug, D. (1981). Microbial degradation of bis(tributyltin) oxide. *Chemosphere*, *10*(10), 1145–1154. https://doi.org/10.1016/0045-6535(81)90185-5

Bates, M.E., Fox-Lent, C., Seymour, L., Wender, B.A., & Linkov, I. (2015). Life cycle assessment for dredged sediment placement strategies. *The Science of the Total Environment*, *511*, 309–318. https://doi.org/10.1016/j.scitotenv.2014.11.003

Birkett, J.W., Noreng, J.M.K., & Lester, J.N. (2002). Spatial distribution of mercury in the sediments and riparian environment of the River Yare, Norfolk, UK. *Environmental Pollution*, *116*(1), 65–74. https://doi.org/10.1016/S0269-7491(01)00121-X

Bridges, T.S., Gustavson, K.E., Schroeder, P., Ells, S.J., Hayes, D., Nadeau, S.C., & Patmont, C. (2010). Dredging processes and remedy effectiveness: relationship to the 4 Rs of environmental dredging. *Integrated Environmental Assessment and Management*, *6*(4), 619–630. https://doi.org/10.1002/ieam.71

Buruaem, L.M., Hortellani, M.A., Sarkis, J.E., Costa-Lotufo, L.V., & Abessa, D.M. S. (2012). Contamination of port zone sediments by metals from Large Marine Ecosystems of Brazil. *Marine Pollution Bulletin*, *64*(3), 479–488. https://doi.org/10.1016/j.marpolbul.2012.01.017

Choi, Y., Thompson, J.M., Lin, D., Cho, Y.-M., Ismail, N.S., Hsieh, C.-H., & Luthy, R.G. (2016). Secondary environmental impacts of remedial alternatives for sediment contaminated with hydrophobic organic contaminants. *Journal of Hazardous Materials*, *304*, 352–359. https://doi.org/10.1016/j.jhazmat.2015.09.069

Cooper, K.M. (2013). Setting limits for acceptable change in sediment particle size composition: testing a new approach to managing marine aggregate dredging. *Marine Pollution Bulletin*, *73*(1), 86–97. https://doi.org/10.1016/j.marpolbul.2013.05.034

Cooper, K.M., Curtis, M., Wan Hussin, W.M.R., Barrio Froján, C.R.S., Defew, E.C., Nye, V., & Paterson, D.M. (2011). Implications of dredging induced changes in sediment particle size composition for the structure and function of marine benthic macrofaunal communities. *Marine Pollution Bulletin, 62*(10), 2087–2094. https://doi.org/10.1016/j.marpolbul.2011.07.021

Cruz, A., Anselmo, A.M., Suzuki, S., & Mendo, S. (2015). Tributyltin (TBT): a review on microbial resistance and degradation. *Critical Reviews in Environmental Science and Technology, 45*(9), 970–1006. https://doi.org/10.1080/10643389.2014.924181

Cruz, A., Caetano, T., Suzuki, S., & Mendo, S. (2007). *Aeromonas veronii*, a tributyltin (TBT)-degrading bacterium isolated from an estuarine environment, Ria de Aveiro in Portugal. *Marine Environmental Research, 64*(5), 639–650. https://doi.org/10.1016/j.marenvres.2007.06.006

Cutroneo, L., Massa, F., Castellano, M., Canepa, G., Costa, S., Povero, P., & Capello, M. (2014). Technical and public approaches to involve dredging stakeholders and citizens in the development of a port area. *Environmental Earth Sciences, 72*(8), 3159–3171. https://doi.org/10.1007/s12665-014-3222-9

Du, J., Chadalavada, S., Chen, Z., & Naidu, R. (2014). Environmental remediation techniques of tributyltin contamination in soil and water: a review. *Chemical Engineering Journal, 235*, 141–150. https://doi.org/10.1016/j.cej.2013.09.044

Ebah, E., Ichor, T., & Okpokwasili, G.C. (2016). Isolation and biological characterization of tributyltin degrading bacterial from Onne Port sediment. *Open Journal of Marine Science, 06*(02), 193. https://doi.org/10.4236/ojms.2016.62015

Eggleton, J., & Thomas, K.V. (2004). A review of factors affecting the release and bioavailability of contaminants during sediment disturbance events. *Environment International, 30*(7), 973–980. https://doi.org/10.1016/j.envint.2004.03.001

Erftemeijer, P.L.A., & Lewis, R.R.R. (2006). Environmental impacts of dredging on seagrasses: a review. *Marine Pollution Bulletin, 52*(12), 1553–1572. https://doi.org/10.1016/j.marpolbul.2006.09.006

Erftemeijer, P.L.A., Riegl, B., Hoeksema, B.W., & Todd, P.A. (2012). Environmental impacts of dredging and other sediment disturbances on corals: a review. *Marine Pollution Bulletin, 64*(9), 1737–1765. https://doi.org/10.1016/j.marpolbul.2012.05.008

Fang, L., Xu, C., Li, J., Borggaard, O.K., & Wang, D. (2017). The importance of environmental factors and matrices in the adsorption, desorption, and toxicity of butyltins: a review. *Environmental Science and Pollution Research International, 24*(10), 9159–9173. https://doi.org/10.1007/s11356-017-8449-z

Hamburger, P. (2002). In defence of dredging: increasing public awareness. *N.S.W.: Institution of Engineers*, 416–427.

Hedge, L.H., Knott, N.A., & Johnston, E.L. (2009). Dredging related metal bioaccumulation in oysters. *Marine Pollution Bulletin, 58*(6), 832–840. https://doi.org/10.1016/j.marpolbul.2009.01.020

IADC. (2016). Dredging in Figures 2016. Retrieved 24 January 2018, from http://iadc-profile-registration.dev.occhio.nl/en/62/news/dredging-in-figures-2016/?id=537

IMO. (1972). Convention on the Prevention of Marine Pollution by Dumping of Wastes and Other Matter.

Kenny, A.J., & Rees, H.L. (1996). The effects of marine gravel extraction on the macrobenthos: results 2 years post-dredging. *Marine Pollution Bulletin, 32*(8), 615–622. https://doi.org/10.1016/0025-326X(96)00024-0

Khanolkar, D.S., Naik, M.M., & Dubey, S.K. (2015). Biotransformation of Tributyltin chloride by *Pseudomonas stutzeri* strain DN2. *Brazilian Journal of Microbiology, 45*(4), 1239–1245.

Kirby, R. (2011). Minimising harbour siltation – findings of PIANC Working Group 43. *Ocean Dynamics, 61*(2–3), 233–244. https://doi.org/10.1007/s10236-010-0336-9

Kirby, R. (2013). Managing industrialised coastal fine sediment systems. *Ocean & Coastal Management, 79*, 2–9. https://doi.org/10.1016/j.ocecoaman.2012.05.011

Kirby, R., & Parker, W.R. (1974). Seabed density measurements related to echosounder records. *Dock & Harbour Authority, 54*, 423–424.

Kirby, R., Wurpts, R., & Greiser, N. (2008). Chapter 1 Emerging concepts for managing fine cohesive sediment. In Kusuda, T., Yamanishi, H., Spearman, J. & Gailani, J. Z. (Eds.), *Proceedings in Marine Science* (Volume 9, pp. 1–15). Elsevier. Retrieved from www.sciencedirect.com/science/article/pii/S1568269208800034

Knight, & Lacey. (1843). History of the dredging machine. *Mechanics' Magazine and Journal of Science, Arts, and Manufactures*, 39, 307–311.

Langston, W.J., Pope, N.D., Davey, M., Langston, K. M., O' Hara, S.C.M., Gibbs, P.E., & Pascoe, P.L. (2015). Recovery from TBT pollution in English Channel environments: a problem solved? *Marine Pollution Bulletin*, 95(2), 551–564. https://doi.org/10.1016/j.marpolbul.2014.12.011

Lee, S.E., Chung, J.W., Won, H.S., Lee, D.S., & Lee, Y.-W. (2012). Removal of methylmercury and tributyltin (TBT) using marine microorganisms. *Bulletin of Environmental Contamination and Toxicology, 88*(2), 239–244. https://doi.org/10.1007/s00128-011-0501-y

Manap, N., & Voulvoulis, N. (2014). Risk-based decision-making framework for the selection of sediment dredging option. *The Science of the Total Environment, 49*(6), 607–623. https://doi.org/10.1016/j.scitotenv.2014.07.009

Manap, N., & Voulvoulis, N. (2015). Environmental management for dredging sediments - the requirement of developing nations. *Journal of Environmental Management, 147*, 338–348. https://doi.org/10.1016/j.jenvman.2014.09.024

Martins, M., Costa, P.M., Raimundo, J., Vale, C., Ferreira, A.M., & Costa, M.H. (2012). Impact of remobilized contaminants in *Mytilus edulis* during dredging operations in a harbour area: bioaccumulation and biomarker responses. *Ecotoxicology and Environmental Safety, 85*, 96–103. https://doi.org/10.1016/j.ecoenv.2012.08.008

McAnally, W.H., Kirby, R., Hodge, S.H., Welp, T L., Greiser, N., Shrestha, P., ... & Turnipseed, P. (2016). Nautical depth for u.s. navigable waterways: a review. *Journal of Waterway, Port, Coastal, and Ocean Engineering, 142*(2), 04015014. https://doi.org/10.1061/(ASCE)WW.1943-5460.0000301

McAnally, W.H., Friedrichs, C., Hamilton, D., Hayter, E., Shrestha, P., Rodriguez, H., ... Kirby, R. (2007). Management of fluid mud in estuaries, bays, and lakes. i: present state of understanding on character and behavior. *Journal of Hydraulic Engineering, 133*(1), 9–22. https://doi.org/10.1061/(ASCE)0733-9429(2007)133:1(9)

Munawar, M., Norwood, W.P., McCarthy, L.H., & Mayfield, C.I. (1989). In situ bioassessment of dredging and disposal activities in a contaminated ecosystem: Toronto Harbour. In *Environmental Bioassay Techniques and their Application* (pp. 601–618). Springer, Dordrecht. Retrieved from https://link.springer.com/chapter/10.1007/978-94-009-1896-2_62

Murthy, R.K., Cabral, L., Vidya, R., & Dubey, S.K. (2007). Isolation and biological characterization of a tributyltin chloride degrading marine bacterium, *Vibrio sp* from Bombay High Oil Field, India. *Current Science, 93*, 1073–1074.

Nasr, S.M., Okbah, M., & Kasem, S.M. (2006). Environmental assessment of heavy metal pollution in bottom sediments of Aden Port, Yemen. *International Journal of Oceans and Oceanography, 1*, 99–109.

Novak, J., & Trapp, S. (2005). Growth of plants on TBT-contaminated harbour sludge and effect on TBT removal. *Environmental Science and Pollution Research International*, *12*(6), 332–341.

OSPAR. (2004). Convention on the Protection of the Marine Environment of the North-East Atlantic: Revised Guidelines for the Management of Dredged Material.

Palermo, M., Schroeder, P., Estes, T., Francingues, N., Gustavson, K., Bridges, T., & Ells, S. (2008). USACE Technical Guidelines for Environmental Dredging of Contaminated Sediments. Scribd.

Pang, Q.X., Han, P.P., Zhang, R.B., & Wen, C.P. (2018). delaying effect of extracellular polymer substances on fluid mud consolidation and application for nautical depth. *Journal of Waterway, Port, Coastal, and Ocean Engineering*, *144*(3), 04018001. https://doi.org/10.1061/(ASCE)WW.1943-5460.0000441

Ponti, M., Pasteris, A., Guerra, R., & Abbiati, M. (2009). Impacts of maintenance channel dredging in a northern Adriatic coastal lagoon. II: effects on macrobenthic assemblages in channels and ponds. *Estuarine, Coastal and Shelf Science*, *85*(1), 143–150. https://doi.org/10.1016/j.ecss.2009.06.027

Pourabadehei, M., & Mulligan, C.N. (2016). Resuspension of sediment, a new approach for remediation of contaminated sediment. *Environmental Pollution (Barking, Essex: 1987)*, *213*, 63–75. https://doi.org/10.1016/j.envpol.2016.01.082

Prexl, E. (2002). Prolonged experience of natural TBT degradation in harbour sediments in storage fields. Unpublished Report to Emden Port Authority (in German).

Roberts, D.A. (2012). Causes and ecological effects of resuspended contaminated sediments (RCS) in marine environments. *Environment International*, *40*, 230–243. https://doi.org/10.1016/j.envint.2011.11.013

Sakultantimetha, A., Keenan, H.E., Beattie, T.K., Aspray, T.J., Bangkedphol, S., & Songsasen, A. (2010). Acceleration of tributyltin biodegradation by sediment microorganisms under optimized environmental conditions. *International Biodeterioration & Biodegradation*, *64*(6), 467–473. https://doi.org/10.1016/j.ibiod.2010.05.007

Sakultantimetha, A., Keenan, H.E., Dyer, M., Beattie, T.K., Bangkedphol, S., & Songsasen, A. (2009). Isolation of tributyltin-degrading bacteria *Citrobacter braakii* and *Enterobacter cloacae* from butyltin-polluted sediment. *Journal of ASTM International*, *6*(6), 1–6. https://doi.org/10.1520/JAI102120

Sany, S.B.T., Salleh, A., Rezayi, M., Saadati, N., Narimany, L., & Tehrani, G.M. (2013). Distribution and contamination of heavy metal in the coastal sediments of Port Klang, Selangor, Malaysia. *Water, Air, & Soil Pollution*, *224*(4), 1476. https://doi.org/10.1007/s11270-013-1476-6

Sonak, S., Pangam, P., Giriyan, A., & Hawaldar, K. (2009). Implications of the ban on organotins for protection of global coastal and marine ecology. *Journal of Environmental Management*, *90*(S1), S96–108. https://doi.org/10.1016/j.jenvman.2008.08.017

Suehiro, F., Kobayashi, T., Nonaka, L., Tuyen, B.C., & Suzuki, S. (2006). Degradation of tributyltin in microcosm using Mekong River sediment. *Microbial Ecology*, *52*(1), 19–25. https://doi.org/10.1007/s00248-006-9079-z

Tessier, E., Amouroux, D., Morin, A., Christian, L., Thybaud, E., Vindimian, E., & Donard, O.F.X. (2007). (Tri)Butyltin biotic degradation rates and pathways in different compartments of a freshwater model ecosystem. *Science of The Total Environment*, *388*(1), 214–233. https://doi.org/10.1016/j.scitotenv.2007.08.047

United States. (1999). Introduction to contaminated sediments. Washington, DC: US Environmental Protection Agency, Office of Science and Technology. Retrieved from https://catalog.hathitrust.org/Record/007405996

Welp, T.L., & Tubman, M.W. (2017). Present Practice of Using Nautical Depth to Manage Navigation Channels in the Presence of Fluid Mud. US Army Corps of Engineers Vicksburg United States, US Army Corps of Engineers Vicksburg United States. Retrieved from www.dtic.mil/docs/citations/AD1037438

Winger, P.V., Lasier, P.J., White, D.H., & Seginak, J.T. (2000). Effects of contaminants in dredge material from the lower Savannah River. *Archives of Environmental Contamination and Toxicology*, *38*(1), 128–136.

Wu, J., Yang, L., Zhong, F., & Cheng, S. (2014). A field study on phytoremediation of dredged sediment contaminated by heavy metals and nutrients: the impacts of sediment aeration. *Environmental Science and Pollution Research International*, *21*(23), 13452–13460. https://doi.org/10.1007/s11356-014-3275-z

Wurpts, R.W. (2005). 15 years experience with fluid mud: definition of the nautical bottom with rheological parameters. *Terra et Aqua*, 22–32.

17 The human geographies of coastal sustainability transitions

Stephen Axon

1 Introduction: towards coastal resilience and sustainability

The coastal zone is rapidly changing as a result of changing environmental conditions, human development pressures and the governance of sustainability transitions. Yet the coastal zone also represents the frontline in addressing climate change and resilience building. Given the increasing pressures on coastal ecosystems and communities, there is a need for innovative low-carbon developments that address the series of multi-faceted coastal issues including economic growth, human development, climate change, energy generation, environmental management and tourism development. In many societies, the coastline is a place of significance and meaning, and is a site of leisure, tourism, health and well-being. Furthermore, many coastal communities depend on the coast for economic growth and development, energy generation, food production, trade and tourism. Consequently, the coast itself is something that communities, cultures and economies identify strongly with. Furthermore, coastal communities can also be conceptualised as marginal places: economically, politically, socially and culturally. For example, coastal areas are marginal given their distinctive landscapes, distance from centres of economic activity and socio-economic characteristics of local residents and businesses. Given these marginal characteristics, coastal areas present distinct problems particularly for the development, and governance, of sustainability transitions and low-carbon developments.

As a result, the normal practice of establishing sustainability innovations is exacerbated in coastal areas given considerations of place, identity and marginality. To address these issues, a clear, wide-ranging and effective stakeholder and public engagement approach should target such considerations and contextualise coastal sustainability transitions as part of meeting the needs of communities. Yet there is often little public engagement with sustainability transitions and other low-carbon innovations overall, evidenced by the little attention paid to identifying public perspectives and meaningfully involving communities. The results from such top-down decision-making are clear. Limited social acceptability often results in protest and little understanding of how low-carbon developments benefit individual members of the public, the community, the local economy or the environment. With limited understanding and acceptability, coastal communities cannot be invested, or participate, in sustainability-related developments. This therefore results in a significant implication: the division between the principles and practices of sustainability. What is needed now is a more enriched understanding of the human geographies of coastal sustainability transitions to address this disconnect.

This chapter, with reference to two case studies (one from the UK and one from the US), therefore serves to bridge this gap in understanding between developing coastal sustainability transitions in theory and in practice. Given the diverse nature of the field of sustainability transitions, there has been little research into how key concepts from the literature can be applied to coastal zones in practice. Historically, social science research has also neglected to draw attention to coastal and marine issues. In recent years, there has been a scholarly turn towards the ocean and coast (Connery, 2006), with greater attention paid towards issues related to placefulness, placelessness, identity and relationships with coastscapes and seascapes. Furthermore, given the challenges of over-development, climate change and natural habitat destruction that will increasingly influence coastal communities and ecosystems, there is surprisingly little research that explores how the particular characteristics of these areas shape coastal sustainability transitions. Therefore, the knowledge base and understandings around effective implementation approaches of coastal sustainability transitions remains rather scarce. Clearly identifying effective approaches to tackling barriers to implementing coastal sustainability transitions fits within the overall call for more societally relevant research and action on addressing vulnerability and adaptation to climate change (Moser, 2010).

This chapter is structured as follows. Firstly, the challenges facing the coastal zone are outlined, followed by a description of how sustainability transitions can address these challenges. Secondly, issues concerning identity and marginality are illustrated and I discuss how a better understanding of these concepts can address a deficit of knowledge relating to the specific characteristics of coastal communities. This is followed by an overview of public engagement, its importance and effective approaches to meaningfully involve the public with coastal sustainability transitions. Thirdly, this chapter presents two case studies, Guernsey, British Channel Islands and Cape Wind, Massachusetts. These illustrate how concepts of identity, marginality and engagement are essential to the development of coastal sustainability transitions. The chapter then concludes with implications arising from the case studies and suggests that to successfully, and effectively, implement coastal sustainability transitions an understanding of the specific characteristics of local coastal places and tailored public engagement approaches are essential.

2 Addressing coastal zone challenges: towards coastal sustainability transitions

2.1 Coastal zone challenges

The coastal zone is home to around 2.6 billion people, over 40% of the world's population (Barbier *et al.*, 2011; Sale *et al.*, 2014). Given the trajectory for global population increase, this number is expected to double over the next 30 years. This places a substantial emphasis on coastal areas and the resources needed to sustain an expanding population that is increasingly urbanised along the coast. The value of ecosystem services from the coastal zone is estimated to be more than one-third of the global total (Barbier *et al.*, 2011; Barragan & de Andres, 2015). However, coastal ecosystems are increasingly at risk as a result of intensive resource use, increasing population growth in coastal areas and substantive coastal development (Duxbury & Dickinson, 2007). While it is well understood that coastal ecosystems provide valuable services – such as carbon sequestration, control of coastal erosion and recreation activities – changes to marine conditions, if continued to be poorly managed, will have substantial impacts on the economic activities of coastal communities (Bradley *et al.*, 2015).

The exponential demand for space and resources required by ever-increasing population growth, coupled with poor coastal management such as over-development and habitat destruction, coastal areas are project to become particularly and increasingly susceptible to the impacts of climate change (Lloyd *et al.*, 2013; Sale *et al.*, 2014). Furthermore, coastal areas are becoming more vulnerable to storm surges as a result of sea level changes (Wang *et al.*, 2008; Androulidakis *et al.*, 2015). The development of large coastal cities, their high population densities and associated economic activities is leading to declining coastal natural habitats and resources being over-exploited (Mee, 2012; Mavrommati *et al.*, 2013). What these coastal zone challenges illustrate is that changes in marine conditions have the potential to substantially alter major socio-economic activities. Increasing over-development and human activity in coastal areas combined with the impacts of climate change are likely to make economic growth in coastal areas challenging, if not increasingly impossible to sustain, should a radical transition towards coastal resilience and sustainability not be undertaken (Day *et al.*, 2014). To address these challenges, long-term solutions are required to develop coastal resilience and sustainability transitions.

2.2 Sustainability transitions in the coastal zone

The concept of 'transition' has become increasingly central to futures-orientated thinking (Feola & Nunes, 2014). This body of literature argues that deeply embedded socio-ecological problems such as global environmental change urgently require radical and inclusive solutions with long-term trajectories (Geels, 2012). The sustainability transitions literature has highlighted the need for multi-dimensional shifts, essential for new models of sustainable production and consumption (Geels, 2002; Foxon *et al.*, 2010; Markard *et al.*,

2012). The application of this socio-technical approach illustrates complex co-evolutionary interactions between policy, markets, industry, technology, civil society and culture (Geels, 2012). These interactions are framed within a systems perspective and exemplified as being part of the multi-level perspective (MLP) nested hierarchy that illustrates how transitions become accepted to replace the dominant unsustainable regime (Geels, 2002; Geels & Schot, 2007; Seyfang & Smith, 2007; Foxon *et al.*, 2010). The MLP suggests that transitions result through interactions between processes at three levels: niche-innovation level; socio-technical regime; and landscape level. Firstly, niche-innovations afford space for new ideas to be tested and developed. Secondly, changes at the landscape level create pressure on the regime. Thirdly, destabilisation of the regime creates windows of opportunity for niche innovations to emerge. Geels & Schot (2007) suggest that the alignment of these processes enables the breakthrough of novelties in mainstream markets where they compete with the existing regime.

These 'niches of innovation' are ideas, technologies, practices and organisational models that have the potential for wider social transformation should they be suitable for wider uptake and diffusion (Geels, 2002; Geels & Schot, 2007; Seyfang & Smith, 2007; Seyfang, 2010). Radical technologies and practices break out of the niche level when a 'window of opportunity' is created that allows such innovations to become integral to the regime (Geels & Schot, 2007). However, challenges exist to scaling-up innovative technologies and practices such as uptake, societal support and investment. The challenge for coastal sustainability transitions is that coastal sustainability projects are often developed, and participated in, by relatively small networks of dedicated actors (Alexander *et al.*, 2007; Geels & Schot, 2007). While this ensures that sustainability in coastal areas is acted upon, gathering wider support can be challenging within broader discourses framed around unsustainable modes of production and consumption. While the sustainability transitions literature has identified a number of important lessons about the actors, networks and diffusion of niche innovations, this research predominately analyses past energy transitions through an historical analysis approach (Fouquet & Pearson, 2012; Grubler, 2012). To date, limited research has studied these processes in coastal areas and how coastal communities are engaged with sustainability transitions. There is a clear knowledge deficit on the particular, situated and place-based characteristics of coastal communities (e.g., identity and marginality), which might influence sustainability projects, and a lack of understanding of how coastal communities are engaged with sustainability transitions.

3 Understanding coastal sustainability transitions: the role of identity, marginality and public engagement

3.1 Identity and marginality

Coastal areas are often conceptualised as marginal in a number of ways. Firstly, coastal areas are frequently marginal in economic terms. The distinctive landscapes of the coastal zone (such as cliffs, strong tidal currents, eroding coastlines or frequent flooding) often put constraints on the types of economic activity that can take place there. Additionally, coastal areas suffer from poor connectivity to other economic centres and to major transport links and as a result suffer from poor accessibility (Rickey & Houghton, 2009). Traditionally, coastal areas have been the centre for limited and small-scale economic activities such as shipbuilding and fishing. This changed dramatically with the emergence of trade and tourism. However, tourism itself is highly seasonal and does not support year-round income or employment (Dwyer *et al.*, 2003). Secondly, the economically marginal nature of the coast means that coastal zones have distinct demographic and social characteristics. Coastal areas in the UK, for example, are characterised by high rates of inward migration of older people for retirement purposes (Office for National Statistics, 2014). At the same time, there is a high rate of outward migration of younger people due to limited local opportunities for employment, with their place being taken over by migrant workers in recent years. Consequently, lower income levels and high levels of unemployment result in many coastal areas being associated with higher than average levels of social deprivation and exclusion (Ward, 2015).

Thirdly, coastal areas are frequently marginal in political terms. They represent the edges or borders of countries that are traditionally of lower concern to policy-makers in urban 'centres'. Given the lower levels of economic activity in coastal areas, they have not been a priority for policy-makers. Additionally, government policies frequently focus on larger population centres (particularly industrial cities), tending to neglect coastal communities. Fourthly, many coastal areas are marginal in cultural terms. That is to say, in terms of the meanings attached to them and associated behaviours. With reference to the concept of liminality – referring to times and spaces that intermediate – coastal areas can represent states of transition or 'in-between' spaces. The beach is an appropriate example of this liminality, indicating an indeterminate transitional zone between land and sea that is often associated with escape and freedom (Shields, 1991). These feelings are often temporary and, as such, visitors to the coast often take advantage of the opportunity to invert everyday rules and routines, e.g., behaving in ways that would otherwise be less acceptable at home (Webb, 2005; Clisby, 2009).

3.2 Public engagement with coastal sustainability transitions

Long-term sustainability requires gradual, and radical, readjustments to science, technology, policy, industry, markets and civil society. Consequently, there is a need for substantial engagement with sustainability transitions processes amongst stakeholders and publics. As such, there is a need for acceptance of, and engagement with, sustainability strategies and projects to progress towards a low-carbon sustainable future. The well-established notion of 'having people on board' is not a new concept (Arnstein, 1969; Wolf & Moser, 2010; Axon, 2016). However, the majority of public participation processes revolve around consultation events rather than meaningful and inclusive citizen involvement (Morrison & Dearden, 2013). This completely fails to connect theory with practice. While there may be a deficit in understanding relating to how to engage the public with (coastal) sustainability transitions, there is an emerging body of literature attempting to address this gap. This literature has much to offer academics, policy-makers and practitioners; specifically in two key areas: 1) identifying how individuals engage with sustainability-related projects, and 2) how projects can improve citizen participation (Peters *et al.*, 2012; Whitmarsh *et al.*, 2013; Axon, 2016). The outcomes of this research present a synergy between what individuals want to do to live sustainably and how they want to achieve this with various interventions, innovations technologies and practices. That said, there has been very little research to date that has considered these issues in the specific context of coastal areas.

Let's start with the first lesson of successful sustainability initiatives: public engagement is integral to the success of projects. Individuals and communities have been identified as contributing to numerous environmental challenges, yet they are also essential to implement solutions to such challenges. Consequently, engaging the public with sustainability transitions is an imperative rather than an option (Wolf & Moser, 2011). This is especially important in coastal areas given that substantial leisure, tourism, energy generation, transport, food production and consumption practices are located in these environmentally sensitive areas of economic importance (Day *et al.*, 2014). To engage the public with coastal sustainability transitions, individuals and communities need to be aware of projects, their aims and activities; feel excited and interested in them; and be able to participate in their operation and success. Therefore, engagement has three key components: cognitive (knowledge and understanding); affective (emotions, interests and concerns); and behavioural (actions) (Axon, 2016). This implies that individuals need to care about sustainability, be motivated by it and be able to take action rather than simply knowing about the issues. To be clear, although 'engagement' is an often-cited and ambiguous term, for the purposes of our understanding it is defined as a "personal state of connection" in contrast to "engagement solely as a process of public participation in policy making" (Wolf & Moser, 2011, p. 550). To be at all successful and meaningful to local communities, successful engagement needs to acknowledge specific issues of economic and social marginality in coastal areas, and respond to these issues within their projects.

Lesson number two: public engagements on sustainability issues are not fixed; rather they are subject to temporal variation and are affected by particular changes and contextual issues. They need to be facilitated in the short-term and sustained in the long-term. This brings us to the largest challenge faced by all sustainability projects: how to effectively turn initial excitement in a project to sustained engagement (Alexander *et al.*, 2007; Peters *et al.*, 2013; Axon, 2016). Coastal projects, however, can use existing infrastructures, resources and facilities as well as new interventions to improve understandings, increase interest, and illustrate how to participate in such projects. Common interventions used to improve knowledge of, and participation in, sustainability strategies often employ information raising, which may lead to active involvement in only very limited terms, if at all (Abrahamse *et al.*, 2005; Verplanken & Roy, 2016; Axon, 2017). By contrast, more involved and more considered strategies may produce better outcomes. For instance, leisure activities, particularly water-based sports and activities, can be used to increase interest and attract additional revenue that could address issues of economic marginality and contribute to the 'blue economy', specifically 'blue tourism' (Ehlers, 2016), while raising awareness of natural habitats and biodiversity. Recruiting local residents to decision-making panels can also support the extent to which local communities are 'invested', rather than simply being consulted, with coastal sustainability transition projects, such as local renewable energy initiatives. These examples illustrate how to engage diverse publics cognitively, affectively and behaviourally.

Lesson number three: 'tick-box' consultation approaches and other forms of tokenistic involvement with local communities is *not* true engagement. While technical decisions and financial arrangements may not directly involve stakeholders and the public, implementing projects in a top-down manner risks marginalising communities. While coastal areas have become increasingly central to sustainable energy transitions as well as being of greater economic importance, decision-making on coastal projects continues to be dominated by top-down approaches. Public engagement in these projects reflects little more than being involved in consultative events. For example, stakeholder engagement as part of Navitus Bay Wind Farm project proposal applied public consultation events, and failed to allay public concerns about the installation of an offshore wind farm, which led to mistrust amongst local residents and failure of the project (Axon *et al.*, 2017). Such tokenistic involvement marginalises the values and needs of stakeholders and users of coastal areas (e.g., residents and tourists, etc.). This raises questions not only about the level of engagement, but also its quality. Individuals can be actively or passively engaged in sustainability projects (Alexander *et al.*, 2007; Axon, 2016). Active engagement reflects substantive cognitive, affective and behavioural engagement. This includes whether people understand the project and its aims and develop an emotional attachment to it. Active behavioural engagement with coastal sustainability transitions will determine whether people follow, and respond to, initiatives (Rogers *et al.*, 2008; Peters *et al.*, 2013; Axon, 2016). Passive engagement illustrates that people have limited understanding of the schemes implemented beyond identifying their existence, have limited emotional connection, and do not contribute towards the project or are not influenced by it, often seeking to remove themselves from its influence. Examples include mass media campaigns such as the "Are you doing your bit" project that result in limited public engagement (Hinchliffe, 1996; Ockwell *et al.*, 2009). With best-practice inclusive and creative approaches, coastal communities should not be faced with tokenistic involvement in sustainability strategies; but rather be engaged as citizens that co-produce, co-govern and co-deliver coastal resilience and sustainability.

4 Practical implications with implementing coastal sustainability transitions

4.1 Case study 1: Guernsey, British Channel Islands

Guernsey is one of the Channel Islands – a group of British islands near France – located around 115 km south of England. The 64.5 km^2 island is home to 63,000 people, with the largest settlements and populations being the capital, St. Peter Port, and in the northeastern part of the island, whereas the western and southern parts of the island are much less densely populated. The jurisdiction of Guernsey also includes several smaller islands including Herm (popular for day trips) and Lihou (an unpopulated nature reserve). Guernsey itself is not part of the United Kingdom, but rather a Crown Dependency (similar to the Isle of Man and Jersey). These are not recognised internationally as sovereign states, but as self-governing dependencies of the British Crown (responsible for their own administrative, legal and fiscal systems). Despite this, the culture of Guernsey can be characterised as being British in terms of language, entertainment and cuisine. Guernsey's energy system can be characterised as very independent. Wiserma (2016) states that given that there is only one electricity company (owned by the States of Guernsey), one single oil-fuelled power station, and one state-owned oil supply ship. Guernsey has no wind turbines and very few solar panels; which is unsurprising given the lack of investment in renewable energy subsidies. Therefore, renewable energy currently plays only a marginal role in Guernsey's energy portfolio – despite substantial resource availability for tidal, wave, wind and solar energy – and there are currently no specific proposals for offshore renewable energy developments under consideration (Wiserma, 2016).

For his Ph.D., Bouke Wiserma explored how representations of place, specific characteristics of the coast and how offshore renewable energy technologies are accepted to 'fit' within coastscapes with reference to the development of hypothetical offshore wind, tidal and wave projects around the coast of Guernsey. Through an auto-photographic approach, residents of Guernsey strongly represent the coastal environment as a 'place' to be used and explored; in opposition to this, coast was viewed in an environmental sense as a 'landscape' which is predominately valued in a visual way. This difference in terminology is important. 'Landscapes' are inherently visual, they are 'looked at' and appreciated whereas 'places' are "very much things to be inside of" (Cresswell, 2004, p. 10), indicating usability and meaning in an otherwise visually orientated narrative. As a result, representations of the coast and sea of Guernsey were underpinned by themes relating to local distinctiveness, natural beauty, wildlife, utilising the coast, leisure activities, and a place to spend time with family and friends (Wiserma, 2016). Consequently, Guernsey's coasts were represented in meaningful ways, with each coastline ascribed different identities. For example, the west coast was represented as a sociable place for leisure and sunsets; the south coast portrayed as a place for exploration and quietness, and the north coast as a less-loved industrial place fundamentally at odds with the rest of the island (Wiserma, 2016). While the existing energy infrastructure in Guernsey was considered to be over-expensive and dependent on outside imports, offshore renewable energy acceptance varied by technology. Wind energy was portrayed as obtrusive, which did not fit necessarily fit the image and identity of Guernsey's coast, while wave and tidal energy were positively framed and discussed during interviews. These findings illustrated that many Guernsey residents believed that wind energy would disrupt the visual, natural beauty of the island, whereas tidal and wave energy were not considered to be visually 'out of place' but also exploiting the significant tidal energy resource and move towards a more self-sufficient energy system (Wiserma, 2016).

These issues were further explored by Wiserma (2016) in engaging residents with siting hypothetical offshore renewable energy technologies through employing the notion of 'place-technology fit' to represent specific areas of Guernsey's coast and sea as (un)acceptable for the development of coastal sustainability transitions. With reference to developing offshore wind-energy projects, there was opposition to development in areas where visual narratives of the coast featured enjoyment of views and sunsets. This was framed in collective terms, arguing in favour of siting offshore wind energy in places used less by the wider community. Specifically the north coast was identified as an industrial and disliked area. Applying a deliberative focus group methodology for public engagement, Wiserma (2016) investigated preferences for siting (hypothetical) local energy options. The first option included 10–15 wind turbines that would provide 25–30% of Guernsey's electricity demand which would be developed by 2020, while the second option was a relatively large-scale tidal energy project of between 20 and 100 turbines, unlikely to happen by 2025 and with projected high cost. Provided with these two options, residents identified offshore renewable energy projects that 'fit' with place-related narratives and the identities ascribed to particular coastlines. As such, support was conditional upon local offshore renewable energy projects being locally owned; predominantly for the benefit of Guernsey; not resulting in substantial increases in energy prices; maintaining or enhancing Guernsey's distinctiveness; and being sited in the 'right' place (Wiserma, 2016).

It was found that residents indicated the 'right place' for siting offshore wind projects was along the north(west) coastline while the east, south and west coasts were identified as unacceptable places. Knapp & Ladenburg (2015) state that people often favour offshore wind to be placed at the greatest possible difference from the shore, yet Wiserma (2016) found that Guernsey residents identified their preference for offshore wind energy with locations relatively close to the shore (within 6 nautical miles of the coast). For tidal energy, Guernsey residents identified that the north and southwest coast were acceptable sites while the east, south and west coasts were suggested as being unacceptable. The deliberative engagement process applied clearly identifies the importance of place identity, particularly the identities that residents ascribe to coastal areas and the meanings underpinning them. This explains why residents found particular areas to be unacceptable and used strong 'place-protective' arguments (Devine-Wright, 2009). Findings suggest that individuals feel more strongly about places that are unacceptable for offshore renewable energy development more so than those that are relatively acceptable, emphasising affective engagements with renewable energy technologies and emotional values ascribed to the coast.

This research found Guernsey and its coast and sea to be meaningful to local residents in numerous ways and at different scales – with a coast that is valued for its quietness, tides, wildlife, leisure opportunities, space for exploration and natural beauty (Wiserma, 2016). Public understandings of tidal and wave energy as a local energy option were highly diverse, and as a result some (although not all) offshore renewable energy options were considered to 'fit' specific place-related meanings. This demonstrates how Guernsey residents view the least intrusive energy option that corresponds with their representation vision of the coast. Specific renewable energy projects were also found to enhance, or disrupt, Guernsey's local distinctiveness. For example, tidal energy projects were represented as enhancing this distinctiveness while offshore wind energy was instead portrayed as making Guernsey more like everywhere else (Wiserma, 2016). Acceptance of renewable energy options was dependent upon the selected site of the project, the extent of the technology chosen, and how the project is interpreted relationally within a context of wider energy systems and policies (Wiserma, 2016).

4.2 Case study 2: Cape Wind, Massachusetts, USA

Offshore wind energy in America is virtually synonymous with the Cape Wind Project. Despite the numerous successes of developing thousands of offshore wind turbines across Europe and the rapid expansion of onshore wind energy in America, the United States has still to develop extensive offshore wind farms. Within the United States, offshore wind energy is a controversial subject with many Americans' perspectives influenced by the Cape Wind Project (Huffington Post, 2013). The Cape Wind Project is an approved offshore wind farm on Horseshoe Shoal in Nantucket Sound off Cape Cod, Massachusetts. The offshore wind farm, proposed by private developer Cape Wind Associates, is estimated to generate 1,500 GW hour of electricity each year. The first proposal for Cape Wind (in 2001) involved building 170 wind turbines in Nantucket Sound – although this was later changed to 130. The project, if completed, was estimated to become one of the largest offshore wind farms in the world and would significantly contribute to reducing 730,000 tonnes of greenhouse gas emissions per year (the equivalent of taking 175,000 cars off the road each year) (Kimmell & Stalenhoef, 2011). The $2.6 billion project was planned with 440-ft-tall intended turbines and projected electricity generation was to be transmitted to the mainland of Cape Cord via cables buried beneath the seabed and would generate 468MW of power, supplying 75% of the electricity needs of Cape Cod, Nantucket Island and Martha's Vineyard – or roughly 200,000 homes (Kimmell & Stalenhoef, 2011).

Cape Wind has spent most of the years since its initial consideration in planning struggles and litigation. Despite the Nantucket Sound site being ideal in wind resource availability, opponents to the development of the wind farm outline that the area is important for recreational boating, commercial fishing and air and ferry traffic (Boston Globe, 2016). The wind speeds in Nantucket Sound are high, averaging 19.75 miles per hour, which is considered "outstanding" from a technical perspective, and the wind blows strongest in Nantucket Sound at times of peak energy demand (Kimmell & Stalenhoef, 2011). Additionally, the proposed site is 5 miles from Cape Cod and 9 miles from Martha's Vineyard. Despite the (technical) advantages of this site, opposition to the project has been fierce. A non-profit entity named the Alliance to Protect Nantucket Sound, funded by Bill Koch and others, spent over $15 million in ten years (between 2001 and 2011) to oppose the project in numerous administrative venues and filed ten different lawsuits to prevent the project from succeeding (Kimmell & Stalenhoef, 2011). Among one of the many legal challenges the project has faced was a lawsuit by struggling fishermen from Martha's Vineyard who argued that the massive wind farm threatened their livelihood, although the lawsuit was withdrawn and Cape Wind offered the complainants an undisclosed settlement (Boston Globe, 2015). As of 2016, Cape Wind has defended against more than 20 lawsuits, and it has won almost all of them, with several still pending (Boston Globe, 2016).

Additionally, 59% of respondents to a Cape Cod Times online poll in January 2015 suggested they were happy that the Cape Wind Project would fail (Boston Globe, 2015). The Alliance to Protect Nantucket Sound stated that the development of Cape Wind directly conflicted with the fishing industry, recreational boating, air and ferry traffic as well as endangering a fragile habitat (Huffington Post, 2013). Furthermore, the Alliance argued that the vistas of Nantucket Sound are worthy of preservation as other symbolic landmarks that defined the coastal towns as "tranquil" (Huffington Post, 2013). The notion that the Cape Wind Project would directly compromise such industries and vistas specifically relates to place identity, the meanings ascribed to coastscapes and the economic marginality of coastal areas. Moreover, while developers promised cheap, clean energy, it was revealed that 77.5% of the power would be sold to Northeast Utilities and National Grid for some two times the average cost of power generated by US suppliers, and the contracted price of 18.7% per kilowatt hour was estimated to rise 3.5% every year of the 15-year contract (Boston

Globe, 2015). To address arguments of loss of jobs, the developers stated that good manufacturing jobs would result. However, the commissioning contract to buy the turbines and offshore transformer went in favour of German company Siemens (Boston Globe, 2015). Despite arguments of economic marginality being central to opponents of the project, ultimately the developers failed to take this into consideration and further marginalised local residents.

In January 2015, National Grid and Northeast Utilities terminated their power purchase agreements making it more difficult for the project to obtain financial support. The future of the Cape Wind Project is consequently in doubt (Boston Globe, 2016). If the project is not constructed and ultimately fails – either because of its large up-front cost or because of the aesthetic concerns of tenacious beachfront property-owners who oppose the project – the United States' commitment to doing its part to address climate change will be further questioned (Kimmell & Stalenhoef, 2011)[1]. Rather than having the mantle of being the first offshore wind farm in the United States – this accolade belongs to the Block Island Wind Farm, Rhode Island commissioned in 2016 (Klain *et al.*, 2017) – the Cape Wind Project has demonstrated that the opposition to coastal sustainability transitions (as well as energy transitions overall) is real and present. Yet the legacy of Cape Wind is far more wide-ranging, even if the project fails. The project highlights issues of siting offshore wind energy facilities in areas that are used for recreation activities and fishing, within a popular tourist area (Kimmell & Stalenhoef, 2011). This demonstrates that considerations of siting local offshore energy projects should take into account the diverse activities and uses of the coast. Part of the argument against the siting of Cape Wind in Nantucket Sound was predicated on notions of economic marginality and identity, and that if successful the project would jeopardise the local tourist and fishing industries. By exploring preferences for siting coastal sustainability-related projects through deliberative democracy and public engagement exercises, as illustrated in the case study in Guernsey by Wiserma (2016), opportunities for allowing local residents to choose acceptable sites for implementing coastal sustainability transitions can be identified. Without doing so, there is a risk that substantial opposition will form against such projects.

5 Concluding discussion: towards human geographies of coastal sustainability transitions

While there has been considerable debate about sustainability transitions, much research to date has neglected coastal areas. Additionally, there appears to be little recognition of specific characteristics of coastal areas that can play a significant role in shaping the responses to, and engagement with, coastal sustainability transitions-related projects. With reference to the case studies of Guernsey and Cape Wind, it is clear that a deeper understanding of the human geographies of coastal sustainability transitions is needed should such projects be successful and effective. The characteristics of this human geography of coastal sustainability transitions starts with understandings of place identity, marginality and public engagement. Given that the literature exploring local energy acceptability has predominantly focused on understanding local opposition to single (wind) energy projects, it has relatively little to say about the construction of support for such projects (Wiserma, 2016). Furthermore, focusing on oppositional responses purely as a quantitative exercise also fails to understand the importance of the coast to different stakeholders and publics. It is clear from the case studies in this chapter that these elements constitute deeper understandings of responses to coastal sustainability transitions and the construction of support or opposition. Therefore, as a research agenda, the human geographies of coastal sustainability transitions can be of substantial value to the practical application of sustainability.

Given that engagement with sustainability is about a personal state of cognitive, affective and behavioural connection, for coastal communities this connection may need to be about much more than issues of sustainability. As the case study of Guernsey outlines, coastal sustainability transition-related public engagement should seek to enhance the local distinctiveness of coastal areas and 'fit within the wider narrative of place identity and how individuals feel attached to places that are ascribed with specific meanings'. Furthermore, as the case study of Cape Wind illustrates, coastal sustainability transitions need to connect with local communities in terms of addressing immediate social and economic concerns. This chapter demonstrates that effective public engagement and deliberative democracy principles are key to successful project delivery for the wider success of coastal sustainability transitions processes. It is essential to learn from both good, and bad, practices in implementing coastal sustainability transitions. There is as much to learn from the failures of sustainability projects as there is to learn from the successes. These understandings need to go hand-in-hand with a comprehensive strategy indicating how stakeholders and the public have been, and will be, meaningfully engaged with coastal sustainability transitions in the short-, medium-, and long-term.

Note

1 This positioning is already being questioned as a result of President Donald Trump's commitment to reinvigorate the coal industry and the withdrawal of the USA from the Paris Climate Change Agreement (Zhang *et al.*, 2017).

Bibliography

Abrahamse, W., Steg, L., Vlek, C., & Rothengatter, T. (2005). A review of intervention studies aimed at household energy conservation. *Journal of Environmental Psychology*, *25*, 273–291.

Alexander, R., Hope, M., & Degg, M. (2007). Mainstreaming sustainable development – a case study: Ashton Hayes is going carbon neutral. *Local Economy*, *22*, 62–74.

Androulidakis, Y.S., Kombiadou, K.D., Makris, C.V., Baltikas, V.N., & Krestenitis, Y.N. (2015). Storm surges in the Mediterranean Sea: variability and trends under future climatic conditions. *Dynamics of Atmospheres and Oceans*, *71*, 56–82.

Arnstein, S.R. (1969). A ladder of citizen participation. *Journal of the American Institute of Planners*, *35*, 216–224.

Axon, S. (2016). "The Good Life": engaging the public with community-based carbon reduction strategies. *Environmental Science and Policy, 66*, 82–92.

Axon, S. (2017). "Keeping the ball rolling": addressing the enablers for, and barriers to, sustainable lifestyles. *Journal of Environmental Psychology*, *52*, 11–25.

Axon, S., Chapman, A., & Light, D. (2017) Stakeholder engagement in coastal sustainability transitions: an emerging research agenda. *Regions*, *308*, 20–22.

Barbier, E.B., Hacker, S.D., Kennedy, C., Koch, E.W., Stier, A.C., & Silliman, B.R., (2011). The value of estuarine and coastal ecosystem services. *Ecological Monographs*, *81*, 169–193.

Barragán, J.M., & de Andrés, M. (2015). Analysis and trends of the world's coastal cities and agglomerations. *Ocean and Coastal Management, 114*, 11–20.

Boston Globe (2015) What really toppled Cape Wind's plans for Nantucket Sound. Retrieved 11 November 2017, from www.bostonglobe.com/magazine/2015/01/30/what-really-toppled-cape-wind-plans-for-nantucket-sound/mGJnw0PbCdfzZHtITxq1aN/story.html

Boston Globe (2016). Offshore wind may finally take off with big projects, none named Cape Wind. Retrieved 11 November 2017, from www.bostonglobe.com/magazine/2016/03/23/offshore-wind-may-finally-take-off-with-big-projects-none-named-cape-wind/FZ9Ng715HYkgKFFoNtlZHN/story.html

Bradley, M., van Putten, I., & Sheaves, M. (2015). The pace and progress of adaptation: marine climate change preparedness in Australia's coastal communities. *Marine Policy*, *53*, 13–20.

Clayton, S., Devine-Wright, P., Stern, P.C., Whitmarsh, L., Carrico, A., Steg, L., Swim, J., & Bonnes, M. (2015). Psychological research and global climate change. *Nature Climate Change*, *5*, 640–646.

Clisby, S. (2009). 'Summer sex: youth, desire and the carnivalesque at the English seaside'. In Donnan, H., & Magowan, F. (Eds.), *Transgressive Sex: Subversion and Control in Erotic Encounters*. Oxford: Berghahn, pp.47–68.

Connery, C. (2006). There was no more sea: the suppression of the ocean, from the bible to cyberspace, *Journal of Historical Geography*, *32*, 494–511.

Cresswell, T. (2004). *Place: A Short Introduction*. Oxford, UK: Blackwell.

Day, J.W., Moerschbaecher, M., Pimentel, D., Hall, C., & Yáñez-Arancibia, A. (2014). Sustainability and place: how emerging mega-trends of the 21st century will affect humans and nature at the landscape level. *Ecological Engineering, 65*, 33–48.

Devine-Wright, P. (2009). Rethinking NIMBYism: the role of place attachment and pace identity in explaining place-protective action. *Journal of Community and Applied Social Psychology*, *19*, 426–441.

Duxbury, J., & Dickinson, S. (2007). Principles for sustainable governance of the coastal zone: in the context of coastal disasters. *Ecological Economics, 63*, 319–330.

Dwyer, L., Forsyth, P., Spurr, R., & VanHo, T. (2003) Tourism's contribution to a state economy: a multi-regional equilibrium. *Tourism Economics*, *9*, 431–448.

Ehlers, P. (2016) Blue growth and ocean governance – how to balance the use and the protection of the seas. *WMU Journal of Maritime Affairs*, *15*, 187–203.

Feola, G., & Nunes, R. (2014). Success and failure of grassroots innovations for addressing climate change: the case of the transition movement. *Global Environmental Change*, *24*, 232–250.

Fouquet, R., & Pearson, P.J.G. (2012). Past and prospective energy transitions: insights from history. *Energy Policy*, *50*, 1–7.

Foxon, T.J., Hammond, G.P., & Pearson, P.J.G. (2010). Developing transition pathways for a low carbon electricity system in the UK. *Technological Forecasting and Social Change*, *77*, 1203–1213.

Geels, F.W. (2002). Technological transitions as evolutionary reconfiguration processes: a multi-level perspective and a case study. *Research Policy*, *31*, 1257–1274.

Geels, F.W. (2012). A socio-technical analysis of low-carbon transitions: introducing the multi-level perspective and a case study. *Journal of Transport Geography*, *24*, 471–482.

Geels, F.W., & Schot, J. (2007). Typology of socio-technical transition pathways. *Research Policy*, *36*, 399–417.

Grubler, A. (2012). Energy transitions research: insights and cautionary tales. *Energy Policy*, *50*, 8–18.

Hinchliffe, S. (1996). Helping the earth begins at home. *Global Environmental Change*, *6*, 53–62.

Huffington Post (2013). Cape Wind: Regulation, Litigation and the power struggle to develop offshore wind power in the US. Retrieved 11 November 2017, from www.huffingtonpost.co.uk/entry/cape-wind-regulation-liti_n_2736008

Kimmell, K., & Stalenhoef, D.S. (2011) The Cape Wind offshore wind energy project: a case study of the difficult transition to renewable energy. *Golden Gate University Environmental Law Journal*, *5*, 197–225.

Klain, S.C., Satterfield, T., MacDonald, S., Battista, N., & Chan, K.M.A. (2017). Will communities "open-up" to offshore wind? Lessons learned from New England islands in the United States. *Energy Research and Social Science*, 34, 13–26.

Knapp, L., & Ladenburg, J. (2015). How spatial relationships influence economic preferences for wind power – a review. *Energies*, *8*, 6177–6201.

Lloyd, M.G., Peel, D., & Duck, R.W. (2013). Towards a social-ecological resilience framework for coastal planning. *Land Use Policy*, *30*(1), 925–933. http://doi.org/10.1016/j.landusepol.2012.06.012

Markard, J., Raven, R., & Truffer, B. (2012). Sustainability transitions: an emerging field of research and its prospects. *Research Policy*, *41*, 323–340.

Mavrommati, G., Bithas, K., & Panayiotidis, P. (2013). Operationalizing sustainability in urban coastal systems: a system dynamics analysis. *Water Research*, *47*, 7235–7250.

Mee, L. (2012). Between the devil and the deep blue sea: the coastal zone in an era of globalisation. *Estuarine, Coastal and Shelf Science*, *96*, 1–8.

Morrison, C., & Dearden, A. (2013). Beyond tokenistic participation: using representational artefacts to enable meaningful public participation in health service design. *Health Policy, 112*, 179–186.

Moser, S.C. (2010). Now more than ever: the need for more societally-relevant research on vulnerability and adaptation to climate change. *Applied Geography*, *30*(4), 464–474.

Ockwell, D., Whitmarsh, L., & O'Neill, S. (2009) Reorienting climate change communication for effective mitigation: forcing people to be green or fostering grassroots engagement. *Science Communication*, *30*, 305–327.

Office for National Statistics (2014). *2011 Census: Coastal Communities.* Newport: Office for National Statistics.

Peters, M., Fudge, S., & Hoffman, S. (2013). The persistent challenge of encouraging public participation in the low-carbon transition. *Carbon Management, 4*, 373–375.

Peters, M., Fudge, S., Hoffman, S., & High-Pippert, A. (2012). Carbon management, local governance and community engagement. *Carbon Management, 3*, 357–368.

Rickey, B., & Houghton, J. (2009). Solving the riddle of the sands: regenerating England's seaside towns. *Journal of Urban Regeneration and Renewal, 3*(1), 47–55.

Rogers, J.C., Simmons, E.A., Convery, I., & Weatherall, A. (2008). Public perceptions of opportunities for community-based renewable energy projects. *Energy Policy, 36*, 4217–4226.

Sale, P.F., Agardy, T., Ainsworth, C.H., Feist, B.E., Bell, J.D., Christie, P., ... Sheppard, C.R.C. (2014). Transforming management of tropical coastal seas to cope with challenges of the 21st century. *Marine Pollution Bulletin, 85*(1), 8–23. http://doi.org/10.1016/j.marpolbul.2014.06.005

Seyfang, G. (2010). Community action for sustainable housing: building a low-carbon future. *Energy Policy, 38*, 7624–7633.

Seyfang, G., & Smith, A. (2007). Grassroots innovations for sustainable development: towards a new research and policy agenda. *Environmental Politics, 16*, 584–603.

Shields, R. (1991). *Places on the Margin: Alternative Geographies of Modernity*. London: Routledge.

Sutton-Grier, A.E., Wowk, K., & Bamford, H. (2015). Future of our coasts: the potential for natural and hybrid infrastructure to enhance the resilience of our coastal communities, economies and ecosystems. *Environmental Science & Policy, 51*, 137–148.

Verplanken, B., & Roy, D. (2016). Empowering interventions to promote sustainable lifestyles: testing the habit discontinuity hypothesis in a field experiment, *Journal of Environmental Psychology, 45*, 127–134.

Wang, S., McGrath, R., Hanafin, J., Lynch, P., Semmler, T., & Nolan, P. (2008). The impact of climate change on storm surges over Irish waters. *Ocean Modelling, 25*, 83–94.

Ward, K.J. (2015). Geographies of exclusion: seaside towns and houses in multiple occupation. *Journal of Rural Studies, 37*, 96–107.

Webb, D. (2005). Bakhtin at the seaside: utopia, modernity and the carnivalesque. *Theory, Culture & Society, 22*(3), 121–138.

Whitmarsh, L., O'Neill, S., & Lorenzoni, I. (2013). Public engagement with climate change: what do we know and where do we go from here? *International Journal of Media & Cultural Politics, 9*, 7–25.

Wiserma, B. (2016). Public acceptability of offshore renewable energy in Guernsey: Using visual methods to investigate local energy deliberations. Unpublished Ph.D. thesis. University of Exeter, UK.

Wolf, J., & Moser, S.C. (2011). Individual understandings, perceptions, and engagement with climate change: insight from in-depth studies across the world. *WIREs Climate Change, 2*, 547–569.

Zhang, Y-X., Chao, Q-C., Zheng, Q-H., & Huang, L. (2017). The withdrawal of the US from the Paris Agreement and its impact on global climate change governance. *Advances in Climate Change Research*, In Press.

Part IV

Social and environmental justice

18 Coastal environmental vulnerability

Sustainability and fisher livelihoods in Mumbai, India

Hemantkumar A. Chouhan, D. Parthasarathy and Sarmishtha Pattanaik

1 Introduction

Marine-coastal ecosystems and coastal communities are poorly represented in public debates on India's social and environmental problems. Coastal and marine ecosystems are the backbone of a fisheries economy that supports livelihoods of millions directly and several hundreds of thousands more indirectly (Rodriguez, 2010). Fisheries are an important source of employment and livelihood for millions of people worldwide (Parthasarathy, 2011). Fishing is one of the significant livelihood and economic activities of India, having deep roots in the indigenous traditions and practices known as the 'sunrise' economy, and has played a crucial role in food supply and food security, generating job opportunities and earning foreign exchange (Karmarkar, 2012). There are about 3.52 million fishers occupying the 3,202 fishing villages spread across the Indian coastline (Vivekanandan, 2007). According to one estimation there were 39 million people working as fish farmers as a full-time or part-time occupation in 2000, and 41 million in 2004 (Sharma, 2010). However, there are problems in the enumeration process, which leads to conflict among fish workers, as well as leading to encroachment on catchment areas.

The coastal zone space is an area from the low tide line (LTL) seawater side till the point seawater reaches during high tides (Singh, 2016). The coasts of India comprise a wide variety of diverse habitats and ecosystems – from estuaries, coral reefs, seagrass beds, mangrove swamps, creeks, backwaters and lagoons to bays, cliffs, sandy and rocky beaches. Coastal zones provide and sustain diverse ecosystems which offer significant public goods and services. In the past, India's coastal areas and resources were managed within a framework of traditional knowledge accumulated over centuries. The modern state impacted the customary practices of these communities and without communitarian controls (Rodriguez, 2010). It is important to note that small-scale fisheries are relatively more sustainable, given the diversity of the equipment employed depending on the season and the species targeted, ensuring that minimal catch is generated and that less energy is consumed per unit of fish output (Vivekanandan, 2007). Fishing communities in India have struggled for greater control over the seas and resource management, struggles which have been directed both inward, that is, between fishing groups, as well as against the state. As a result, developmental activities have encouraged disparity in social and economic status of various fishing groups, with the non-mechanised sector, especially the traditional fishing community, falling to the bottom of the spectrum. Even though the sector is rapidly changing today and is relatively more technology- and capital-intensive, small-scale fisheries remain viable for fishery-dependent livelihoods and are likely to do so well into the future. Artisanal/traditional fisher folk may have adapted to new social relations of production by beginning to move their production towards larger markets or in working as wage labourers in the mechanised sector, but their negotiation with the latter has not been without protest. The history of change in maritime societies has witnessed conflict between small fishing communities and the large mechanised sector, especially with respect to competition over resources or, in other words, to the 'common' space of the sea (Ram, 1991). Fishing communities, particularly the artisanal communities, find themselves most impacted by this alienation, as the development drive has left them marginalised (Vivekanandan, 2007). The Indian coastline is a contested space. In recent years, the country's coastal stretches have become pressure points for random and unsustainable development pressures (Rodriguez, 2010). In the name of development, however, a wide range of projects are appearing in coastal areas that impose additional and intense environmental, spatial and livelihood pressures in coastal zones (Patil, 2001).

The fishers in India thus face severe environmental and ecological challenges. Fishing villages have had historical relationships with environmental systems which have eroded over time. Even today, various fishing villages have unique relationships with the city and the landscape. Today these areas, being situated in prime locations, are facing real-estate pressures, inviting fishing villages to redevelop. The drastic environmental and ecological changes in fishing villages have brought a greater impact to fishing communities. The natural beauty of fishing sites has attracted tourists and many areas are declared as ecotourism zones for the urban population. As a result, these sites are largely in demand by an elite population for their private sea-facing apartments. Various slum development schemes have led to the destruction of mangroves without replacement of the cut mangrove trees. Lack of employment opportunities and a hike in property rates represent major temptations for the fishing community to sell their land and surrounding areas. The CRZ is an area which provides livelihoods to local inhabitants by providing goods and services. In terms of resource-dependent communities such as the fishing populace, coasts are like a natural common property, which provides free ecosystem services benefiting the poorest of the poor who strive by various means to survive and eke out livelihoods in close proximity to nature. Some outsiders consider these ecosystem services as an asset on which they can 'free-ride' (Costanza *et al.*, 1997).

Using structured and unstructured interviews at household level as well as meetings with co-operative society leaders, environmentalists and experts in focus group discussions among different age groups, this study attempted to comprehend changes in livelihood opportunities and the impacts of CRZ violations on the environment as well as fishing activities in the Mumbai region. The following case studies were selected to assess the environmental as well as livelihood impacts from CRZ violations across the region: Cuffe Parade Collaba, Sewri creek, Thane-Mulund creek and Uran Koliwada[1]. These areas were selected on the basis of their different environmental and social characteristics and on the level of intensity of CRZ violations. These cases brought out issues facing fishing communities, including conflicts, their resistance, contestations and impacts on livelihood as well as sustainability of coastal ecology.

2 Coastal environmental vulnerability: sustainability and fisher livelihoods

Coastal areas are assuming greater importance in recent years, with increasing population size and density, urbanisation, and accelerated economic activities. Many economic, infrastructure and developmental projects are located near the seashore across coastal states in India. These have become new centres of attraction for the developmental state, builder lobbies, real-estate firms, recreational needs, and infrastructure sectors – exercising new coastal claims. Against a background where industrialisation, including industrial fisheries, appear to be a 'natural' form of economic progression, the small-scale fisheries model, which is labour-intensive and scaled for subsistence and local markets, is seen as irrational and inefficient, and therefore it is viewed as an anomaly and problem by the developmental state. In the name of modernisation and development of the fisheries sector, policy-makers have encouraged industrial fishing to replace small-scale and traditional fish catch. Yet evidence from the field study shows that small-scale fisheries appears to offer a variety of ecological and economic benefits which sustain marine resources. Modernised fisheries have brought intended and unintended consequences that produce conflicts at different levels, such as conflict between states, conflict between state and market, and conflict between one section of natural resource users and state (Kurien, 2008). The overpowering demand-pull for fish has led to over-exploitation of natural resources. It has transformed the fishing industry suddenly because of high demand, which has led to overfishing (Kurien, 2008). The export-orientated thrust, started after the implementation of neo-liberal policies in 1991, which has come to define the sector, was aided and expedited by the country's wider attempts to boost foreign exchange earnings (Kurien, 2008). Overfishing has led to scarcity issues for small-scale fisher folk as it has skewed distribution of the benefits and cost of the fish economy, which have larger socio-political implications for society (Kurien, 1992).

The coastal environment is increasingly pressurised to maintain regular flow of goods and services, and doing this in a sustainable manner is a challenge of another level (Ramesh *et al.* 2013). Coastal areas are under pressure from environmental harm, climate change, development-related encroachments and degradation of marine resources, which are leading to loss of livelihoods for fishing communities. Research on small-scale fisheries typically emphasises two assumptions. First, poor fisher folk are poor because they are fishers; this leads to a conclusion that these people should do something else (like factory workers) and that fishing should be entirely industrial. Instead, it turns out that people fish because they are poor; in other words, fishing provides unique opportunities for alleviating poverty (rather than making it worse) and small-scale fisheries should be encouraged rather than undermined (Allison & Horemans, 2006; Bene, 2003; Bene *et al.,* 2007). A second assumption is that because of their poverty, poor fishers have no other choice but to deplete fisheries to the point of overfishing. Instead, small-scale fisheries are turning out to be an important model for the future of fishing, because they are more efficient and less degrading than industrial fishing; in fact, they manage fisheries resources very effectively (Allison & Ellis, 2001; Dyer & McGoodwin, 1994; Pauly 2007). Research reveals that small-scale fisheries with small boats do not lead to overfishing, but rather these people are harmed by resource depletion that they do not themselves create (Mansfield, 2011). Traditional fishing technologies (nets, tackle and methods of fishing) are suited particularly for specific ecological contexts, especially the selective nature of fishing nets and the 'passive' nature of fishing operations (i.e., allowing fish to get entangled in the net rather than catching them by disturbing their location).

In addition, elite groups are trying to push fishing communities from their existing places due to the impacts of such activities on, for example, local land prices (e.g., the smell of fishing activities). The coastal lands of fishing communities are highly expensive and the developer lobby is increasingly trying to capture this land for real-estate business and redevelopment of these areas. The fishing community typically does not have land deeds since their local fisheries are traditionally commonly owned and managed. Such communities are facing threats of eviction from builders and the state. After the introduction of trawler and purse seine nets, many fishermen took loans due to a livelihood crisis and have been forced into bonded labour, making them vulnerable to hyper-exploitation. In Budhwar Park, Bhandup, Shevri creek, many fishers are adopting other modes of occupation. These diverse problems offer a glimpse into the political economy verses political ecology debates. New coastal claims and encroachments involve contestations among elite and small-scale fisher folk.

From a political perspective, there is little representation of the fishing community in the political sphere, and this leads to political disempowerment, resulting in unfavourable fisheries policies that restrict fishers' access and entitlement as well as creating conditions for class conflict. Due to this, the issues of the fishing community are not taken seriously in policy frameworks. This has pushed traditional fisheries into a state of extreme powerlessness. Encroachments over coastal spaces are also leading to power conflicts among fishing communities, and smaller fishermen are becoming marginalised and more vulnerable in a gradual, but ever-increasing manner.

The introduction of large trawlers, catching fish primarily for export, led to major changes in the ecology and economy of fisheries. A rapid increase in fish landing in the early years of trawling was followed by stagnation and relative decline. This conflict gave rise to a widespread movement – involving strikes, processions and violent clashes with trawler owners – in which small fishers pressed for restrictions on the operations of trawlers. Fishing communities, whose livelihoods have been seriously undermined through a combination of resource flows biased against them and growing deterioration of the environment, resisted to the best of their ability. The origins of these conflicts lie in the process of development itself (Kurien & Thankappan, 1990). The CRZ protects and sustains livelihood as much as the coastal ecosystems do. The need for livelihood security gets linked to conservation in Para 1 of CRZ Notification in subsection (I) and clause (V) of subsection (2) of the Environment and Protection Act (1986). Through this, the central government is given responsibility for ensuring sustainable livelihood security to fishing communities and other local communities living in coastal areas. In setting these, the CRZ legitimately imposes many land-use restrictions such as on the setting up or expansion of industries, operations or processes and manufacture, and handling or storage or disposal of hazardous substances. In some cases the Supreme Court has set aside high court permissions for constructions which violate the CRZ guidelines for land-use diversions or reclassifications, and appointed expert committees to prepare an Integrated Coastal Zone Management Plan (ICZMP) for sustainable development of these particular regions (Singh, 2016).

Some fundamental research questions emerge from the case studies undertaken for this chapter: What are the different types of resources under stress from new coastal claims and developments? What are the new claims on coastal zones – both internal and external to coastal fishing communities? What is the role of environmental legislation, its violation, amendments, and the state's failure to implement these laws, in environmental degradation, and marginalisation of resource dependent groups?

Like conflicts over land and in the workplace, conflicts over natural resources typically pit against each other two unequal antagonists. While forests, water and other natural resources are diverted to produce energy and commodities for the rich, the poor are made to bear the social and environmental cost of economic development, whether in the form of declining availability of natural resources, a more polluted environment, or in an increasing physical displacement (Gadgil & Guha, 1994).

In the case of the fishing community of the Mumbai region, the smaller fisher folk, due to the encroachment process of bigger trawler operators, are becoming alienated from their fishing activities, while simultaneously urban elites are capturing their land and pushing the small-scale fishing community away from their spaces and traditional locations for fishing activities. In Budhwar Park, Bhandup, Shevri creek, and several other coastal areas, fisher folk are moving away from their fishing activities to other forms of livelihood. In the case of men who were totally dependent on fishing, they have moved to other jobs such as wage labour in urban construction sites, in many cases earning less than typical incomes from their fish trade. Others are engaged as security guards and auto-rickshaw drivers. Fisher women have taken up jobs as household help. They clean pots, wash clothes, sweep and mop floors at these households and earn a monthly income.

3 CRZ violations and challenges for sustainability: reflections from Mumbai's coastal areas

3.1 Cuffe Parade Colaba

Budhwar Park (population 4,000–5000) from the Machchimar Colony, Colaba, located in the southern part of Mumbai, is one of the case study sites.

Figure 18.1 Map showing the case study areas (Uran koliwada is 63 km from Mumbai, thus not included)

Source: BMC (Bombay Municipal Corporation) records, 2016, after Chouhan *et al.* (2016)

Colaba is one of the busiest and most affluent areas of Mumbai. The families of the fishing community are not the only residents of the Cuffe Parade area. Garib Janta Nagar, Mahatma Phule Nagar and Dhobi Ghat are their immediate neighbours. The residents include migrants who are not dependent on fish catching, though they stay very close to this community and the sea. However, there is a clear indication of encroachment on the sea due to their proximity. Currently, the sea has become a dumping ground for debris, plastic bags and other waste material of this community. This pollutes the sea water and thereby affects breeding of fish in this area.

3.1.1 The impact of encroachment and CRZ violation on fishing livelihoods

Due to the dumping of plastic bags and other waste material, small fish cannot live in the polluted water. This adversely affects the breeding of fish. The people of the fishing community in earlier times were able to catch crabs near the seashore without using a net. However, due to pollution of the seawater, the fishing community is finding it hard to catch not only crabs but also other species of fish such as corambi, buimawla and domi[2]. Before this encroachment and CRZ violation in this area, small fishermen were catching about one tonne of fish per day. This number has now reduced to less than 15 kg per day. Sometimes they even come back empty-handed. The encroachments have led to an increase in population density and unavailability of space (land) for the construction of jetties. There are buildings with a higher than 2.5 FSI (floor space index), properties such as the Adarsh building, which is not permitted in CRZ-I zoned area. As a result, there are not enough jetties, although there is provision to construct them in the CRZ[3].

3.2 Sewri Creek

The Sewri area, locally known as Shivdi, is a small hamlet on the eastern shore of Mumbai. It belongs to the CRZ-1 because of its mangrove cover. The fishing community in these parts has been established here since the British period. The community is dependent on small-scale fishing for their livelihood. According to the local respondents, the state government has reclaimed this area after independence. Initially, the fishermen were catching fish from the seashore near the Bombay Port Trust. Large parts of Sewri currently belong to the Bombay Port Trust including the adjacent harbour facilities (Chouhan *et al.*, 2016). The mangrove swamps of Sewri were declared a protected ecology. An example of this is the Sewri Mangrove Park, which the Bombay Port Trust created on 15 January 1996[4]. This park consists of 15 acres of mangroves in the mudflats between Sewri and Trombay. These mudflats are near the Sewri jetty, reached by a 20-minute walk from the Sewri railway station on the harbour line. There are a number of well-known housing societies like Sewri Koliwada, BDD Chawls, Dnyaneshwar Nagar, Shivaji Nagar, Gulmohar Society, Labor Camp and Bhatwadi in the area. The market area is called 'Sewri Naka'. The area is of ecological importance as flamingoes from other parts of India come to these mangroves to breed. They arrive at the mudflats between October and March every year.

In 2007, more than a decade after being declared protected, this wetland habitat was in danger of being wiped out by the planned Mumbai-Nava Sheva road link. In recent years a lot of construction activity has taken place in this area due to the opening up of the mill lands in the Mumbai Port Trust area. The Sewri Nhava-Sheva trans-harbour link planned by the Mumbai Metropolitan Region Development Authority (MMRDA) threatens the habitat of the migratory flamingoes in this area. The discharges from the Bombay Port Trust (BPT) include coal and oil which are directly dumped into the seawater. Mangroves are essential to the ecology of the coast and the island. They provide fertile ground for the fish to feed and breed and nurture a large variety of birds. Seven species of mangroves have been identified in this area. But now these mangroves are being destroyed as large amounts of waste material are being dumped in this area by the Port Trust in order to increase their activities. Due to this, the mangroves are being destroyed. Water pollution has also led to a decline in the fishing areas.

To keep the mangroves from extinction, the local fishermen brought small mangrove species from other areas and planted them here. BPT and other agencies are championing construction activities in a major way in this area due to the opening up of the mill lands and the Sewri-Nhava-Sheva trans-harbour link. The local fishing community considers this as the leading cause of pollution of the seashore due to dumping of debris, that in turn has led the community to incur heavy business losses.

In this area, there are about 35 families who depend on fish catching. They now get to catch only crabs in contrast to earlier times when their catch included other species of fish like boi, bangda, tarla, tingala, pale, sewad, kolambi, pamplet, goli, rawas, toli and wam. There are 30 small boats and three machine boats owned by the fishers here. Overall, fishing communities are in a minority in the city. They do not have the capacity to fight with the state. Earlier there were more families catching fish, but now their numbers have dwindled. They have shifted to work other than fishing. Some are driving cars, some are working as coolies (porters), and some are involved in *majuri* or low wage labour. Due to the encroachment of BPT, there are no jetties available to park boats and no land available for weaving nets or other activities.

3.3 Thane-Mulund Creek (Bhandup Village)

The Thane-Muland belt falls under CRZ-I, although there is a dispute now as to whether it belongs to CRZ-I or CRZ-III. The dispute arose around 2005 as a Special Economic Zone (SEZ) was earmarked for a 134-acre mangrove belt between Mulund and Thane. According to CRZ 1991 and 2011, a SEZ is not allowed on a CRZ-I area. An additional problem with this area is that it falls under the jurisdiction of two municipal corporations – Mumbai (BMC) and Thane (TMC). According to BMC records, the Mulund side of this plot is under CRZ-III, while the area under Thane was still under CRZ-I. Therefore, the Ministry of Environment and Forests (MoEF) changed the CRZ classification of the Thane side of the plot from CRZ-I to CRZ-III in 2006. This was done in order to implement the proposed SEZ project. The MoEF had at first questioned the irregularity in the records but gave its consent on the recommendation of the state government that in turn had been approached by the SEZ developer.

The proposed land for the SEZ project is barely 200 metres from Thane Creek. The land was surrounded by a thick cover of mangroves on three sides. Some of these mangroves have been cleared to free up the land. Inter-tidal waters can be seen in and around the plot. Environmentalist D. Stalin[5] and others filed a complaint to the then Union Environment Minister Mrs. Jayanti Natarajan on the issue of environmental clearance of mangroves in the area. State Environment Secretary Mrs. Valsa Nair Singh said that she could not comment specifically on the Mulund-Thane SEZ project until after checking departmental records. However, on a general note, she stated that "as per CRZ 2011 rules, SEZs are prohibited on CRZ area. Under old rules, they were permissible in CRZ II and CRZ III areas". According to D. Stalin, the planned environmental clearance is bound to affect the mangroves which will in turn cause water logging in the future. He also stated that "raising the height of the land to build the SEZ will expose Kopri village to flooding. Flooding is already a problem for Patilwadi area". In addition to flooding, this SEZ would adversely hurt local marine life, thereby depriving small fishermen of their livelihood.

In addition to this, private industries are discharging their untreated waste water and chemicals along the seashore; sewerage pipelines are also meeting the sea as is observed in many parts of the city. Water in the area is totally polluted, which is leading to an increase in the high presence of mosquitoes, which in turn prevents fishermen from venturing into these spaces. In the opinion of one of the fishermen, the "mosquitoes are multiplying and not fish species" (Chouhan *et al.*, 2016).

3.4 Uran Koliwada

The coastal environment of Uran has been under considerable stress since the onset of other industries and Jawaharlal Nehru Port Trust (JNPT) in 1989. The intense activities of container freight stations, urbanisation, industrialisation and reclamations in the stretch of creek around Uran have resulted in a loss of mangrove biodiversity. In fact, the Jawaharlal Nehru Port Trust itself is located on a major reclaimed area. This reclamation can perhaps be justified on the grounds of infrastructure needs, but environmental and social costs are high, and it can be strongly argued that these are too high. For instance, about 7,000 hectares of land is proposed to be reclaimed for the new Special Economic Zone. Under these proposals, no buffer space is left for the surplus water of the sea to flow. The sea has no alternative but to hit the land. Several incidences of coastal pollution have occurred because of leakage/discharge of transporting materials along with industrial effluents. The disposal of domestic wastes and untreated or partially treated industrial effluents in the coastal region of Uran, Navi Mumbai has depleted coastal resources, risking public health and causing a considerable loss of coastal marine biodiversity. These impacts have severely affected fisher livelihoods.

3.4.1 Uran mangrove destruction

The dumping ground for Uran town has been established among the mangroves in the area. The mangroves on 1.21 hectares of land have already been destroyed. In addition to the dumping yard, there is more destruction of mangroves around Uran.

A number of roads have been constructed by CIDCO, the JNPT and ONGC in this area. Many of them are located on the creek opening sites. If these roads do not pass the creek openings over properly constructed bridges, the tidal inflow is hampered and the mangrove ecosystems dependent on it are at risk. The mangroves at stake cover a total area of 301.2 ha (202.8 ha in area 1; 35.7 ha in area 2; 22.5 ha in area 3; and 40.2 ha in area 4). These are very large tracts of mangroves and need to be protected.

Currently, 17.9 ha of mangroves in one area and 20.9 ha in another area are damaged. These patches are located along the same creek. Construction and development in the area have blocked the entry of saltwater into the creek and as a result the mangroves have dried up. The area with mangroves has lost its typical pattern and texture. The bright tone of vegetation in the area suggests dumping and associated adverse impacts on biodiversity. As discussed earlier, these areas are likely to be used soon for residential and commercial construction.

Due to the dumping of sand and debris in surrounding areas, the flow of the creek water leads to coastal land erosion. It adversely affects the surrounding fishing hamlets and their livelihoods. It gives rise to flood situations in the surrounding villages that are deeper within the coastal area. And there is no way for water to flow back and meet the creek, so it affects the deepest area. CIDCO dumping its waste in the creek mouth from every possible site also leads to a fluctuation in creek flows.

4 Conclusion

Fishing villages located in urban areas are legally classified as Koliwadas and Gaothans, which are 'no development zones'. These areas are used for different purposes such as landing of boats, fishing, drying of fish, drying and weaving nets and for residential purposes. Today, these Gaothans and

Koliwadas in the Mumbai Metropolitan Region (MMR), despite legal protection, face huge problems of encroachment, reclamations and reclassification from 'no development or no construction zones' to 'commercial zones'. Various projects and SEZs are emerging in these locations, which are further aggravated by encroachment by working-class migrants on the livelihood and residential spaces of the fishers. Encroachment leads to a decline in the availability of land in areas where there is already a restriction on space available; as their families grow bigger, two or three generations of fisher households may now reside in one small room because of non-availability of space, and problems of expansion. A significant problem is the case of newly married couples who find it difficult to have separate rooms constructed for them. At the same time, as the infamous Adarsh housing scam in the CRZ zone reveals, the elite perpetuate CRZ violations and access expensive and upmarket housing, while the indigenous fishing community suffers from a housing crisis (Chouhan *at el.*, 2016).

In Uran in the Navi Mumbai part of the MMR, JNPT, CIDCO and ONGC have encroached upon thousands of acres of land. In Cuffe Parade in south Mumbai, rich residents have encroached heavily on fishing lands. These encroachments and reclamations tend to accelerate environmental damage and have led to scarcity of land for fishing and ancillary activities. Some fishing villages are encroached upon as part of beautification projects, as in the case of Cuffe Parade, where 60 metres of dense mangrove was wiped out to construct a garden for the local elite residents.

It appears that the main purpose of the beautification project is to drive away the current fishing population from the area, in order to create exclusive spaces for the local elites. According to the local fishing community, the plan to integrate the existing fishing facilities into the project to improve their living and environmental conditions is a fraud. They say that it is not feasible to incorporate all of the proposed facilities within this small stretch of land. They suspect that the main objective of the project is to shift them out of this area, as once they are marginalised into certain spaces their access to the seashore would be severely restricted thereafter. This will lead to a significant loss of livelihood. They claim that the state is also party to this conspiracy. Since fisher households do not have land deeds, and have mostly owned and accessed land as commons, they find it difficult to resist such elitist projects. Gradually trawlers/commercial fishing and builders are encroaching upon traditional fisher spaces, and making the fishing community economically fragile.

In Sewri, wetland habitats are in danger of being wiped out by the planned Mumbai-Nava Sheva road link. Also, in recent years a lot of construction activity has been taking place in this area due to the opening up of mill lands and the Mumbai Port Trust area for real estate. In Uran, mangrove destruction is ongoing at an alarmingly large scale and rate due to infrastructure development for the ports, oil refineries and upcoming airport. This has changed hydrological flows, affecting fish breeding and catch.

The MMR is a densely populated area, and rubbish disposal has emerged as a key source of pollution and marginalisation for fishers. This has been exacerbated by CRZ violations, for example by creating dumping grounds in wetlands by reclassifying them as 'waste land', in the process destroying flora and fauna in the coastal belt. In many such spaces, apart from encroaching on fishing spaces, new developments have also either taken away or made it impossible for other service provisions or amenities to be effective. These include primary health centres, schools, and playgrounds. As Banerjee-Guha (2002) argues, inside global and globalising cities a new geography of centrality and marginality emerges. Uran, Cuffe Parade, Thane-Mulund Creeks and several other sites are a good example of this. Essentially the metropolitan space under globalisation tends to become a contradictory space, characterised by contestation and internal differentiation. New coastal claims on land have resulted in the CRZ being seen as an opportunity for coastal grab; Koliwadas were always under surveillance for 'land grabbing' behaviour by the state, market and illegal elements, but CRZ has enabled the process by creating a legalised screen for such land alienation. The Koliwadas, as Warhaft notes, "happen to occupy some of the most expensive real estate in the world pursuing an occupation many consider defiling" (Warhaft, 2001). From an environmental justice and equity perspective, this is a problem of 'geographic equity'. Inequality is enabled and supported by the CRZ, through relaxations to CRZ notifications, whereby several sites are reclassified from CRZ I (ecologically sensitive area) where commercial constructions cannot take place to CRZ II or CRZ III, where constructions are allowed in the name of growth and development. Since the economic reforms of the early 1990s, a huge transformation has taken place in fishing livelihoods. In 1990, fishers in the region were catching fish twice a day close to the coast, and in the mangrove areas. Due to blockage in flow of coastal waters, heavy siltation of the creeks, and pollution, this is no longer the case. Several species of fish and crustaceans that are the source of livelihoods and food are disappearing, including tiger prawn, mudskipper and flier crab. Another source of fisher marginalisation in the MMR is sand mining in rivers along the coast.

All of these acts of deprivation and marginalisation have forced younger members of the traditional fishing community to abandon fishing and to shift to new occupations. In Bhandup village, about 250 families have abandoned fishing for other jobs. The same situation is evident in Cuffe Parade and Sewri. Some are engaged as security guards and others as auto-rickshaw drivers. Women have had to take up jobs as household help, losing autonomy, choice and agency in the process. There is also an aspiration among men for white-collar jobs, which however are not accessible to them due to educational constraints. The future situation for fisher spaces, ecosystems and livelihoods is grim, unless urgent steps are taken to reverse the damage caused by inadequate and improper use, violation and implementation of CRZ norms.

Notes

1 Koliwada literally means the habitat of the Kolis. The Kolis are a fishing community that lived in the seven islands of the Arabian Sea that subsequently merged to form the city of Mumbai. There are several Koliwadas scattered all over the city.
2 Local names of fish species.
3 CRZ Notification, 2011.
4 According to the board erected by the BPT in the area.
5 Project Director of Vanashakti. Vanashakti is an NGO which aims at environmental conservation and to instil a sense of responsibility among citizens (www.vanashakti.in/)

Bibliography

Allison, E.H., & Horemans, B. (2006). Putting the principles of the sustainable livelihoods approach into fisheries development policy and practice. *Marine Policy, 30*, 757–766.

Allison, E.H., & Ellis, F. (2001) The livelihoods approach and management of small-scale fisheries. *Marine Policy, 25*, 377–388.

Banergee-Guha, S. (2002) Shifting cities: urban restructuring in Mumbai. *Economic and Political Weekly, 37*(2), 121–128.

Bene, C. (2003) When fishery rhymes with poverty: a first step beyond the old paradigm on poverty in small-scale fisheries. *World Development, 31*(6), 949–975.http://dx.doi.org/10.1016/S0305-750X(03), 45–47.

Bene, C., Macfadyen, G., & Allison, E.H. (2007). Increasing the contribution of small scale fisheries to poverty alleviation and food security. *Food and Agriculture Organization Fisheries Technical Paper No. 481.* FAO, Rome, Italy.

Costanza, R., d'Arge, R., deGroot, R., Farber, S., Grasso, M., Hannon, B., Limburg, K., ... van denBelt, M. (1997) Total value of the world's ecosystem services and natural capital. *Nature, 387,* 253–260.

Chouhan Hemantkumar, A., Parthasarathy, D. & Pattanaik, S. (2016) Coastal ecology and fishing community in Mumbai: CRZ policy sustainability and livelihoods. *Economic and Political Weekly, 51*(39), 48–57.

Dyer, C.L., & McGoodwin, J.R. (1994). *Folk Management in the World's Fisheries.* University Press of Colorado.

Gadgil, M., & Guha, R. (1992). *This Fissured Land: An Ecological History of India.* Delhi: Oxford University Press.

Gadgil, M., & Guha, R. (1994) *Ecology and Equity: Steps towards an Economy of Permanence.* Geneva, UNRISD.

Karmarkar, D. (2012). *Fishy Spaces: Globalisation and Livelihood of Indigenous Fishermen A Case of Mumbai.* LAP LAMBERT Academic Publishing.

Kurien, J., & Achari, T. (1990) Overfishing along Kerala coast: causes and consequences. *Economic and Political Weekly, 25*(35/36): 2011–2018.

Kurien, J. (1992). Ruining the commons and the response of the commoners: coastal overfishing and fish workers action in Kerala State, India. In *Grassroots Environmental Action: People's Participation In Sustainable Development,* London, Routledge.

Kurien, J. (2008) State, modernization and conflict in fisheries revisiting the fish workers movement in Kerala. In Narayanan, N.C. (Ed.), *State, Natural Resource*

Conflicts and Challenges to Governance, Where do we go from here. Academic Foundation in Association with the Institute of Rural Management (IRMA), Anand, Gujarat, India.

Mansfield, B. (2011) Modern industrial fisheries and the crises of overfishing. In Peet, R., Robbins, P. & M. Watts (Eds.), *Global Political Ecology*, New York, Taylor and Francis, 84–99.

Parthasarathy, D. (2011). Hunters, gatherers and foragers in a metropolis: communizing the private and public in Mumbai. *Economic and Political Weekly, 46*(50), 54–63.

Patil, R. (2001). Coastal zone conflicts in Maharashtra. International Collective in Support of Fishworkers (ICSF), International Ocean Institute (IOI), India, 156–157.

Pauly, D. (2007) The sea around us project: documenting and communicating global fisheries impacts on marine ecosystems. *AMBIO: a Journal of the Human Environment, 34*(4), 290–295.

Ram, K. (1991). *Mukkuvar Women, Gender, Hegemony and Capitalist Transformation in a South Indian Fishing Community.* Allen and Unwinpty Ltd.

Ramesh, D.A., Senthil, V.A., Paul, T., & Wadekar, V. (2013) Relationship of socio-economic parameters in coastal litigations of coastal states of India. *International Journal of Humanities and Social Sciences Invention, 2*(6), 27–32.

Rodriguez, S. (2010). Claims for survival: coastal land rights of fishing communities. Dakshin Foundation, Bangalore, 42.

Sharma, C. (2010). Report of the Workshop. UNDP, New Delhi.

Singh Kuldeep, J. (2016) *Supreme Court of India Judgment*, Chawla Publication (P) Ltd.

Vivekanandan, V. (2007).Changing climate of the livelihood and rights of fishermen on the coast. Presentation made at the workshop on ‘Combating Coastal Challenges, organised by Citizen Consumer and Civic Action Group (CAG) Chennai, 7–8.

Warhaft, S. (2001) No parking at the bunder: fisher people and survival in capitalist Mumbai, South Asia. *Journal of South Asian Studies, 24*(2): 213–223.

19 Creative and constrained hybridisations in subarctic Inuit communities

Communal fishery development in Nunatsiavut, Canada

Paul Foley, Charles Mather, Natalya Dawe and Jamie Snook

1 Introduction

1.1 Access to fish resources for Indigenous groups in settler societies

Indigenous groups in several settler societies have successfully secured access to fish resources for commercial gain (Davis & Jentoft, 2001; Durette, 2007; Capistrano & Charles, 2012; Lalancette, 2016; van der Porten *et al.*, 2016). These achievements by Indigenous groups in the United States, Canada and Australasia have not involved the transfer of sovereign fishing rights. Instead, Indigenous rights to fish have been recognised through separate licensing and quota allocation mechanisms, which remain part of existing, state-controlled, fisheries management systems (Coates, 2000; Davis & Jentoft, 2001). These state-controlled access arrangements have tended to take a particular form: they involve allocating licences and quotas to Indigenous *groups or organisations*, rather than to *individuals* in the hope that these allocations will have a strong redistributive impact within Indigenous communities. In Alaska, for example, the Community Development Quota (CDQ) system has been in place since the early 1990s and it allocates significant fish quota to poor and mostly Indigenous communities (Ginter, 1995; Carothers, 2011; Haynie, 2014). Indigenous groups that receive CDQs in Alaska do not always fish these quotas themselves, but they are allowed to trade these allocations to owners of licensed commercial vessels in return for royalties, which are in turn used to support local economic development initiatives (Mansfield, 2007). A similar licensing system was developed in Canada in the early 1990s to allocate fish resources to Indigenous groups. The Aboriginal Communal Fisheries Licence (ACFL) policy has provided a mechanism for the Canadian state to allocate communal licences to Indigenous groups for commercial fishing purposes (Harris & Millerd, 2010; Krause & Ramos, 2015). Once allocated to Indigenous organisations, qualified Indigenous fishers are "designated" or permitted to fish under the authority of communal licences for commercial purposes (Allain & Frechette, 1993; Stanbury, 2003). The Canadian system is different from the CDQ in Alaska in that communal licences are used to empower individual Indigenous fishers to become directly involved in the commercial fish sector. Nonetheless, both systems share a common goal of allocating licences and quotas to Indigenous groups in an effort to address social objectives of poverty alleviation, economic empowerment and general restitution for colonial legacies.

Drawing on the case of the CDQ system in Alaska, Mansfield (2007, 492) argues that communal allocations for Indigenous groups should be seen as a form of property that allows the state to "unite neoliberal and social justice approaches" to fisheries access. In other words, these communal allocations allow the state to enclose fishing rights, but for specific social purposes. There is, however, evidence, especially from New Zealand, that the redistributive role of these communal rights can dissipate over time. Fiona McCormack (2016a, p. 229) has recently written that resource allocations to Indigenous groups "can lead to new and more permanent forms of loss, as the assets and resources returned are increasingly entangled with capitalist markets". De Alessi's (2012, p. 408) work on New Zealand points to additional problems that go beyond the loss of resources: he argues that the transfer of fishing rights to Māori groups has "had a profound effect on Māori social relations and identity". He goes on to suggest that these fishing rights for Indigenous groups have "led to the capitalist penetration of Māori fishing practices and social organization on a scale unheard of with community quotas in places such as Alaska" (De Alessi, 2012, p. 391). Although there is limited writing on the experience of Indigenous commercial fishers in Canada, the research that does exist also points to the erosion of social objectives in communal fishing allocations. Writing from the context of the Canadian Maritimes, where Indigenous groups have secured access to communal fishing licences, Wiber & Miley argue that "Native communities have not been able to maintain aboriginal values in the harvesting of fish, nor in the distribution of benefits. In fact, many signatory communities are experiencing sharp debt as a result of the 'right' to fish commercially" (Wiber & Miley 2007, p. 184).

The evidence from New Zealand and Canada suggests that the social objectives of Indigenous commercial fishing rights are being eroded as they become 'entangled' in capitalist markets. But this is not the only finding on how Indigenous rights for fish have played out in practice. McCormack's (2016a) work in New Zealand also points to a more complex outcome that involves Indigenous groups struggling to maintain a balance between social objectives and economic outcomes (also see Wiber & Miley, 2007, p. 184; McCormack, 2016b). McCormack writes that Indigenous groups face the challenge of trying to balance the "oft-conflicting demands of customary obligation and market engagement" and must weigh up "the opportunities for wealth creation with customary distributional economies and kin obligation" (McCormack, 2016a, p. 192). Her argument is that the outcome of these struggles cannot be known in advance, but may result in what she calls 'creative hybridizations', a term that reflects this effort to balance distributional economies and market engagement. McCormack suggests that these creative hybridisations may increasingly characterise how Indigenous rights to fish and other resources play out in practice.

1.2 Creative and constrained hybridisations

This chapter aims to contribute to these debates on Indigenous claims to fish and other resources that become entangled in capitalist markets. These debates are pressing and urgent because they occur in a context where Indigenous coastal communities are struggling to overcome the disastrous legacy of colonial and post-colonial policies. We do so through an analysis of the implementation of Government of Canada communal licences by the Labrador Inuit Association (LIA) and, since 2005, the Nunatsiavut Government within Nunatsiavut. Nunatsiavut is the self-governing territory of the Inuit established in 2005 through the Labrador Inuit Land Claim Agreement covering the northern coastal region of Labrador (Figure 19.1). Since the early 2000s, the LIA/Nunatsiavut Government has designated some Inuit fish harvesters to fish under its communal licences for northern shrimp, snow crab and turbot. Our analysis is not primarily about the erosion of social objectives in these allocations, an issue that has been a central theme in the literature on Indigenous fishing rights in a number of different contexts. Instead, we reveal a more complex process of change that provides new ways of thinking with the concept of 'creative and constrained hybridisations', which modifies McCormack's concept by incorporating a notion of constraints that can capture state and capitalist barriers. In this way our work contributes to ongoing debates on resource allocations to Indigenous coastal groups that become entangled in capitalist markets (c.f. Egan & Place, 2013) by recognising both agency (creativity) and structure (constraints).

Our research on recent claims by Indigenous groups in Nunatsiavut for fish resources in a remote Canadian coastal zone represents an important contribution to recent efforts to rethink and reframe the concept of sustainability transitions (Lawhon & Murphy 2012; Truffer *et al.*, 2015; Morrissey & Heidkamp, this volume). A key concern in this work has been to stress the importance of the impact of geography on sustainability transitions, specifically through the idea of socio-spatial embedding (Murphy, 2015). Socio-spatial embedding, as Truffer *et al.* (2015) argue, demands an explicit focus on the "cultures, institutions, political systems and networks of capital stocks" have been used to support sustainability transitions in specific sites. It also requires that we play close attention to the "coalitions of actors use to advance socioeconomic, political, and environmental agendas" (Murphy 2015, p. 88). Our work on the 'creative and constrained' efforts to establish independent inshore harvesters in Nunatsiavut confirms the significance of place: we reveal the crucial role of situated historical and contemporary political and social dynamics that are central to understanding the efforts of government and other groups in working towards a sustainability transition in this remote coastal zone.

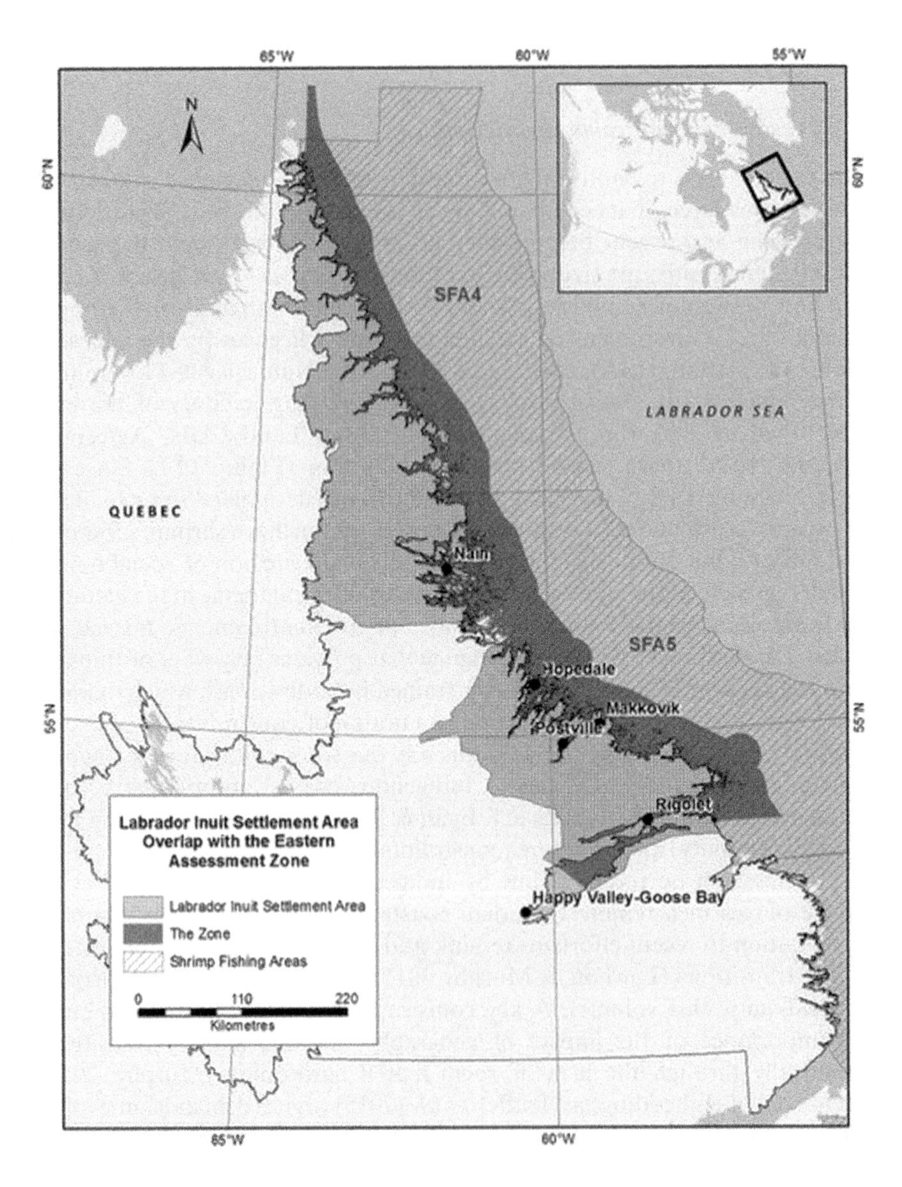

Figure 19.1 Nunatsiavut and key coastal settlements
Source: Bryn Wood, GIS Analyst, Torngat Wildlife Plants and Fisheries Secretariat

This chapter is part of a long-term project on the resource politics of the northern shrimp sector in Atlantic Canada (Foley *et al.*, 2015; Foley & Mather, 2016; Foley *et al.*, 2017). Our research for this chapter draws on two years of work in the specific context of Nunatsiavut, where we have established research collaborations with local government and non-governmental organisations involved in fishery development in this region of Canada's north (Foley *et al.*, 2017). We draw on secondary sources and key informant interviews with public and private organisations such as the Torngat Fish Producers Co-operative Society Ltd., the Nunatsiavut Government, the Torngat Joint Fisheries Board and the Torngat Wildlife, Plants and Fisheries Secretariat (Snook *et al.*, 2018), as well as with Inuit commercial harvesters. These interviews with key informants, representatives of public organisations and Inuit commercial harvesters were conducted between 2015 and 2017 in St John's, Happy Valley Goose Bay and Makkovik, Labrador. We also draw on material collected at two fisheries workshops in northern Labrador hosted by the Torngat Joint Fisheries Board in 2015 and 2016.[1]

The remainder of this chapter is structured into three sections. In the first section we provide background to the northern shrimp fishery and the allocations that groups in Nunatsiavut secured from the Canadian state. We describe state policies that led to the allocation of fish resources to Indigenous groups in this region, but we also emphasise the active role of historically marginalised Inuit peoples of Labrador in securing and managing these important resources for coastal communities in Nunatsiavut. In the next section, we outline how these resources were creatively used to support Nunatsiavut's inshore fishery, including more recent efforts to establish independent fish harvesters through a federal communal licence system. In the final section we explore the growth of this inshore fish sector and the constraints and tensions that are emerging as these harvesters attempt to create livelihoods opportunities with relatively small amounts of quota, and become more entangled in capitalist markets. We conclude by exploring the concept of 'creative and constrained hybridisations' and its utility in capturing the dynamics that emerge from resource allocations to Indigenous groups.

2 Northern shrimp: enclosure and resource allocations to Nunatsiavut groups

Atlantic Canada's northern shrimp sector, which emerged in the wake of Canada's decision to declare a 200-mile Exclusive Economic Zone (EEZ), has been crucial to Indigenous fishery development in Nunatsiavut. The declaration of Canada's EEZ in 1977 effectively enclosed northern shrimp stocks for Canadian fishing interests and excluded northern European offshore boats, which had previously fished northern shrimp in what were now Canadian waters (Foley *et al.*, 2015). Canada's Department of Fisheries and Oceans (DFO) issued 11 offshore shrimp licences in 1978 and another licence in 1979, for a combined total quota of 8,100 tonnes. The initial allocation of offshore licences was strongly influenced by a progressive Minister of Fisheries, Romeo LeBlanc, who was committed to ensuring that coastal communities benefited significantly from the fish resources within Canada's newly established 200-mile EEZ (Foley & Mather, 2016). To this end, LeBlanc allocated offshore licences for northern shrimp across Atlantic Canada to fishing co-operatives, Indigenous groups, as well as several companies with access to factory freezer trawlers and with processing facilities in coastal communities. Significantly, allocations to Indigenous groups in Canada pre-date by many years subsequent efforts to transfer fishing rights through programmes like the Aboriginal Fisheries Strategy.

While a progressive minister was an important factor in enabling Nunatsiavut access to shrimp, the allocation of significant northern shrimp resources to Nunatsiavut interests in the form of licences and special allocations must also be understood in the context of Inuit political mobilisation since the early 1970s in response to colonial and post-colonial resource extraction practices. Indeed, the period leading up to the allocation of northern shrimp resources to Nunatsiavut groups was characterised by vigorous political mobilisation by Inuit, which would eventually lead to a successful land claim and the establishment of the self-governing territory of Nunatsiavut. The Labrador Inuit Association (LIA) played a key role in mobilising Indigenous cultural, social and economic interests in the region, including claims for northern shrimp. It was also responsible for setting up the Labrador Resources Advisory Council (LRAC) in 1976 to promote Labrador interests with regard to oil, gas and other resources including fish.

When the announcement was made that a licence would be granted to fishing interests in Nunatsiavut, the LRAC demanded that the resource be used in a way that benefited local residents and the northern Labrador fishery. As they argued, the allocation of northern shrimp resources, should not "repeat the familiar and very bitterly-resented pattern which already governs the exploitation of every other resource in Labrador. Whether it is fish, iron ore, pulpwood or hydro power, wealth flows out of Labrador in a form which ensures that the main benefits will be felt somewhere else" (LRAC, 1978, p. 1). The top priority, they pointed out, should be "to rebuild the Labrador inshore fishery, not to help already prosperous firms and thriving communities fatten themselves on our future" (LRAC, 1978, p. 1). In other words, the use of the northern shrimp licence should break with past patterns of resource exploitation that brought few benefits to Inuit communities in Nunatsiavut. The LRAC articulated additional demands associated with the northern shrimp licence including the training of Inuit fishermen as crewmen on offshore boats into senior positions, processing of shrimp onshore as a way of generating processing jobs for Inuit coastal communities, and the use of resources from the shrimp licence to bolster the inshore fleet and the processing plants that they served.

Northern Labrador/Nunatsiavut interests benefited significantly from these early northern shrimp allocations.[2] The Torngat Fish Producers Co-operative Society Ltd. (hereafter Torngat Co-operative), established in 1980, was the first organisation in Nunatsiavut to hold an offshore shrimp licence. The LRAC's demands for how the benefits of northern shrimp should be used applied to the licence allocated to the Torngat Co-operative, but also to the ones allocated to companies that fished for northern shrimp in waters adjacent to Nunatsiavut. The Torngat Co-operative is not formally recognised as an Indigenous organisation, and it has never been linked administratively to the Labrador Inuit Association or the Nunatsiavut Government. Nonetheless, its membership is mainly Inuit with a small number of First Nation Innu people and some settlers living in small and isolated towns on Nunatsiavut's coastline (Figure 19.1).

A second offshore licence was granted to Pikalujak Fisheries of Labrador in 1987, which was a joint venture between the Labrador Inuit Association (LIA), the organisation that represented Inuit interests up until the establishment of the Nunatsiavut government, and a southern fishing company. The LIA was established in 1973 with a mandate to strengthen Inuit culture and to motivate for a formal land claim on the basis of Indigenous rights (Brantenberg & Brantenberg, 1984). The organisation played a key role in the negotiations that led to a successful land claim for the Inuit of northern Labrador in 2005, which established the self-governing territory of Nunatsiavut.

From the mid-1990s, stock assessments showed that the northern shrimp resource was growing very rapidly. In response, Canada's Department of Fisheries and Oceans not only increased quotas for existing offshore licence holders but also began granting permits/licences to inshore harvesters in NL and 'special allocations' to community-based organisations and Indigenous groups across Atlantic Canada. Nunatsiavut interests were again beneficiaries through the LIA which received special allocations of northern shrimp in 1997 and again in 2003. The LIA also benefited from additional northern shrimp quotas through its membership of the Northern Coalition, which supports the interests of northern and Indigenous northern shrimp licence and special allocation holders in Canada. In late 1990s, the Northern Coalition secured a very large increase in northern shrimp quotas, which is shared equally between the six members of the organisation, including the Torngat Co-operative and the LIA. In spite of these significant gains in access to northern shrimp, the Nunatsiavut Government has frequently pointed out that these gains do not match the commitment made in the Labrador Inuit Land Claims Agreement signed in 2005, an agreement that carries constitutional authority (TJFB, 2016). Nonetheless, the ability of Nunatsiavut interests to secure access to northern shrimp through offshore licences and special allocations illustrates early dimensions of the "creative hybridizations" emerging as Indigenous groups interact with state-owned resources.

3 Using allocations for inshore development: from royalty charters to communal licences

A key focus of our ongoing research on northern shrimp in Canada has been on how offshore licences and special allocations of northern shrimp to co-operatives, social enterprises and Indigenous groups tended to be exchanged for royalty payments that have benefited remote coastal communities, inshore fish harvesters and processing workers (Foley & Mather, 2016; Foley *et al.*, 2015, 2017). Our research in Nunatsiavut has focused on the activities of the Torngat Co-operative and the LIA/Nunatsiavut Government, and this research has revealed a complex and dynamic history of not only the creative use of royalties for the local fishing sector but also the creative integration of allocations under state-sanctioned communal licences.

The Torngat Co-operative was established in 1980 and was the first offshore licence holder in Nunatsiavut. The Torngat Co-operative has always used its royalty payments to support inshore fishery development. In the first decade after it was awarded its offshore licence, the Torngat Co-operative embarked on an ambitious effort to reinvigorate inshore harvesting and processing. It used royalty earnings to upgrade fish-handling units that supported Inuit small-scale cod and salmon harvesters, it arranged new training opportunities for processing workers at fish plants in Makkovik and Nain, and it provided equipment and supplies to fish harvesters along the coast. Through its royalty agreements it also provided new and very lucrative employment opportunities for its members on offshore factory freezer trawlers. Employment provided cash income, and allowed members of the Torngat Co-operative to access unemployment insurance during the winter season when fishing inshore was not possible. Allocations of northern shrimp to the LIA and the Nunatsiavut Government were used in similar ways to support inshore fishery development, but these resources were also used to support other non-fishery economic development projects that created new opportunities for Inuit in Nunatsiavut.

Up until the early 2000s, northern shrimp allocations to groups in Nunatsiavut were used mainly to generate royalties to support onshore fish-processing facilities and employment opportunities. This changed in the early 2000s when Nunatsiavut fishery interests began to explore the idea of integrating their special allocations under communal licences, which would allow it to expand inshore fishing capacity for northern shrimp, crab and turbot and in turn enhance employment opportunities in onshore processing facilities.

The communal licence mechanism had been in place in Canada since the early 1990s. In the late 1980s and early 1990s, Indigenous groups on the west and east coasts of the country succeeded in winning several landmark court cases where they claimed the historical right to fish resources. Canada's Department of Fisheries and Oceans responded to these court cases in 1992 by launching the Aboriginal Fisheries Strategy (AFS), which aimed to provide Indigenous groups with greater access to fish resources for both food and ceremonial purposes as well as for commercial purposes. A special licensing system, called the Aboriginal Communal Fishing Licence Regulation (ACFLR), was introduced under the broad umbrella of the AFS to provide communal licences to Indigenous groups, who then 'designated' individual Indigenous harvesters (known in Nunatsiavut as 'designates') to fish under the authority of the licence.

The first communal licences in Nunatsiavut were for crab, but communal licences were soon also established for northern shrimp (caught inshore and processed onshore) and turbot, which allowed designates to fish commercially for more than one species. These communal licences were administered by the LIA and, after 2005, by the Nunatsiavut Group of Companies, which designated individuals with experience in the commercial fishing sector across the coastal communities Nunatsiavut (Figure 19.1). Nunatsiavut designates who did not own their own inshore vessels leased vessels from boat-owners in southern Labrador and Newfoundland. The leasing agreements typically involved the southern boat-owner paying the Inuit designate a very small portion of the value of the catch, typically in the range of 15%.

The communal licensing programme was a significant new development for Nunatsiavut. Rather than simply trading quotas for royalties, the Nunatsiavut Government began supporting independent Inuit fishers to become independent commercial inshore harvesters for northern shrimp, crab and turbot. The communal licence designate programme by the Nunatsiavut Government was designed to complement ongoing Torngat Co-operative initiatives, including building inshore harvesting and processing capacity and opportunities. Despite the promise of this new arrangement, research participants suggested to us that it "languished" for a decade or more. A significant proportion of the quotas under the communal licence system were not used in the early years of the programme and had to be re-allocated in the same season to other commercial interests in return for royalties. More recently, however, for reasons that we discuss below, the Nunatsiavut Government's designate programme has overcome some of the constraints of participation and attracted growing attention amongst Inuit harvesters in Nunatsiavut. We now turn to examine the recent growth in the communal licence and designate programme and the dynamics that are shaping its ongoing development.

4 Constraints of communal licences

In the previous section we noted that the effort to establish independent inshore harvesters in Nunatsiavut appeared to languish until the mid-2010s. Since then, changes in the northern shrimp resource and allocation policy shifts have created more favourable conditions for the programme.

Northern shrimp stocks began to decline from the late 2000s, for reasons that are indeterminate, but are likely related to a complex interplay between fishing pressure and climate change factors. The changing biomass of northern shrimp had two important effects that relate to the Nunatsiavut Government's communal licence and designate programme. First, it has led to dramatic cuts in northern shrimp allocations along the northeast coast of Newfoundland as the shrimp biomass weakened in those areas. Inshore and offshore allocations adjacent to Newfoundland and southern Labrador have been slashed from the highs of the late 2010s. At the same time, because northern shrimp stocks are stronger in more northerly zones, northern and Indigenous allocation holders, including those in Nunatsiavut, were not as negatively affected. More recently Nunatsiavut interests have gained additional, albeit modest, quotas of northern shrimp.

The decline in northern shrimp stocks has had a second important impact on the industry, with implications for the Nunatsiavut Government's communal licence and designate programme. The decline in northern shrimp stocks has led to higher prices for this shellfish species which is exported to international markets in Europe and the United States. Canada is by far the largest global supplier of wild-caught cold-water shrimp and decreases in supply have pushed prices to record levels.

For existing and potential new designates in northern Labrador, these shifts in biomass that have led to slightly larger quotas, in a context of overall resource decline, combined with much higher prices, have made the communal licence system a much more attractive proposition. Communal licence designates can now gain access to relatively larger quotas, and the prices they receive through leasehold agreements are now higher. From a financial perspective, accessing quotas linked to communal licences through the designate programme has become a more lucrative proposition. At the same time, lower quotas for southern inshore harvesters means that there is excess capacity to fish for northern shrimp, which has encouraged boat-owners from southern Labrador and Newfoundland to enter into leasehold agreements with Nunatsiavut designates. One designate told us, for example, that they are now contacted directly by inshore shrimp boat-owners from Newfoundland with attractive offers that significantly exceed the standard 15% of the value of the catch.

In this context, it is not surprising that the number of applications for designate licences from Nunatsiavut residents has increased rapidly. In 2015–16, for example, the Nunatsiavut Government designated up to 14 beneficiaries who operate between seven and nine vessels and who were required to hire a minimum of one additional crew, resulting in as many as 32 people employed annually (Foley *et al.*, 2017). Indeed, the number of applications for communal licences regularly exceeded the existing quotas in recent years, which has required the Nunatsiavut Government to establish a new framework for allocating communal licences. These developments within the communal licensing programme have also generated new tensions and contradictions for the Nunatsiavut Government, as some Inuit have been excluded from the designated programme while others participating in the programme are pressing for multi-year licences as a vehicle to accumulate capital.

For Inuit designates in Nunatsiavut, a multi-year designate system holds the obvious attraction of having some certainty with regard to future fishing seasons. There is, however, a far more important consideration at play that relates to the role that fishing licences play in Canada's broader commercial fishery sector. Under Canada's Fisheries Act, individual fishing licences allocated outside the Aboriginal Fisheries Strategy policy are allocated to individuals or enterprises and these are considered to be 'privileges' that are granted on an annual basis. The Department of Fisheries and Oceans has stressed that licences should not be considered a form of property. Yet in practice, fishing licences in Canada are traded for very large sums of money. Indeed, the value of a licence usually far exceeds the value of the boat that is used to fish the quota (Allen, 2018). Harvesters also use fishing licences as collateral to secure loans from financial institutions. These loans have become very important for harvesters wanting to purchase vessels or upgrade existing fishing boats or otherwise invest in their fishing enterprises. In strict legal terms, fishing licences in Canada are not property, but in practice they are traded for large sums and used as collateral to secure loans from financial institutions.

Individual designates fishing under the authority of communal licences do not have the same financial privileges, however. While individual fishing licences in Canada can be used as collateral to secure loans from banks and other financial institutions, communal licences cannot be used in the same way. This is because licences allocated under the ACFLR are allocated to Indigenous groups rather than to individuals. Communal licences cannot be traded between groups or between harvesters, and because they are allocated on an annual basis, they are not considered to be a legitimate form of collateral by financial institutions. Nunatsiavut designates who fish commercially under communal licences are, therefore, unable to use their access to quotas to secure loans to purchase or upgrade vessels, as is the case for settler harvesters in other parts of Canada.

These constraints of communal licences or, rather, the relationship between communal licences and capitalist institutions, was a key theme at both the 2015 and 2016 fisheries workshops in Nunatsiavut (Whalen *et al.*, 2015).[3] Inuit harvesters voiced their concern that the terms of the licence did not allow them to purchase boats and become truly independent enterprise owners. One harvester compared his situation to "Canadian citizens" who were able to take their fishing licences to the bank and use them as collateral to secure loans. This communal licence holder went so far as to suggest that he "just wanted to be like a normal Canadian citizen, to operate in the fishery like other Canadians". Another harvester challenged the policy of communal licences arguing that these were "our licences, they should have been given to us, not to the Nunatsiavut government". In other words, the constraints faced by individual designates facilitated their perception that their designate quotas should not have been allocated under communal licences, but rather should have been allocated to individual Inuit harvesters who could then use them to secure loans to buy a boat. Another designate claimed that her progress as a fish harvester was being compromised through the communal licence system:

> I am being held back right now. Even funding and grant agencies won't look at me. The banks won't look at me. Investors are having trouble investing money in me. Why? Because you don't even own a quota. You don't know whether you are going to have that licence next year.
>
> (Interviews)

5 Creative and constrained hybridisations

At first glance, the Nunatsiavut Government's communal licence experience appears to mirror the findings from research that has examined state allocations of fish resources to Indigenous groups in settler societies. As we noted in our introduction, a key finding of the research in countries like New Zealand and other parts of Canada is that the social objectives of these allocations tend to erode as licence holders become more entangled in capitalist markets. With the recent growth of the Nunatsiavut communal licence system, the initial evidence we have presented appears to suggest that a similar process is under way. For some Nunatsiavut fish harvesters, the communal licence system represents an obstacle to their individual commercial aspirations, which they articulate through an identity politics that appears to question Indigenous belonging and attachment. Yet a more careful analysis of this dynamic period of change points to a more complex process of transformation that can usefully understood through the idea of 'creative and constrained hybridisations'.

Creative hybridisations take a number of forms. For example, while a number of fish harvesters in Nunatsiavut expressed frustration with the communal licence system, and called for changes that will allow them to become owners of boats and formal fishing enterprises, other designates appear to be more risk-averse and are uninterested in the financial risks and uncertainties in borrowing large sums of money from a bank to buy a boat or a fishing enterprise. Another example is the ongoing experimentation of the Nunatsiavut Government to creatively overcome constraints, such as the initiative to explore whether the communal licence designate system could be extended beyond a single year to allow designates to secure loans to purchase a vessel.

The Nunatsiavut Government is exploring a range of different options, with associated financial risks, including the current model that involves entering into lease agreements with boat owners in southern Labrador and Newfoundland (Whynot, 2016). The diversity of options reflects the fact that some communal licence holders are satisfied with the existing system that allows them to earn some income without significant risk, and which also allows them time to pursue traditional activities. As one Nunatsiavut Government official explained:

> There is definitely some interest from our designates to be vessel owners themselves. But a lot of our designates like the fact that they walk onto a vessel and five or six weeks later they are done for the year and they can then pursue their traditional activities like hunting and going to the cabin. That is very appealing to them,
>
> (Interviews)

He also pointed out that contracts from southern boat-owners are far more generous, which has reduced the incentive to buy a boat as a way of earning a higher income:

> There is a lot of competition and boat owners in the south are offering higher shares because they don't have resource to fish like they used to have. A common sharing arrangement even three or four years ago you were lucky if you got in excess of 15 percent, and now designates from what we are hearing are getting up to 40 percent. The designate share has gone up. And so if you are getting a lot of money from a lease, what's the incentive to get a boat?
>
> (Interviews)

Commercial fishing for significant financial reward, for some designates, does not require boat-ownership, and can, significantly, be achieved without compromising meaningful traditional activities on the land. Thus, creative but constrained hybridisations are not only reflected in the communal licence designate system and its relationship to the Canadian state but also in the diversity of relationships and initiatives among Inuit designates. There is a second complexity in the debates and discussions on how communal licences might be used by some designates to secure loans, buy boats and build commercial fishing enterprises. The Nunatsiavut Government is actively exploring ways that these communal licences might be used as collateral with financial institutions. One option involves allocating quotas to individual designates for several years at a time, a move that may convince financial institutions to use these as collateral against a loan. Yet there are those within the Nunatsiavut Government who are concerned that this multi-year allocation system will compromise the original objectives of these communal licences:

> We have more of a social conscience than a business conscience. If we are looking at a multi-year designation system and we are going through a process where a designate becomes a vessel owner, my feeling is that the decision-making process would be more business minded than social minded.
>
> (Interviews)

In other words, a multi-year designate programme might constrain the ability of the Nunatsiavut Government to use communal licences for social and redistributive purposes. As he noted below, communal licences are not owned by individual designates, and they should be used for the benefit of designates and the beneficiaries of Nunatsiavut's hard-fought Land Claim agreement.

> The assets that the designates fish under, don't belong to them. We as the government own the quotas... We provide maximum benefits to our designates, to our beneficiaries.
>
> (Interviews)

The final complex hybridity we consider has to do with how these developments in the communal licence system intersect with identity politics. We noted earlier that during the discussions around the communal licence system, several designates contrasted the constraints they face within the communal license system to the situation of 'Canadian citizens' who are able use their individual licences to secure loans to buy or upgrade fishing vessels. One interpretation of this statement would be that the communal licence system is causing this Indigenous fisher to distance himself from Indigenous identity in favour of a Canadian one that will allow them to achieve their economic aspirations, which is consistent with De Alessi's (2012) argument that Māori identity and social formations have been transformed and weakened through commercial allocations of fish resources. In the case of Nunatsiavut, while the communal licence system may be interacting with identity politics, it is premature to see it as causing a clear *erosion* of Inuit identity. Indeed, one economically ambitious communal licence holder suggests that economic success, Inuit identity and the self-governing territory of Nunatsiavut are not mutually exclusive:

> I want to have the Nunatsiavut flag flying. I want a white boat with the Labrador colours on it – the white, blue and green. I want us to be proud. That when we pull into Charlottetown (southern Labrador shrimp plant) people see that's a Nunatsiavut boat with a Nunatsiavut quota being caught by a beneficiary landing their own stuff. That's something for us to be proud of – not only me, but Nunatsiavut. Otherwise the Nova Scotians and the Newfoundlanders are going to continue to catch our quota and the money is going to leave as well.
>
> (Interviews)

In this designate's view, Indigenous identity and individual economic success are compatible, and provide a way of challenging long-standing issues associated with fish resource politics in this region of northern Canada. It is too simple to suggest that the tension that has emerged with the communal licence system is reducible to tensions between Indigenous identity and economic aspirations.

6 Conclusion: towards an understanding of 'creative and constrained hybridisations'

This chapter provides empirical and theoretical insight into specific processes of socio-spatial embedding (Foley *et al.*, 2015) that have been identified as significant aspects in sustainability transitions (Truffer *et al.*, 2015). To theorise coalitions of actors seeking to advance new socio-economic, political and environmental agendas that can be understood as supporting sustainability transitions (Murphy, 2015, p. 88), we have used the concept of creative and constrained hybridisations to examine a settler state's allocation of fish resource rights to Indigenous groups for commercial purposes. Our case is the self-governing territory of Nunatsiavut where communal licences and designate programme have been used to support a small but vibrant group of inshore harvesters. We have explored tensions within the communal licence system as some designates have begun to press for changes that will allow them to accumulate individual access rights as a less constrained path to commercial success. Our argument is that these developments do not represent a clear case of the erosion of Indigenous identity and the social objectives that are associated with communal fishing rights. Instead, we have pointed to both the creativity and constraints in complex and fluid processes of social and environmental change. While some communal licence designates have experienced the communal licence programme as a constraint, others experience it as a less risky opportunity for maintaining diverse livelihoods that involve both seasonal commercial fishing and traditional, offseason livelihood and culturally significant pursuits. The Nunatsiavut Government, moreover, is also unwilling to give up formal resource rights to individual harvesters as they continue to see communal licences as integral to its responsibility for supporting collective social development. Finally, we argue that Indigenous identity intersects with individual commercial aspirations in complex and interesting ways.

How does the Nunatsiavut case help us extend the idea of creative hybridisations or, as we push the concept further, creative and constrained hybridisations? The concept of creative hybridisations is helpful in that it points to more complex outcomes associated with resource use by Indigenous groups for commercial gain. While the existing literature tends to focus on the erosion of Indigenous identities and the loss of social objectives, the idea of creative hybridisations provides a way of revealing unexpected outcomes associated with Indigenous resource economies. These unpredictable outcomes cannot be determined in advance and are likely to vary from site to site. However, we acknowledge the contributions that highlight the structural influence of capitalist markets and processes. In this way, we believe that creative and constrained hybridisations provides a concept recognising both agency and structure in understanding and explaining how Indigenous groups use resources for commercial gain within settler-colonial and capitalist contexts. In providing a counter to the literature that tends to see only negative outcomes from Indigenous resource use for commercial gain, we are not arguing that creative and constrained hybridisations result in a 'balance' of social objectives, Indigenous identities and market engagement. On the contrary, our case suggests that these different contradictory priorities will shift and change over time in particular social contexts as Indigenous groups continue to gain access to resources.

Acknowledgements

This research was supported through a Leslie Harris Centre Applied Research Fund grant, Memorial University of Newfoundland and through support from the Torngat Secretariat, Happy Valley-Goose Bay, Labrador.

Notes

1 We received ethical clearance for this research from Grenfell Campus Research Ethics Board at Memorial University as well as research clearance from the Nunatsiavut Government's research office.
2 For consistency and clarity, from this point on we use the term Nunatsiavut to refer to developments in the northern shrimp sector in what was northern Labrador before 2005 and what since then is the self-governing territory of Nunatsiavut.
3 These annual fisheries workshops involve participation from key stakeholders associated with the Nunatsiavut fishery including Inuit harvesters, fish processors (i.e., the Torngat Co-operative), the Nunatsiavut Government, Canada's Department of Fisheries and Oceans, the Newfoundland and Labrador Department of Fisheries and Aquaculture, and the Torngat Secretariat. The Torngat Secretariat was established in 2005 following the Labrador Inuit Land Claims Agreement. It is directly related to fisheries through the role it plays as the implementation agent of the Torngat Joint Fisheries Board, which advises the Nunatsiavut Government on fisheries issues.

Bibliography

Allain, J., & Frechette, J-D., (1993). *The Aboriginal Fisheries and the Sparrow Decision*. Background paper. Research Branch, Library of Parliament, Ottawa.

Allen, S. (2018). 'For sale: one fishing enterprise including boat, gear, licence and quota': mapping property rights in Newfoundland's commercial fishery. Ph.D. thesis. Department of Geography, Memorial University.

Brantenberg, A., & Brantenberg, T. (1984). Coastal Northern Labrador after 1950. In Sturtevant, W.C. (Ed.), *Handbook of North American Indians*, 5, Smithsonian Institution, Washington D.C., pp. 689–699.

Capistrano, R.C.G., & Charles, A.T. (2012). Indigenous rights and coastal fisheries: a framework of livelihoods, rights and equity. *Ocean and Coastal Management, 69,* 200–209.

Carothers, C. (2011). Equity and access to fishing rights: exploring the community quota program in the Gulf of Alaska. *Human Organization, 70*, 213–223.

Coates, K. (2000). *The Marshall Decision and Native Rights.* McGill-Queen's Press, Kingston Ontario.

Davis, A., & Jentoft, S. (2001). The challenge and the promise of indigenous peoples' fishing rights: from dependency to agency. *Marine Policy*, *25*, 223–237.

De Alessi, M. (2012). The political economy of fishing rights and claims: the Māori experience in New Zealand. *Agrarian Change, 12*(2–3), 390–412.

Durette, M. (2007). *Indigenous Property Rights in Commercial Fisheries: Canada, New Zealand and Australia Compared.* Centre for Aboriginal Economic Policy Research.

Egan, B., & Place, J. (2013). Minding the gaps: property, geography, and indigenous peoples in Canada. *Geoforum*, *44*, 129–138.

Foley, P., & Mather, C. (2016). Making space for community use rights: insights from "community economies" in Newfoundland and Labrador. *Society & Natural Resources*, *29*(8), 965–980.

Foley, P., Mather, C., & Neis, B. (2015). Governing enclosure for coastal communities: social embeddedness in a Canadian shrimp fishery. *Marine Policy*, *61*, 390–400.

Foley, P., Mather, C., Morris, R., & Snook, J. (2017). Shrimp Allocation Policies and Regional Development Under Conditions of Environmental Change: Insights for Nunatsiavutimmuit. Report for the Harris Centre, Memorial University, available at http://research.library.mun.ca/12788/.

Ginter, J.J. (1995). The Alaska community development quota fisheries management program. *Ocean & Coastal Management*, *28*(1-3), 147–163.

Harris, D.C., & Millerd, P. (2010). Food fish, commercial fish, and fish to support a moderate livelihood: characterizing Aboriginal and Treaty rights to Canadian fisheries. *Arctic Review on Law and Politics, 1*, 82–107.

Haynie, A.C. (2014). Changing usage and value in the Western Alaska Community Development Quota (CDQ) program. *Fisheries Science*, *80*(2), 181–191.

Krause, C., & Ramos, H. (2015). Sharing the same waters. *British Journal of Canadian Studies*, *28*(1), 23–41.

Labrador Inuit Land Claims Agreement. (2005). *Land Claims Agreement Between the Inuit of Labrador and Her Majesty the Queen In Right Of Newfoundland and Labrador and Her Majesty the Queen In Right Of Canada.* Newfoundland and Labrador. Retrieved from www.nunatsiavut.com/wp-content/uploads/2014/07/Labrador-Inuit-Land-Claims-Agreement.pdf

Lalancette, A. (2016). Creeping in? Neoliberalism, indigenous realities and tropical rock lobster (kaiar) management in Torres Strait, Australia. *Marine Policy*, *80*, 47–59.

Lawhon, M., & Murphy, J.T., (2012) Socio-technical regimes and sustainability transitions insights from political ecology. *Progress in Human Geography, 36*(3), 354–378.

LRAC. (1978). Position paper on the Labrador Shrimp Fishery. Labrador Resources Advisory Council, Goose Bay, Labrador.

Mansfield, B. (2007). Property, markets, and dispossession: the Western Alaska Community Development Quota as neoliberalism, social justice, both, and neither. *Antipode, 39*, 479–499.

McCormack, F. (2016a). Indigenous claims: hearings, settlements, and neoliberal silencing. *Political and Legal Anthropology Review*, *39*(2), 226–243.

McCormack, F. (2016b). Sustainability in New Zealand's quota management system: a convenient story. *Marine Policy*, *80*, 35–46.

Murphy, T., (2015). Human geography and socio-technical transitions studies: promising intersections. *Environmental Innovation and Sustainability Transitions*, *17*, 73–91.

Snook, J., Cunsolo, A., & Morris, R. (2018). A half century in the making: governing commercial fisheries through indigenous marine co-management and the Torngat Joint Fisheries Board. In Vestergaard, N., Kaiser, B. A., Fernandez, L., & Nymand Larsen, J. (Eds.), *Arctic Marine Resource Governance and Development* (pp. 53–73). Cham: Springer International Publishing.

Stanbury, W.T. (2003). Accountability for subordinate legislation: the case of the Aboriginal Communal Fishing Licence Regulations. Fraser Institute Digital Publication, Fraser Institute, Calgary.

TJFB. (2016). Torngat Joint Fisheries Board (TJFB) submission to the Ministerial Advisory Panel for the LIFO Review, Ministerial Advisory Panel for the Review of Last-In-First-Out policy. Department of Fisheries and Oceans, Ottawa.

Truffer, B., Murphy, J.T., & Raven, B. (2015) The geography of sustainability transitions: contours of an emerging theme. *Environmental Innovation and Societal Transitions*, *17*, 63–72.

von der Porten, S., Lepofsky, D., McGregor, D., & Silver, J. (2016). Recommendations for marine herring policy change in Canada: aligning with Indigenous legal and inherent rights. *Marine Policy*, *74*, 68–76.

Whalen, J., Snook, J., White, B., & Wood, B., (2015). Fisheries Workshop 2015. Torngat Wildlife, Plants and Fisheries Secretariat, Happy Valley-Goose Bay, Labrador. Available at www.torngatsecretariat.ca/home/files/cat2/2015-fisheries_workshop_report.pdf

Whynot, Z. (2016). Development of a multi-year allocation policy. Presentation to the Joint Fisheries Board Conference, 24 November, Happy Valley-Goose Bay.

Wiber, M., & Milley, C. (2007). After Marshall: implementation of aboriginal fishing rights in Atlantic Canada. *The Journal of Legal Pluralism and Unofficial Law*, *39*(55), 163–186.

20 Tourism in the coastal zone: livelihoods and opportunity for youth in Dominica

Kristin Weis, Catherine Chambers and Patrick J. Holladay

1 Introduction

On 18 September 2017, Hurricane Maria passed directly over Dominica, resulting in 31 fatalities and 37 people missing (United Nations, 2017a). In about a day and a half, it quickly intensified offshore from a tropical storm to a category 5 hurricane and then made landfall in Dominica. The island sustained extensive damage, including ripped roofs and buildings, destroyed crops, disrupted power and water supplies, landslides, flooding and washed-out roads and bridges. Within a few months, most schools had reopened and economic activities had resumed, such as businesses and shops reopening and markets offering a wide selection of food products (United Nations, 2018). School attendance, however, had not returned to pre-hurricane levels, 80% of the population had not regained electricity, the majority of housing had not regained adequate roofing and many people lost their source of livelihood, especially livelihoods within tourism and agriculture (United Nations, 2017b, 2018).

These factors are salient to the economic landscape of Dominica as more than a quarter of the island's 70,000 residents are in extreme poverty (Central Intelligence Agency, 2018). This poverty ties into vulnerabilities and related economic impact of natural disasters that are compounded in Small Island Developing States (SIDS) such as Dominica. Of particular interest in this study is the connection of youth to livelihood opportunities – farming, fishing, and tourism – in the face of such difficulties. What are youth perceptions of employment alternatives, the access of young people to the alternatives, and the future of those livelihoods? This regional economic geography approach concerns itself with youth prospects, sustainability, resilience, and the island's future. The overarching goal of this research, therefore, was to develop an understanding of youth perceptions of livelihoods in Dominica's coastal communities. This chapter focuses in particular on youth and youth engagement in tourism as a livelihood option and factors that enable or hinder youth access to the opportunities and benefits provided by tourism.

1.1 SIDS and tourism

SIDS are physically remote, underdeveloped nations with limited land mass, natural resources and socio-economic development (Sharpley & Telfer, 2015; van der Velde, *et al.*, 2007). These characteristics contribute to economic, social and environmental vulnerabilities, and much of SIDS literature focuses on their disproportionate vulnerability and decreased resilience (Briguglio, 1995, 2003; Pelling & Uitto, 2001). SIDS challenges can be understood as issues of limited access (e.g., biophysical and socio-economic resources), including limited physical access to the international community and limited economic access to goods and services (Briguglio, 1995). SIDS are perceived as highly dependent on foreign aid, investment or development, and SIDS economies are characterised as "open and heavily dependent on trade" (Sharpley & Telfer, 2015, p. 2). The literature often links this dependency to colonial histories, noting that much of SIDS trade is with their former colonial powers (Briguglio, 2003; Scheyvens & Momsen, 2008a).

SIDS are commonly perceived as dependent on tourism (Sharpley & Telfer, 2015) and many scholars understand tourism in SIDS to be a form of neo-colonialism (Scheyvens & Momsen, 2008a). This dependency perception stems from the fact that tourism is often a major revenue source for SIDS and the only source of regular economic growth (Scheyvens & Momsen, 2008b). While tourism is commonly considered to be a poverty-reduction strategy among international development actors, Scheyvens & Momsen (2008b) argue that tourism is only effective at reducing poverty when it is part of more holistic policies that address social sustainability along with the traditional environmental and economic sustainability focus. Tourism is a major global industry and is "widely recognized as one of the world's most significant forms of economic activity" (Hall & Page, 2014, p. 1). While tourism has been viewed as a positive tool for development by both academics and policy-makers, some scholars caution that the positive association between tourism and development requires further and expansive study and note that tourism is not immune to economic, political or environmental uncertainties and surprises (Sharpley & Telfer, 2015). Regardless, the tourism sector continues to expand and the number of international tourists is estimated to grow: roughly doubling over 20 years, from approximately 1.186 billion in 2010 to 1.8 billion in 2030, to increase about 72 times the amount in 1950 (25 million) (UNWTO, 2016).

Coastal zones and tropical islands of SIDS are now some of the most popular tourism destinations (Student, *et al.*, 2016; Weaver, 2001, 2017), and tourism in coastal and marine areas is experiencing an ongoing trend of rapid expansion and increased popularity (Gillis, 2012; Hall, 2001; Hall & Page, 2014). Especially since the mid-20th century, tourism in coastal areas has increased as traditional coastal industries have declined (e.g., fishing), and coastal tourism in SIDS is a prime example of this development. It has become common practice for many coastal nations to incorporate tourism into their development and economic policies as tourism is viewed to be a more reliable revenue source and more resilient to external pressures (e.g., economic crises, terrorism, climate change) (Sharpley & Telfer, 2015).

1.2 Dominica in context

For many years, Dominica benefited from the European Union's preferential treatment of Caribbean banana exports (Slinger, 2002). In 1993, the United States and South American multinational corporations successfully challenged this trade relationship and engaged in a "banana trade war" against Caribbean banana exporters and the European Union (Holladay & Powell, 2016). The resulting World Trade Organization (WTO) ruling effectively ended the "green gold" era for Dominica: banana exports decreased by about two-thirds (in both tonnage and USD) between 1993 and 2001 and the number of banana farmers decreased by 92% from the 1980s to 2009 (Holladay & Powell, 2016). Current poverty rates are linked to the legacy of Dominica's past dependence on banana exports and the related WTO ruling (UNDP, 2016). Dominica has received a large amount of investment from the European Union to develop community-based tourism as a strategy for enhancing dimensions of resilience; however, an understanding of specifics related to social-ecological resilience remains undertheorised (Holladay & Powell, 2013). Dominica has a primarily coastal tourism model as most of its target tourist activities take place within or near the coastal zone (e.g., coastal walks, coastal community festivals, coastal restaurants) and it has many marine tourism activities (e.g., whale watching, snorkelling, scuba diving). Unlike its Caribbean neighbours, Dominica did not develop 3S (Sea-Sun-Sand, Emmanuel, 2014) coastal tourism because it lacks 3S-style white sand beaches, and access to the island is limited by Dominica's geographical location and topography (Holladay & Powell, 2013; Weaver, 1991). As a result of being unable to effectively compete with 3S markets, Dominica developed a niche tourism model and established itself as *The Nature Island* of the Caribbean (Discover Dominica Authority, 2013; Miller & Henthorne, 2007). Dominica's *Nature Island* brand attracts international tourists to the island by providing a competitive alternative to 3S tourism options in the region (Miller & Henthorne, 2007). Dominica has leveraged tourism as a key development tool for its population and its tourism model appears to rely on unique features such as the island's distinct culture and volcanic activity (Edgell & Swanson, 2013).

Dominica is often characterised as rich in essential resources (Allen & Lines, 2001; Hypolite *et al.*, 2002), where small-scale fishing and agriculture allow for a highly localised subsistence lifestyle (Peteru *et al.*, 2010). However, it is also viewed as wild and potentially volatile due to its volcanic activity, steep terrain, exposure to large amounts of rainfall and hurricanes (the most recent being Hurricane Maria in 2017), and the severe damage that such conditions can cause to modern development, infrastructure and agriculture (e.g., washed-out roads from swollen rivers, landslides, wind-damaged crops). Coupled with international economic instability, reduced tourism income, and diminished Foreign Direct Investment (FDI), Dominica's economy has weakened (Holladay & Powell, 2016, p. 1). Essentially, unexpected costs for reconstruction have accumulated while revenues have decreased; Dominica's public debt is now more than 85% of its GDP and its ability to borrow is thus limited (IMF, 2016). The Caribbean Development Bank (CDB), of which Dominica is a borrowing member, estimated that Dominica had 6.4% negative growth and lost 225% of its GDP as a result of Hurricane Maria in 2017 (Government Information Service, 2018). The CDB cited successes for the region as a general decrease in government debt and an increase in stayover visitors; however, challenges of "high debt, unemployment, low productivity and competitiveness and environmental vulnerability" remain for the region (Government Information Service, 2018, para. 4).

As Dominica rebuilds after Hurricane Maria, the CDB has projected that Dominica's economy will grow by 6.9% in 2018 and the Dominican Government has announced its aim to become the world's first climate-resilient nation (Government Information Service, 2018). There is concern about the ability to rebuild and grow Dominica's economy in the face of potential future storm damage during upcoming annual hurricane seasons: phrases such as "time is not on our side" and Dominica is on "the frontline of the war on climate change" by Dominica's Foreign Minister indicate an urgency felt by the government to recover before the next storm damage occurs (Reuters, 2018, paras. 1–2). One example of adapting to future severe storm damage is a plan to increase the island's amateur radio operator network (e.g., training, equipment) after the island was completely reliant on a few amateur radio operators and satellite phones for all communication for about three days after Hurricane Maria (ARRL, 2018; Government of the Commonwealth of Dominica, 2017). In 2016, estimated incoming remittances to Dominica were $24 million, with the most coming from the United States ($12 million), France ($3 million), US Virgin Islands ($2 million), and the United Kingdom ($2 million); outgoing remittances were an estimated $5 million with the most also going to the United States ($1 million) (Pew Research Center, 2018). Net migration is estimated at a rate of -1 per 1,000 people from 2010 to 2015, which is a change from -8 per 1,000 people from 1985 to 1990. The projected rate of migration among Dominica's working-age population (ages 15 to 64) is consistently -1 per 1,000 people from 2015 to 2050 (Pew Research Center, 2018). Dominica's primary industries are agriculture and tourism (Payne, 2008; Weaver, 1991), with national and international perceptions that tourism is a more reliable industry and should replace agriculture as the island's primary revenue source (Holladay & Powell, 2016). The EU is Dominica's most significant grant aid donor and it has aimed to support Dominica's recovery as it adapts to the new post-WTO ruling international economy: it has funded "agricultural and economic diversification" efforts, including programmes to develop ecotourism, youth skills, and train former banana farmers for new opportunities (UNDP, 2016, para. 7).

1.3 Social-ecological resilience theory and youth

Tourism destinations are complex in nature and it can be difficult to predict impacts on the interrelated environment and tourism systems (Baggio, 2008). Therefore, there is a need to better understand the dynamics of tourism markets to effectively develop tourism strategies that enhance dimensions of a destination's social-ecological resilience. Resilience literature related to tourism and sustainable development describes four dimensions of resilience: social, political, economic and ecological (Adger, 2000; Holladay & Powell, 2013, 2016). The more resilient a community is, the less vulnerable it will be to sudden or unexpected economic, political, social, and environmental adversities (Holladay & Powell, 2013). Communities with a high level of resilience are more likely to develop effective strategies for responding to unavoidable scenarios, which may challenge the status quo. A resilient community is understood as one that is able to "harness its resources in order to adapt to change" (Bec, McLennan & Moyle, 2016, p. 2). Human interactions, economic forces, and ecological resilience all contribute to a community's level of social-ecological resilience (Maciejewski & Cumming, 2015).

Particularly for youth, access to livelihood opportunities can be an important potential benefit that contributes to overall community resilience, especially in rural areas or communities with limited employment prospects (Sebele, 2010). The literature on youth in resource-dependent communities consistently discusses local employment opportunities, out-migration, and the negative consequences when neither option is available. Within SIDS literature, youth may be briefly discussed as a component of social equity, tourism employment or common out-migration trends. Ecotourism models and adaptation to climate change are also frequent themes. Resource-dependent communities experience at least some similar youth migration dynamics – both in developing countries and in rural areas of developed countries (Bjarnason & Thorlindsson, 2006). Community connections have become less based on location and familial ties and more based on individual identity and preference (Bjarnason & Thorlindsson, 2006). Essentially, if a community has a limited menu of lifestyle and livelihood options – limited opportunities – then an individual seeking a specific lifestyle or livelihood is less likely to remain in that community and will instead seek another community with a more extensive and customizable menu of options. The challenge for resource-dependent communities is that they are unable to compete with other communities and thus their community members may leave, especially youth. The resulting population decline has often led to an increased vulnerability of economic, social, and cultural systems within the community (Bjarnason & Thorlindsson, 2006).

2 Methods

2.1 Study communities

The purpose of this study was to understand trends *within* communities in Dominica, rather than an in-depth comparison *between* communities or to extrapolate findings to all of Dominica. Therefore, the six communities of Roseau, Scotts Head, Soufriere, Layou, Mero and Portsmouth were chosen. Roseau is the capital of Dominica and is an urban hub for government and business, as well as the primary cruiseship port on the island. Portsmouth also has a cruise port which is used by smaller cruiseships. Two communities, Scotts Head and Layou, were fishing villages with underdeveloped tourism activity. The two additional communities, Soufriere and Mero, were neighbouring communities to Scotts Head and Layou, respectively, and had more developed tourism activity. As noted above, Hurricane Maria happened after the completion of this research. All of the communities in this study were affected but information about their sustained damage and subsequent recovery is limited and beyond the scope of this chapter.

2.2 Data collection

2.2.1 Semi-structured interviews with key informants

Semi-structured interviews were conducted with four to eight key informants in each of the six study communities for a combined total of 40 individuals in November 2016. The interviews were part of a larger research project (see Weis, 2018) intended to identify factors that influence residents' access to the benefits of tourism and describe the effect of tourism on social and cultural dynamics. Key informants were identified on the ground as individuals who were self-employed in either the farming, fishing or tourism industries, community leaders, or those who were involved in tourism-related efforts or policy at a mid- to senior-level. Interviews were distributed almost evenly by gender, with 19 males (47.5%) and 21 females (52.5%); 33 informants were native Dominicans (82.5%) and seven were born abroad (17.5%).

2.2.2 Youth focus groups

Four focus groups composed of three to ten individuals between the ages of 14 and 18 provided insight into the perceptions of youth regarding their access to tourism-related opportunities that was absent in semi-structured interviews with key informants. The focus groups intended to elicit information relevant to the connections between youth preferences and aspects of resilience and sustainability, which intrude into the lives of youth on the island (Table 20.1). In Dominica, there are not schools in every community and so students often travel to a secondary school in another town or area. Focus groups therefore do not necessarily reflect perspectives specific to a single community; instead,

Table 20.1 Focus group protocol

Activity	*Description*	*Discussion question(s)*
Ranking livelihood preferences	Present five livelihood opportunities (tourism, fishing, agriculture) to the group and have them rank them in order of preference. Observe the decision-making discussion and participants' rationale behind their rankings	What are some words that you think describe working in tourism here?
Post-activity discussion	After the activity, have discussion to explore the rationale behind the ranking process. Come back together as one group if participants had split up into multiple groups	Why did you choose tourism as X place on list (first, last, etc.)? What are the good and bad things about working in tourism? So it looks like you all are thinking X, so… In last five years, what do you think has changed in tourism? Is it hard to get a job in tourism? What is that process like? How would you describe your community? Tell me some words that describe your town/community.

they reflect youth perspectives in Dominica in general. Participants were selected by the area's secondary school administrators or teachers (based on student availability or interest in tourism).

After some initial introductions (e.g., collecting participant age and school year), students were asked to rank five occupations amongst themselves. The five occupations were farming, fishing and three others (at least one of which was related to tourism). For some of the focus groups, we asked them to choose the remaining three occupations and for others we chose the occupations for them. We then asked the participants to discuss the occupations amongst themselves and rank them by preference. At the end of the session, we asked the group about their experience ranking the options and their reasoning behind the order of preference (e.g., "what makes the top choice preferable?" and "why did you choose taxi driver over fisher?").

2.3 Data analysis

All research followed the informed consent process and interviews and focus group sessions were audio-recorded. Audio recordings from semi-structured interviews and focus groups were transcribed and edited for accuracy.

Transcript data alongside field note data were coded inductively by hand to allow understanding of the data to develop from studying informant answers. Themes were identified by the prominence of occurrence in interviews and focus groups, but also by relevance to the research questions. Selected comments were then identified as either reaching data saturation, a strong theme echoed by many, or a smaller theme only mentioned by a few.

3 Results

3.1 Focus groups: perceptions of youth

Focus group participants felt tourism-related opportunities were accessible, but tourism professions were not necessarily desirable (Figure 20.1). Tourism-related opportunities included more indirect social and economic aspects as well as other contributors to development. Access to social benefits centred on general interaction with "*people from abroad*" and sharing Dominica's culture, "*festivals*" and "*knowledge of the country*" – these activities were seen as both benefits of and skills needed for tourism-related opportunities. Participants often cited interaction skills needed to access social benefits, such as "*communication*", "*patience*" and "*socializing skills*", and indicated that they had access to them – their optimism and confidence implied that

Figure 20.1 Focus group participant responses describing their future; text size indicative of frequency

they either already had these skills or could easily develop them. Access to economic benefits centred on the ability to earn income, where tourism was associated with "*employment*", expected to "*create jobs*" and promote "*local products*" and "*crafts*" made by "*craftsmen*". Access to other benefits centred on "*development*", with "*infrastructure*" and "*physical development*" often balanced with "*natural resources*" and the necessity to "*protect forests*" and "*popular sites*".

While participants consistently associated tourism in Dominica with economic development (e.g., *income, finance, money, jobs, create jobs, employment, economy, products, advertise*) and social and cultural interaction (e.g., *communication, foreign exchange, people from abroad, cooperation, patience, socialising skills, country knowledge*), some participants referred to negative aspects. Two participants mentioned the social challenges of "*language barrier[s]*" and "*divorce*" and two others mentioned economic challenges, associating tourism with "*poor*", "*seasonal development*" and "*seasonal benefits*" – implying that benefits of tourism were not consistently available year-round. These economic challenges were consistent with key informant responses that described tourism activity as seasonal and not consistent or reliable enough to rely on year-round as a single source of income.

Pursuing higher education and income were consistent themes when participants were asked to describe their plans after graduation, e.g., "*college*", "*further education*", "*get a job*", "*get any job*", "*anything available*". Views on tourism-specific opportunities were mixed. Some participants appeared to view tourism prospects as less desirable professions or short-term options while others described entrepreneurial tourism businesses as their ideal profession. This discrepancy could be at least partly due to the numerous tourism professions and individuals were not always discussing the same profession or circumstances. For example, participants who described tourism jobs as undesirable were discussing their future plans more generally and appeared to give longer-term answers about "*college*" rather than referring to tour guiding, which was also described as a starter job for young men to make extra money in high school. These mixed perceptions of tourism opportunities can perhaps be understood from another angle: individual autonomy was also a theme and a few participants specifically described their future as one of "*independence*" and "*freedom*".

3.2 Perceptions **about** *youth*

Informants mentioned that there was a difference between the current younger generation and older generations. Previous generations may have moved abroad (e.g., to the UK) for work, but then returned to Dominica once they were older. However, youth and their parents now aim for the youth to move abroad permanently (e.g., to the United States) for more opportunities. This description could be related to economic development, upward social mobility across generations, and/or new generational expectations. Further, informants described how they thought young people do not do hard farming work and noted that youth prefer to do fishing instead, which is viewed as easier labour when compared to farming; the profit is quicker as the harvest is quicker (i.e., one day to catch fish versus months of waiting for crops to grow). Some informants expressed that they would like to see more tourism-related education in schools so that youth would be more prepared for tourism jobs; more education was also desirable to improve public awareness and common sense about visitors on the island, such as understanding tourist expectations and how to deliver a consistent form of good service.

3.3 Livelihoods and opportunities: "Everything goes into the same pot"

Informants consistently described the need for multiple revenue sources from varying combinations of tourism, farming and fishing opportunities. Many informants who held office jobs participated in either farming, fishing or tourism, or a combination, on weekends and when not performing their office job duties. Some informants noted that they fish in the morning and/or farm at a family's plot on weekends and then earn money from tourism-related income during the week (e.g., as a taxi or tour bus driver). Youth often helped out on family farms or with fishing on the weekends. As one informant noted, in general, "*in Dominica, it's hard to make money*" and one needs many revenue sources to create an income. Informants described farming and fishing as "*up and down*", and that harvest- (from land or sea) and tourism-related opportunities are not constant. As a result, they felt it prudent to have multiple sources of income. Informants described having diverse income portfolios so that, if one source diminishes, other potential income sources remain available. Many informants, adults and youth alike, specifically referred to the banana crash as reasoning for ensuring a diversity of economic options at the individual and household level.

A consistent view expressed by interview informants and youth was that, when choosing between farming, fishing and tourism, the choice did not matter; one simply needs to make money and then one can make decisions about personal preferences. As one informant said, "*everything goes into the same pot at the end of the day*". Individuals seemingly attached little or no identity to the type of work they did. This perception was most strongly expressed that there was no difference between people who worked in tourism and people who did not. As one said, "*everybody likes to make money*" and "*everyone does better*" in the tourism season. Choice was said to be based on which opportunities were available, often described as depending on season, circumstances or day.

Informants also consistently stated that tourism, farming and fishing were interdependent or interconnected. This was also perceived to be intentional and supported by top-down efforts, as one informant said that the Dominica Hotel and Tourism Association (DHTA) – a "*very strong*" public-private organisation – aims "*to blend*" tourism, farming, and fishing. As another explained, "*it's not choosing for me*" because "*you're not choosing agriculture over tourism. You're choosing agriculture as a manifestation of what you'd like to produce in this world and that is linked to agriculture, fisheries, and tourism.*" For example, another said that farming and fishing complement each other, with seafood and agriculture products often being cooked together to create a meal that can be sold to tourists. Farming and fishing were described as having "*endless opportunities*". Farming, fishing and tourism were understood to be "*all linked up*" where farming and fishing provided products to local tourism businesses, such as hotels and restaurants, that the tourists were then able to buy. As one informant said, "*tourism is a six-month thing*" and that, regarding farming and fishing, "*everything has its season*". This was a frequent response among informants, who explained that people do tourism when there's a season for it, and otherwise do farming or fishing when tourism is not an option.

Tourism was frequently described as a young industry and still forming. "*Tourism is still a developing thing here [in Dominica]*" and people are slowly developing tourism ideas. Informants in Portsmouth compared Dominica to surrounding islands, saying that other Caribbean islands have a more developed or more advanced tourism industry, such as convenient marinas for yachters. One theme that emerged was the perception that there are more government programmes and funding available to help fishers than farmers, and that funding for tourism was essentially nonexistent since the industry was still growing. There was a sense that the fishing industry was comparably more organised than the farming industry, including more government support and a clearer guiding strategy. One informant stated that farming efforts needed more government help due to poor road conditions (worsened by severe storm damage, e.g., Tropical Storm Erika), making it difficult for farmers to transport their crops to market. Another informant described fishing as more established, with more government support, and that farming is more disorganised as an industry and is mainly home gardening where produce is grown for one's own consumption. Informants mentioned that tourism workers needed to acquire more training and earn certifications to be more knowledgeable about the island, however, some also cited difficult or limited access to tourism opportunities. Unclear or complicated bureaucratic processes and a difficulty of finding answers to one's questions created the perception that entering into tourism is "*difficult because no one wants to help you*".

4 Discussion

This research set out to explore the relationship of tourism to aspects of socio-ecological resilience and found that for youth in Dominica, opportunities in tourism mainly fall under aspects of economic resilience. Recent storms in Dominica have caused extensive economic losses, such as Hurricanes Dean and Omar in 2007 and 2008 (Holladay & Powell, 2016, p. 1) and Tropical

Storm Erika in 2015 (Stewart, 2016). The most current Hurricane Maria serves as a reminder that research into resilience in coastal communities is complex, ever-changing, and of utmost importance. Assuming unstable climate patterns continue, SIDS like Dominica must develop more reliable revenue streams and a more resilient economy.

Tourism is of particular value to SIDS because it can be used to enhance dimensions of social-ecological resilience and reduce vulnerability to sudden or unexpected social, political, economic, and ecological disturbances (Adger *et al.*, 2005; Bec *et al.*, 2016; Briguglio, 1995). Within the context of social-ecological systems (SESs) as complex dynamic systems, resilient social systems are able to continually develop and adapt in response to changing social or environmental components (Folke, 2006). The economic dimension of resilience involves economic diversity (diversity of income), income stability, and prevents leakage (i.e., earned income is spent within the community, rather than earned in the community and then spent elsewhere) (Holladay & Powell, 2016). Economic dimensions of resilience are similar to political dimensions in that they both involve decentralised, local distribution of key factors, such as income and power, respectively. Highly resilient economic, social and institutional dimensions all require degrees of equity and trust (Adger, 2000). With a steady increase in the number of international tourist arrivals and amount of tourism-generated revenue, tourism can impact communities by increasing access to economic benefits that may also improve quality of life (Andereck *et al.*, 2005; Kim *et al.*, 2013). Potential benefits of tourism include economic diversity from tax revenues, culture and entertainment (e.g., festivals, attractions, restaurants), outdoor recreation, and employment and livelihood opportunities (Andereck *et al.*, 2005; Kim *et al.*, 2013). Individual and community access to benefits of tourism are dependent on the biosphere and specific SESs in which they live (Folke, 2016). Accessing potential benefits of tourism – such as economic diversity – can contribute to individual and community resilience to environmental change, including climate change and related shocks (Folke, 2016; Walker *et al.*, 2004).

As the legacy of colonialism and the subsequent banana crash remain fresh in locals' minds, Dominicans aim to diversify economic activities that simultaneously offer sustained income for individuals and families as well as the opportunity for development and growth at regional and national levels. While tourism is commonly used as a development strategy for SIDS, its effectiveness in reducing poverty or increasing social or economic equality is questionable (Scheyvens & Momsen, 2008b). SIDS are increasingly concerned about byproducts of poverty, such as domestic violence and youth suicide (Scheyvens & Momsen, 2008b). A main benefit of tourism for SIDS communities is the introduction of employment or income-generating opportunities (Shakeela *et al.*, 2010). If employment, education or lifestyle opportunities are unavailable or undesirable, youth are likely to migrate out to communities with more desirable opportunities (Halseth *et al.*, 2009). Communities that lack employment they may become vulnerable to social and economic challenges associated with poverty or weakened cultural or community identities (Halseth *et al.*, 2009). Youth migration is also present in SIDS communities in the Caribbean, and ecotourism had been cited as contributing to community ties, specifically "stronger village organization and social cohesion" (Shah & Dulal, 2015, p. 5).

5 Conclusion

Compared to other Caribbean islands, the tourism industry in Dominica is underdeveloped. As tourism development continues to grow, it must be understood as a geographic process acting across global, regional and local scales. Youth in Dominican communities are bound by the complex interactions of industry, technology, markets, policy and culture. The changing livelihood expectations of youth in Dominica show how youth perceptions can help scholars understand changing livelihood, urbanisation and poverty trends as a geographic process, and one that is often uneven. Anticipating future trends based on youth studies will allow for more response and planning in advance of expected challenges.

For Dominican youth, an independent lifestyle appears to be valued more than the industry through which it is accomplished, meaning that individual identity is not tied to how one earns a living. How one chooses to earn an income is less important to informants than the consistency of income and its income in comparison to other alternatives. For youth, tourism gives opportunities for freedom and individual control over income pathways. At the same time, despite the international push to increase tourism in Dominica, our results show that youth do not view tourism as a viable year-round economic opportunity. Instead, economic diversity through alternative livelihoods to and within tourism can expand the range of income options for youth as future job-seekers; a tourism opportunity can be one option of many. The value of the autonomy given by employment in tourism as a small-business owner is an important driver for youth in their considerations of future career aspects; however, other careers in tourism, such as being a tour guide or hotel staffer, were not as valued. Although tourism can offer independence and expanded livelihood opportunities and is seen as a positive development for youth in Dominica, it is seen as another "seasonal development". As such, tourism does not offer a drastically different livelihood opportunity from other more traditional pathways, but it does increase economic resilience and diversity for youth in Dominican communities as another complimentary income pathway.

Acknowledgments

This research would not have been possible without the assistance and support of research participants as well as Nancy Osler and many other generous individuals who identified potential research participants, made introductions, and gave helpful tips. The chapter was greatly improved by the comments of the external reviewer. The authors also wish to thank Dr. Patrick Heidkamp and Dr. John Morrissey, book editors, for comments on the chapter and feedback on preliminary results presented at the Coastal Transitions Conference on 31March 2016 in New Haven, Connecticut.

Bibliography

Adger, W.N. (2000). Social and ecological resilience: are they related? *Progress in Human Geography*, *24*(3), 347–364. https://doi.org/10.1191/030913200701540465

Adger, W.N., Hughes, T.P., Folke, C., Carpenter, S.R., & Rockström, J. (2005). Social-ecological resilience to coastal disasters. *Science*, *309*(5737), 1036–1039. https://doi.org/10.1126/science.1112122

Allen, B., & Lines, L. (2001). External economic pressures and park planning: a case study from Dominica. In *Harmon D. (Ed.), Crossing Boundaries in Park Management (pp.* 332--335*)*. Hancock: The George Wright Society. Retrieved from www.georgewright.org/57allen.pdf

Andereck, K.L., Valentine, K.M., Knopf, R.C., & Vogt, C.A. (2005). Residents' perceptions of community tourism impacts. *Annals of Tourism Research*, *32*(4), 1056–1076. https://doi.org/10.1016/j.annals.2005.03.001

ARRL. (2018). Dominica post-disaster needs assessment cites amateur radio's role after Maria. Retrieved 16 February 2018, from www.arrl.org/news/dominica-post-disaster-needs-assessment-cites-amateur-radio-s-role-after-maria

Baggio, R. (2008). Symptoms of complexity in a tourism system. ResearchGate. Retrieved from www.researchgate.net/publication/43496732_Symptoms_of_Complexity_in_a_Tourism_System

Bec, A., McLennan, C., & Moyle, B.D. (2016). Community resilience to long-term tourism decline and rejuvenation: a literature review and conceptual model. *Current Issues in Tourism*, *19*(5), 431–457. https://doi.org/10.1080/13683500.2015.1083538

Bjarnason, T., & Thorlindsson, T. (2006). Should I stay or should I go? Migration expectations among youth in Icelandic fishing and farming communities. *Journal of Rural Studies*, *22*(3), 290–300. https://doi.org/10.1016/j.jrurstud.2005.09.004

Briguglio, L. (1995). Small island developing states and their economic vulnerabilities. *World Development*, *23*(9), 1615–1632. https://doi.org/10.1016/0305-750X(95)00065-K

Briguglio, L. (2003). The Vulnerability Index and Small Island Developing States: A Review of Conceptual and Methodological Issues. Presented at the AIMS Regional Preparatory Meeting on the Ten Year Review of the Barbados Programme of Action, Praia, Cape Verde. Retrieved from www.um.edu.mt/__data/assets/pdf_file/0019/44137/vulnerability_paper_sep03.pdf

Central Intelligence Agency. (2018). The World Factbook: Dominica. Retrieved 19 February 2018, from www.cia.gov/library/publications/the-world-factbook/geos/do.html

Discover Dominica Authority. (2013). Discover Dominica, the Nature Island. Retrieved 12 April 2017, from http://dominica.dm/

Edgell, D.L.S., & Swanson, J. (2013). *Tourism Policy and Planning: Yesterday, Today, and Tomorrow*. Routledge.

Emmanuel, K. (2014). Shifting tourism flows in a changing climate: policy implications for the Caribbean. *Worldwide Hospitality and Tourism Themes*, *6*(2), 118–126.

Folke, C. (2006). Resilience: the emergence of a perspective for social–ecological systems analyses. *Global Environmental Change*, *16*(3), 253–267. https://doi.org/10.1016/j.gloenvcha.2006.04.002

Folke, C. (2016). *Resilience* (Republished). http://dx.doi.org/10.5751/ES-09088-210444

Gillis, J.R. (2012). *The Human Shore: Seacoasts in History*. Chicago: University of Chicago Press.

Government Information Service. (2018). Dominica's economy projected to grow by 6.9% in 2018. Retrieved 16 February 2018, from www.news.gov.dm/index.php/news/4526-dominica-s-economy-projected-to-grow-by-6-9-in-2018

Government of the Commonwealth of Dominica. (2017). Post-disaster needs assessment Hurricane Maria September 18, 2017 (p. 143). Retrieved from https://reliefweb.int/sites/reliefweb.int/files/resources/dominica-pdna-maria.pdf

Hall, C.M. (2001). Trends in ocean and coastal tourism: the end of the last frontier? *Ocean & Coastal Management*, *44*(9), 601–618. https://doi.org/10.1016/S0964-5691(01)00071-0

Hall, C.M., & Page, S.J. (2014). *The Geography of Tourism and Recreation: Environment, Place and Space* (Fourth edition). New York: Routledge.

Halseth, G., Markey, S.P., & Bruce, D. (2009). *The Next Rural Economies: Constructing Rural Place in Global Economies*. CABI.

Holladay, P.J., & Powell, R.B. (2013). Resident perceptions of social–ecological resilience and the sustainability of community-based tourism development in the Commonwealth of Dominica. *Journal of Sustainable Tourism*, *21*(8), 1188–1211. https://doi.org/10.1080/09669582.2013.776059

Holladay, P.J., & Powell, R.B. (2016). Social-ecological systems, stakeholders and sustainability: a qualitative inquiry into community-based tourism in the Commonwealth of Dominica. *Caribbean Studies*, *44*(1–2), 3–28.

Hypolite, E., Green, G.C., & Burley, J. (2002). Ecotourism: its potential role in forest resource conservation in the Commonwealth of Dominica, West Indies. *International Forestry Review*, *4*(4), 298–306. https://doi.org/10.1505/ifor.4.4.298.40529

International Monetary Fund. Western Hemisphere Dept. (2016). Dominica: Selected Issues (Country Report No. 16/245 No. 16/245). Retrieved from http://www.imf.org/en/Publications/CR/Issues/2016/12/31/Dominica-Selected-Issues-44108

Kim, K., Uysal, M., & Sirgy, M. J. (2013). How does tourism in a community impact the quality of life of community residents? *Tourism Management*, *36*, 527–540. https://doi.org/10.1016/j.tourman.2012.09.005

Maciejewski, K., & Cumming, G.S. (2015). The relevance of socioeconomic interactions for the resilience of protected area networks. *Ecosphere*, *6*(9), 1–14. https://doi.org/10.1890/ES15-00022.1

Miller, M.M., & Henthorne, T.L. (2007). in search of competitive advantage in Caribbean tourism websites. *Journal of Travel & Tourism Marketing*, *21*(2–3), 49–62. https://doi.org/10.1300/J073v21n02_04

Payne, A. (2008). After bananas: the IMF and the politics of stabilisation and diversification in Dominica. *Bulletin of Latin American Research*, *27*(3), 317–332. https://doi.org/10.1111/j.1470-9856.2008.00272.x

Pelling, M., & Uitto, J.I. (2001). Small island developing states: natural disaster vulnerability and global change. *Global Environmental Change Part B: Environmental Hazards*, *3*(2), 49–62. https://doi.org/10.1016/S1464-2867(01)00018-3

Peteru, S., Regan, S., & Klak, T. (2010). Local vibrancy in a globalizing world: evidence from Dominica, Eastern Caribbean. *Focus on Geography*, *53*(4), 125–133. https://doi.org/10.1111/j.1949-8535.2010.00015.x

Pew Research Center. (2018). Remittance flows worldwide in 2016. Retrieved 16 February 2018, from www.pewglobal.org/interactives/remittance-map/

Reuters. (2018). Hurricane-hit Dominica hurries to prepare for next storm season. Retrieved 16 February 2018, from www.financialexpress.com/world-news/hurricane-hit-dominica-hurries-to-prepare-for-next-storm-season/1038815/

Scheyvens, R., & Momsen, J. (2008a). Tourism in Small Island States: from vulnerability to strengths. *Journal of Sustainable Tourism*, *16*(5), 491–510. https://doi.org/10.1080/09669580802159586

Scheyvens, R., & Momsen, J.H. (2008b). Tourism and poverty reduction: issues for Small Island States. *Tourism Geographies*, *10*(1), 22–41. https://doi.org/10.1080/14616680701825115

Sebele, L.S. (2010). Community-based tourism ventures, benefits and challenges: Khama Rhino Sanctuary Trust, Central District, Botswana. *Tourism Management*, *31*(1), 136–146. https://doi.org/10.1016/j.tourman.2009.01.005

Shah, K.U., & Dulal, H.B. (2015). Household capacity to adapt to climate change and implications for food security in Trinidad and Tobago. *Regional Environmental Change*, *15*(7), 1379–1391. https://doi.org/10.1007/s10113-015-0830-1

Shakeela, A., Breakey, N., & Ruhanen, L. (2010). Dilemma of a paradise destination: tourism education and local employment as contributors to sustainable development. *CAUTHE 2010: Tourism and Hospitality: Challenge the Limits*, 1340.

Sharpley, R., & Telfer, D.J. (2015). *Tourism and Development: Concepts and Issues*. Bristol; Buffalo: Channel View Publications.

Slinger, V.A.V. (2002). *Ecotourism in a Small Caribbean Island: Lessons Learned For Economic Development and Nature Preservation*. Doctoral dissertation, University of Florida, Gainesville, FL. Retrieved from http://uf.catalog.fcla.edu/uf.jsp?st=UF029665854&ix=pm&I=0&V=D&pm=1

Stewart, S.R. (2016). National Hurricane Center Annual Summary: 2015 Atlantic Hurricane Season (p. 16). NOAA. Retrieved from http://www.nhc.noaa.gov/data/tcr/summary_atlc_2015.pdf

Student, J., Amelung, B., & Lamers, M. (2016). Vulnerability is dynamic! conceptualising a dynamic approach to coastal tourism destinations' vulnerability. In Filho, W.L. (Ed.), *Innovation in Climate Change Adaptation* (pp. 31–42). Springer International Publishing. https://doi.org/10.1007/978-3-319-25814-0_3

UNDP. (2016). About The Commonwealth of Dominica. Retrieved 26 October 2017, from www.bb.undp.org/content/barbados/en/home/countryinfo/the_commonwealth_of_dominica.html

United Nations. (2017a). Dominica: Hurricane Maria Situation (Text No. 11) (p. 9). Dominica: United Nations. Retrieved from https://reliefweb.int/report/dominica/dominica-hurricane-maria-situation-report-no-11-16-november-2017

United Nations. (2017b). Dominica: Hurricane Maria Situation Report No. 13 - (as of 14 December 2017). Retrieved 17 February 2018, from https://reliefweb.int/report/dominica/dominica-hurricane-maria-situation-report-no-13-14-december-2017

United Nations. (2018). Dominica: Hurricane Maria - Overview of the Humanitarian Response in 2017 (September-December 2017), 8 February 2018. Retrieved 16 February 2018, from https://reliefweb.int/report/dominica/dominica-hurricane-maria-overview-humanitarian-response-2017-september-december-2017

UNWTO. (2016). UNWTO tourism highlights (p. 16). UNWTO. Retrieved from http://mkt.unwto.org/publication/unwto-tourism-highlights-2016-edition

van der Velde, M., Green, S.R., Vanclooster, M., & Clothier, B.E. (2007). Sustainable development in small island developing states: agricultural intensification, economic development, and freshwater resources management on the coral atoll of Tongatapu. *Ecological Economics*, *61*(2), 456–468. https://doi.org/10.1016/j.ecolecon.2006.03.017

Walker, B., Holling, C.S., Carpenter, S., & Kinzig, A. (2004). Resilience, adaptability and transformability in social–ecological systems. *Ecology and Society*, *9*(2). https://doi.org/10.5751/ES-00650-090205

Weaver, D.B. (1991). Alternative to mass tourism in Dominica. *Annals of Tourism Research*, *18*(3), 414–432. https://doi.org/10.1016/0160-7383(91)90049-H

Weaver, D.B. (2001). *The Encyclopedia of Ecotourism*. Wallingford: CABI Pub.

Weaver, D.B. (2017). Core–periphery relationships and the sustainability paradox of small island tourism. *Tourism Recreation Research*, *42*(1), 11–21. https://doi.org/10.1080/02508281.2016.1228559

Weis, K. (2018). *Social-ecological Resilience and Tourism in The Coastal Zone: A Case Study in Dominica*. Master's Thesis, University Centre of the Westford's, Iceland.

21 Conclusion

Outlook for coastal transitions

C. Patrick Heidkamp and John Morrissey

1 Coastal resources and development pressures

The global sustainability challenge of improving well-being and living standards, in the context of maintaining and protecting the biophysical processes and ecosystem services that underpin well-being to begin with (Dearing *et al.*, 2014), is amplified considerably by the ongoing degradation of ecosystems worldwide[1], by increasing demand on these ecosystems (Alexander and Graziano, Chapter 14) and by pressures from the direct and indirect impacts of climate change (Braun and Bernzen, Chapter 11). A major structural barrier to sustainability has been that decision-making at the economy-wide level has tended to underestimate or outright ignore the fundamental importance of ecosystem services. For example, pollution and adverse ecosystem impacts associated with resource production and consumption are externalities that are not generally accounted for in neo-classical economic models (Day *et al.*, 2014). Associated welfare impacts of coastal recreational activities are also frequently not reflected in market valuations (Ghermandi, 2015). Conventional land-use planning practice has also tended to operate within conditions of assumed economic growth (Lloyd, Peel & Duck, 2013). For coastal communities, coastal ecosystems provide a wide range of ecosystem services including; raw materials, plant and animal habitats, water and air quality regulation and tourism, recreation, education, and research infrastructure (Barbier *et al.*, 2011; Sutton-Grier *et al.*, 2015). Coastal ecosystems also provide invaluable protective services to coastal communities. For example, saltmarshes provide coastal protection, erosion control and carbon sequestration benefits (Barbier *et al.*, 2011). Mangroves constitute a rich habitat for biodiversity and the destruction can lead to increased flood and erosion risks for local communities (Chouhan *et al.*, Chapter 18).

Ecosystem services such as these form the fundamental basis for the global economy (Day *et al.*, 2014). Nowhere is this more evident than in coastal economies. Because changes in marine conditions are closely linked to changes in marine ecosystems, far-ranging impacts are likely from a deterioration of marine conditions on industries such as aquaculture, fishing and tourism (Bradley *et al.*, 2015). For example, changes in primary and secondary productivity and species range shifts will alter the availability and abundance of wild caught marine species (Daw, Adger, Brown & Badjeck., 2009). Changes in marine conditions will increase vulnerability in the aquaculture industry (Bradley *et al.*, 2015), through increased incidence and impact of diseases, site suitability impacts from inundation, fluctuations in salinity and temperature, and increased risk from storm events (DeSilva & Soto, 2009). The livelihoods of many coastal communities is dependent on the sustainability and perhaps even expansion of such industries (Johnson and Hanes, Chapter 10). Coastal livelihoods are therefore also increasingly at risk (Bradley *et al.*, 2015).

Globally, rapid demographic change of coastal urban agglomerations, the speed of urbanisation over time, the overall impact of coastal space occupation as well as the limited capacity to adjust to those changes in a poverty context (Barragán & de Andrés, 2015) represent common themes in coastal zones. Braun and Bernzen, in Chapter 11, profile the challenges for one of the world's poorest and most densely populated countries, Bangladesh. Globally, coastal urbanisation has been very rapid; in 70 years the number of coastal cities and agglomerations increased by a factor of almost five, from 472 in 1945 to 2129 in 2012 (Barragán & de Andrés, 2015). At present, human activities related to agriculture, mining and construction move more earth material than do natural geomorphic processes related to rivers, glaciers, wind and waves (Hooke, 1994). Coastal degradation, especially in tourism-dependent economies, is increasingly reflected in the retreat of the coastline and increasing risk of erosion (Elliott *et al*, 2014). Due to urban development and the ubiquitous emergence of hardened, impervious surfaces (such as roads), flood protection measures or steep gradients in topography, many coastal ecosystems are limited in their ability to migrate inland as sea levels rise, due to climate change (Sutton-Grier *et al.*, 2015). In this context, the anticipated increased exploitation of marine resources and associated competing and conflicting uses (Lloyd *et al.*, 2013) are real concerns for future ecosystem health and protection. However, such concerns need to be moderated in the context of functioning coastal economies. Weis *et al.* (Chapter 20) emphasise the key importance of tourism to Small Island Developing States, for instance, where the sector can be used to enhance dimensions of social-ecological resilience and reduce vulnerability to sudden or unexpected social, political, economic and ecological disturbances.

In the medium- to long-term, ecosystem pressures will increasingly manifest as economic pressures. The continued degradation of ecosystem services, when combined with macro-level pressures such as climate change and energy availability, are likely to make continued economic growth difficult if not impossible (Day *et al.*, 2014). In fact, global resource and environmental problems, such as climate change, require a radical reorientation of production and consumption systems in industrialised countries (Bai *et al.* 2009). Decision-makers face difficult choices, between short-term growth and long-term sustainability, between electing for lower returns from ecological resources in favour of longer-term viability. Transformative new governance approaches, for example the Integration Transition Pathways advocated by

Kelly *et al.* (Chapter 4), new conceptualisations (Lewis, Chapter 6, Le Heron *et al.* Chapter 7) and new technological innovations (Polrot *et al.*, Chapter 16, Graves, Chapter 13) are urgently required. Such transformation needs not only to be cognisant of and accommodating of key local stakeholders, but need to be driven by these actors and groups (Axon, Chapter 17) and in a way that local innovation and ingenuity is fostered and harnessed for sustainable and locally beneficial outcomes (Foley *et al.*, Chapter 19).

2 Towards just transition(s) in the coastal zone

Based on our observation that issues of social and environmental justice will need to be considered in any type of sustainability and resilience planning scenario, we argue for an approach aimed at assuring a *Just Transition* (Swilling & Annecke, 2012) in the Coastal Zone.

The idea of bringing social and environmental justice concerns into the framework of sustainability and transitions is nothing new, and has been expertly argued for by Swilling & Annecke (2012) focused on *Just Transitions* and by Agyeman, Bullard & Evans (2002) and again Agyeman (2013) focused on the concept of *Just Sustainabilities*. While the work by Agyeman is situated largely in the context of the global North, and Swilling & Annecke provide a much-needed perspective from the global South, the arguments overlap considerably and call for the ideas of social, economic and environmental justice as a guiding principles. Both Agyeman (2013) and Swilling & Annecke (2012) argue that persisting – or rather widening – inequality, is the key issue that needs to be addressed when envisioning a transition towards a more sustainable society. A transition towards just and sustainable communities, Agyeman (2013, p. 7) argues, requires four interconnected conditions to be met: "improving quality of life and wellbeing; meeting the needs of both present and future generations (intragenerational and intergenerational equity), justice and equity in terms of recognition, process, procedure and outcome; and living within ecosystem limits (also called 'one planet living')". Therefore, sustainability – and resilience toward climate change – can only be achieved in societies that exhibit significant levels of trust and cohesion, which can lead to more action towards the common good. Inequality, on the other hand, drives competitive consumption, and thus a reduction in the pursuit of the common good (Wilkinson & Pickett, 2010). Additionally, the transition towards more sustainable societies requires innovation, adaptability and creativity, qualities which are also found to be more prevalent in more equal societies (Agyeman, 2013; Wilkinson & Picket, 2010). Knowing that economic growth does not lead to equality, and more often than not leads to a reduction in environmental quality, requires us to rethink, if the prevalent economic growth paradigm can be reconciled with a transition towards sustainability. A consensus emerged that analysing and/or facilitating any such transition requires us to employ a systems perspective that views transitions as processes as long-term (40–50 years), multi-actor co-evolutionary – encompassing technological, social and institutional change – processes that aim to reconfigure the institutional and organisational structures and systems of society (Swilling & Annecke, 2012, p. xvi). Given this, a just transition then becomes an overtly political project, especially considering Swilling & Annecke (2012, p. xiii) statement that "A Just Transition, must be a transition that reconciles sustainable use of resources with a pervasive commitment to what is increasingly referred to as sufficiency (that is, where over-consumers are satisfied with less, so that under-consumers can secure enough, without aspiring for more than their fair share"). Economic geography as a discipline has much to contribute to a discussion about uneven consumption patterns and other uneven development processes; in fact it can be considered the core of the discipline. Or, as Eric Sheppard put it aptly regarding the field of economic geography: "Yet to me to me at its heart has been the goal of accounting for and redressing unequal livelihood possibilities" (2006, p. 11).

Recognising that transitions are long-term processes, significant political and economic changes are, and should be, clearly part of the discussion about how to assure a Just Transitions framework for coastal change and adaptation. Given the increasing urgency of climate change pressures we are, however, in a bit of a conundrum – 40 to 50 years might make quite a bit of difference in the coastal zone. Due to this urgency, we would argue that we will need to move beyond removed theoretical discussions and embrace pragmatism as a guiding principle, which requires increased engagement with stakeholders at all levels (see also Axon, Chapter 17), and a transdisciplinary (Nicolescu, 2014) approach to problem-solving that facilitates co-learning among all stakeholders in order to facilitate a just transition in the coast zone. Such a discussion cannot and will not ever be apolitical and it is past time to engage with the political in a discussion of coastal sustainability and resilience, as already outlined by Murphy (2015). A recent article by Hardy *et al.* (2017) focused on highlighting the apolitical narratives to sea level rise in the United States, and may provide an interesting starting-point for a meaningful discussion. We would hope that the preceding chapters will inspire readers to build on the work presented in this volume and, in Swilling & Anecke's (2012, p. xiii) words, "search for actionable imaginaries", towards a social, economic and environmentally just transition in the coastal zone.

Note

1 Marine environments worldwide are in severe decline, as a result of over- exploitation, pollution and the indirect and direct impacts of climate change (Marques *et al.*, 2009).

Bibliography

Agyeman, J. (2013). *Introducing Just Sustainabilities: Policy, Planning, and Practice.* Zed Books Ltd.

Agyeman, J., Bullard, R.D., & Evans, B. (2002). Exploring the nexus: bringing together sustainability, environmental justice and equity. *Space and Polity*, *6*(1), 77–90.

Barbier, E.B., Hacker, S.D., Kennedy, C., Koch, E.W., Stier, A.C., & Silliman, B.R. (2011). The value of estuarine and coastal ecosystem services. *Ecological Monographs, 81*(2), 169–193. http://doi.org/10.1890/10–1510.1

Bai, X., Wieczorek, A.J., Kaneko, S., Lisson, S., & Contreras, A. (2009). Enabling sustainability transitions in Asia: the importance of vertical and horizontal linkages. *Technological Forecasting and Social Change, 76*(2), 255–266. http://doi.org/10.1016/j.techfore.2008.03.022

Barragán, J.M., & de Andrés, M. (2015). Analysis and trends of the world's coastal cities and agglomerations. *Ocean & Coastal Management, 114*, 11–20. http://doi.org/10.1016/j.ocecoaman.2015.06.004

Bradley, M., van Putten, I., & Sheaves, M. (2015). The pace and progress of adaptation: marine climate change preparedness in Australia's coastal communities. *Marine Policy, 53*, 13–20. http://doi.org/10.1016/j.marpol.2014.11.004

Coenen, L., Benneworth, P., & Truffer, B. (2012). Toward a spatial perspective on sustainability transitions. *Research Policy, 41*(6), 968–979.

Day, J.W., Moerschbaecher, M., Pimentel, D., Hall, C., & Yáñez-Arancibia, A. (2014). Sustainability and place: how emerging mega-trends of the 21st century will affect humans and nature at the landscape level. *Ecological Engineering, 65*, 33–48. http://doi.org/10.1016/j.ecoleng.2013.08.003

Daw, T., Adger, W.N., Brown, K., & Badjeck, M.C. (2009). Climate change and capture fisheries: potential impacts, adaptation and mitigation. *Climate change implications for fisheries and aquaculture: overview of current scientific knowledge. FAO Fisheries and Aquaculture Technical Paper*, *530*, 107–150.

Dearing, J.A., Wang, R., Zhang, K., Dyke, J.G., Haberl, H., Hossain, M.S., ... Poppy, G.M. (2014). Safe and just operating spaces for regional social-ecological systems. *Global Environmental Change, 28*, 227–238. http://doi.org/10.1016/j.gloenvcha.2014.06.012

De Silva, S.S., & Soto, D. (2009). Climate change and aquaculture: potential impacts, adaptation and mitigation. *Climate change implications for fisheries and aquaculture: overview of current scientific knowledge. FAO Fisheries and Aquaculture Technical Paper*, *530*, 151–212.

Elliott, M., Cutts, N.D., & Trono, A. (2014). A typology of marine and estuarine hazards and risks as vectors of change: a review for vulnerable coasts and their management. *Ocean and Coastal Management, 93*, 88–99. http://doi.org/10.1016/j.ocecoaman.2014.03.014

Ghermandi, A. (2015). Benefits of coastal recreation in Europe: identifying trade-offs and priority regions for sustainable management. *Journal of Environmental Management*, *152*, 218–229.

Hardy, R.D., Milligan, R.A., & Heynen, N. (2017). Racial coastal formation: the environmental injustice of colorblind adaptation planning for sea-level rise. *Geoforum*, *87*, 62–72.

Hooke, R.L. (1994). On the efficacy of humans as geomorphic agents. *GSA Today*, *4*(9), 217.

Lloyd, M.G., Peel, D., & Duck, R.W. (2013). Towards a social-ecological resilience framework for coastal planning. *Land Use Policy, 30*(1), 925–933. http://doi.org/10.1016/j.landusepol.2012.06.012

Marques, J.C., Basset, A., Brey, T., & Elliott, M. (2009). The ecological sustainability trigon –a proposed conceptual framework for creating and testing management scenarios. *Marine Pollution Bulletin*, *58*(12), 1773–1779.

Murphy, J.T. (2015). Human geography and socio-technical transition studies: promising intersections. *Environmental Innovation and Societal Transitions, 17*, 73–91.

Nicolescu, B. (2014). Methodology of transdisciplinarity. *World Futures*, *70*(3–4), 186–199.

Sheppard, E. (2006) The Economic Geography Project. In *Economic Geography: Past, Present and Future*. Bagchi-Sen, S., & Lawton-Smith, H. (Eds.). Routledge.

Sutton-Grier, A.E., Wowk, K., & Bamford, H. (2015). Future of our coasts: the potential for natural and hybrid infrastructure to enhance the resilience of our coastal communities, economies and ecosystems. *Environmental Science & Policy, 51*, 137–148. http://doi.org/10.1016/j.envsci.2015.04.006

Swilling, M., & Annecke, E. (2012). *Just Transitions; Explorations of Sustainability in an Unfair World*. Claremont, South Africa. UCT Press.

Wilkinson, R., & Pickett, K. (2010). *The Spirit Level: Why Equality is Better For Everyone*. Penguin UK.

Index

Bold page references indicate relevant **Tables**; *italic* references are to *Figures*.